MW00760203

Lasers and Electro-Optics Research and Technology Series

PHASE TRANSITIONS INDUCED BY SHORT LASER PULSES

LASERS AND ELECTRO-OPTICS RESEARCH AND TECHNOLOGY SERIES

High-Power and Femtosecond Lasers: Properties, Materials and Applications
Paul-Henri Barret and Michael Palmer (Editors)
2009. ISBN: 978-1-60741-009-6

Optical Solitons in Nonlinear Micro Ring Resonators: Unexpected Results and Applications
Nithiroth Pornsuwancharoen, Jalil Ali andPreecha Yupapin
2009. ISBN: 978-1-60741-342-4

From Femto-to Attascience and Beyond
Janina Marciak-Kozlowska and Miroslaw Kozlowski
2009. ISBN: 978-1-60741-164-2

Fiber Lasers: Research, Technology and Applications
Masato Kimura (Editor)
2009. ISBN: 978-1-60692-896-7

Phase Transition Induced by Short Laser Pulses
Geortgy A. Shafeev (Editor)
2009 ISBN: 978-1-60741-590-9

Lasers and Electro-Optics Research and Technology Series

PHASE TRANSITIONS INDUCED BY SHORT LASER PULSES

GEORGY A. SHAFEEV
EDITOR

Nova Science Publishers, Inc.
New York

For permission to use material from this book please contact us:
Telephone 631-231-7269; Fax 631-231-8175
Web Site: http://www.novapublishers.com

LIBRARY OF CONGRESS CATALOGING-IN-PUBLICATION DATA

Phase transitions induced by short laser pulses / [edited by] Georgy A. Shafeev.
p. cm.
Includes index.
ISBN 978-1-60741-590-9 (hardcover)
1. Phase transformations (Statistical physics) 2. Solid-liquid interfaces. 3. Laser pulses, Ultrashort. 4. Laser ablation. I. Shafeev, Georgy A.
QC175.16.P5P485 2009
530.4'14--dc22

2009017372

Published by Nova Science Publishers, Inc. ✝ New York

CONTENTS

PREFACE

This book presents the recent advances in the studies of phase transitions induced by short laser pulses in solids and the interface solid-liquid. The review is presented with previous and new results of experimental and theoretical descriptions of phase boiling phenomena under the action of short laser pulses. Time-resolved characterization of the phase transitions induced by short laser pulses in metals in confined geometry is described in detail, and the results are compared with classical equations of states of the materials in the sub- and super-critical range of parameters. Phase transitions in the films of cryogenic molecular compounds doped with different impurities are thoroughly studied with respect to their use in MALDI and MAPLE processes. Modeling and new experimental data are described on explosive boiling of liquids under laser exposure of its free surface. Novel experimental results are presented on instabilities at the solid-liquid interface exposed to pico- and femtosecond laser pulses. Morphological and optical properties of resulting nanostructure's periodic relief are described. Theoretical modeling of instabilities and structure formation under laser exposure is given on the basis of defect-deformational instability. Modeling results are found to be in good agreement with experimental data. The book is of interest for scientists, post-graduate students and engineers working in the field of laser-matter interaction.

The book contains several chapters written by specialists that currently work in the field of laser-induced phase transitions in condensed media. The content of chapters gives modern insight into both experimental approaches to the problem and to modeling this complex, highly non-equilibrium phenomenon.

The first chapter presents the detailed review of previous and recent results on the interaction of short laser pulses with solid targets near the threshold of liquid-vapor phase transition. It is demonstrated that such phase transitions manifest themselves in different applied problems. For example, the rate dependence of laser holes drilling is interpreted from the viewpoint of sub-critical and supercritical liquid-vapor phase transformations induced on surfaces and in bulk materials under the action of short and ultra-short laser pulses. The validity of semi-empirical equations of states (EOS) for near-critical non-equilibrium phase transitions induced by laser radiation is discussed. A special emphasis is given to the analysis of experimental results on explosive boiling obtained with a large variety of experimental techniques. These techniques, which allow characterization of liquid-vapor phase transitions under laser exposure, are: ultrasonic imaging of the interface, debris re-deposition, removal rates, optical imaging, etc. The detailed analysis of experimental data distinguishes two different situations: surface and bulk near-critical liquid-vapor phase transformations. The contribution of the plasma plume near the target under laser exposure to its heating is thoroughly analyzed and modeled. The estimates show that this contribution can be comparable with the energy to be directly absorbed by the target from the laser beam at the absence of plasma screening. Current challenges and problems in both experimental and theoretical investigations of laser-induced near-critical phase transitions are discussed.

The second chapter describes results on the laser-induced phase transitions in metals, namely, Pb and Hg in the confined geometry at nanosecond pulse duration. This is achieved via laser exposure of corresponding metal with a dielectric medium transparent at the laser wavelength. Real-time measurements of the thermodynamic state of metals are performed using acoustical methods combined with reflectivity measurements and optical pyrometry. The confined geometry allows almost complete suppression of plasma formation, so that the absorbed laser energy is spent for phase transitions without material evaporation. The chapter contains detailed analysis of different factors that may affect the precision of measurements of phase transition induced in metals by laser beams. Distribution of the pressure in adjacent media is modeled on the basis of the solution of heat conduction equation. Temperature dependence of physical parameters of metals under study is carefully taken into account during modeling. The contribution of density change into pressure value during laser-induced melting is also taken into account. It is pointed out that boiling of the metal can be delayed due to increased pressure induced by laser pulse, and the liquid metal can exist in a metastable state of overheated liquid. The estimates made are used for careful processing of experimental data on a phase diagram of either Pb or Hg under the action of a nanosecond laser radiation. Maximal pressure attained in the work is 100 and 750 MPa for Pb and Hg, respectively. The developed approach allowed determination of critical parameters for lead. Simultaneous measurements of pressure and temperature allowed the plotting of a P-T diagram of Hg and to trace its transition to supercritical state.

Chapter 3 is devoted to the investigation of molecular systems subjected to pulsed UV laser radiation. Phase transitions in this case are less pronounced and are therefore less studied than in metals and semiconductors. However, molecular systems are very important for many practical applications, e.g., for Matrix-Assisted-Laser-Desorption-Ionization (MALDI) of biopolymers and macromolecules or Matrix Assisted Pulsed Laser Evaporation (MAPLE). The chapter deals with the contribution of phase transitions induced by UV laser sources to the ablation of molecular cryogenic solids. The basic studied material is solid toluene doped with different organic compounds having various volatilities. Unlike the confined geometry, the confinement of stress inside the laser-irradiated solid is much weaker, and its influence on the spallation of material is not significant. The results of modeling the explosive boiling using the Molecular Dynamics (MD) method are critically analyzed. Experimental data show that clusters of the ablated material leave the solid surface and therefore are not formed during subsequent collisions above the substrate. The formation of bubbles under the surface of laser-exposed organic solid and their coalescence is traced using the scattering of the probe laser beam. Their formation has a threshold-like character. It is found that spinodal decomposition of the solid occurs at a somewhat lower temperature than it follows from MD calculations and known temperatures in equilibrium conditions. The conclusion of the chapter is that promotion of laser-induced phase transformations, e.g., devitrification and bubble formation rely sensitively on the pre-existence of defects.

Chapter 4 presents new results on measurements of photo-acoustical and vaporization pressure induced by laser pulses in condensed media. The laser wavelength of 2.94 μm fits the maximum of absorption by liquid water. The process of laser heating of free water surface by a laser beam is modeled. It is shown that starting from threshold laser fluence the temperature of the liquid exceeds the spinodal line or even its critical temperature. It is discussed that explosive boiling is accompanied by short pressure pulses and therefore can occur several times during one sufficiently long laser pulse. As a role, the acoustical signal that corresponds to explosive boiling is uni-polar. It is suggested that the sub-ns acoustical signal observed in some experiments is an artifact and is related to the construction of a pressure sensor. Carefully designed experimental setup allows distinguishing two components of the pressure signal under the exposure to laser radiation. The pressure pulse exhibits well-pronounced fluence dependence. Namely, after a certain threshold fluence the measured signal contains a short (sub-ns) component at laser pulse duration of 150-200 ns. This is interpreted as the onset of explosive boiling. Numerous uni-polar peaks appear with the increase of laser fluence. Much longer bi-polar pressure signal is attributed to the evaporation pressure from free surface of water. At even higher fluence the explosive boiling and corresponding peaks can be suppressed by large evaporation pressure.

The fifth chapter deals with a relatively new effect that consists in the formation of self-organized nanostructures (NS) on the target surface immersed into a liquid and exposed to sufficiently short laser pulses. The range of laser pulse durations that allows observation of such NS spans from tens of femtosecond to hundreds of picoseconds. The formation of NS is interpreted as the evaporational instability that develops at the interface between two adjacent media under simultaneous phase transitions. Short laser pulses are needed to avoid damping of nano-sized relief by capillary forces. The high cooling rate of the melt on the target allows freezing these periodic structures presented in the expanding vapors of the liquid that surrounds the target. These NS can be considered as the visualization of fragmentation of supercritical compounds that surrounds the solid target during laser exposure. The chapter contains mostly novel experimental data on formation of NS on various metallic (Ag, Au, Ni, Ta, Zn, Al, etc) and non-metallic (e.g., Si) targets under their ablation in liquid environment (water, ethanol.) The size distribution of NS depends on the laser fluence, and at elevated fluences, the distribution function is bimodal. Estimations show that the pressure difference in the expanding vapors may be as high as 400 atm at the scale of hundreds of nanometers, which exceeds by far the critical pressure for water. NS are characterized by specific optical properties. The free electrons confined in the NS demonstrate the plasmon resonance, which stipulates coloration of metallic surfaces with NS. Some potential applications of NS are also discussed, e.g., for Raman plates or medical implants.

The sixth chapter is devoted to the theoretical description of micro- and nanostructures that emerge under the action of laser radiation on solids at fluence level sufficient for phase transitions. The driving force for development of structures is defect-deformational (DD) instability. The interaction between the concentration of defects and stresses induced by inhomogeneous heating provides the positive feedback that may lead to the development of instability. The strong point of this approach is the analytical (vs. numerical) description of this instability based on the closed nonlinear equation of Kuramoto-Sivashinsky type. The approach developed in the chapter allows formulation of cooperative DD mechanism of nanostructures nucleation and growth. The important advantage of this approach is its validity for both laser light exposure and ion beam bombardment of the solids. The model predicts a bimodal distribution function of a structure's size under laser exposure. This conclusion is in good agreement with experimental results carefully reviewed in the chapter. The model also predicts generation of the second harmonics of spatial relief – the effect often observed in laser-matter interactions. This is assigned to non-linear interaction of DD gratings. The influence of external anisotropy on the symmetry of DD gratings and its evolution with laser fluence and laser pulse duration is considered.

Georgy A. Shafeev

editor

In: Phase Transitions Induced by Short Laser Pulses
Editor: Georgy A. Shafeev
ISBN: 978-1-60741-590-9

Chapter 1

LASER-INDUCED SURFACE AND BULK NEAR-CRITICAL LIQUID-VAPOR PHASE TRANSFORMATIONS: BASIC CONCEPTS AND RECENT ADVANCES

Sergey I. Kudryashov*

P.N. Lebedev Physical Institute of the Russian Academy of Sciences

ABSTRACT

Basic concepts, previous and original experimental data, and recent theoretical findings are presented to illustrate thermal surface and bulk explosive near-critical liquid-vapor phase transformations occurring in condensed materials under pulsed laser irradiation. Homogeneous nucleation and spinodal decomposition in bulk superheated molten materials under sub-critical conditions, and continuous liquid-vapor transformations of supercritical fluids are considered as the fundamental liquid-vapor phase change phenomena in their various appearances, while their occurrence on nano-scale spatial and ultra-fast temporal scales through various sequences of extreme, high-temperature and high-pressure thermodynamic states makes their basic experimental studies a formidable task, requiring diverse state-of-the-art approaches. The broad overview of the previously reported and original experimental signatures and indications of laser-induced explosive liquid-vapor transformations points out two distinct laser ablation regimes, which dramatically differ not only in effective laser parameters (mainly, incident laser intensities), but also in ablated depths and morphologies of produced craters, parameters of laser plume (chemical, phase and charge composition, velocities of ablated species, temperature and pressure, emission spectral components), and can be simply associated with either surface, or bulk such phase transformations. Surface liquid-vapor phase transformation in the form of a phase separation wave, accompanied by expulsion and re-deposition of a mixture of vapor and nanoscale droplets around and inside a shallow (sub-micrometer deep) crater, are shown to be driven directly by incident laser flux in superheated sub-critical or supercritical molten surface layers at the proximity of the liquid-vapor spinode curve or beyond the corresponding critical

* sikudr@sci.lebedev.ru

points, respectively. In other terms, bulk liquid-vapor phase transformation phenomena are induced via strong heating of bulk materials until the sub-critical (near-spinodal) or even supercritical temperatures indirectly – by highly penetrating x-ray emissions of hot short-living ablative laser plasmas, being accompanied by microsecond long delayed forward expulsions of fast micrometer-sized droplets of the molten materials and formation of multi-micron deep ablation craters. Diverse appearances and practical implications of surface and bulk liquid-vapor phase transformations induced by short and ultra-short lasers, as well as current challenges in their experimental and theoretical exploration, are comprehensively discussed.

I. Introduction

A. History of the Topic and Basic Definitions: Many Faces of Liquid-Vapor Phase Transformation

Recent molecular dynamics (MD) simulations, being up to date the only way to provide enlightening visualization of short- and ultrashort-pulse laser ablation phenomena in condensed matter, have demonstrated an increasing initial energy density and a number of characteristic ablation pathways related to thermo-mechanical spallation (rupture) of a surface liquid layer under tensile stress, phase explosion of the superheated liquid layer in the sub-critical region, and fragmentation of the supercritical fluid into its vapor-droplet mixture, respectively [1-6]. Though some of these ablation pathway names have historical analogues (e.g., impact-driven spallation of solid targets [7,8], or "vapor explosion" of rapidly heated liquid [9]), the overall appearance of a vapor-droplet mixture along all of them makes distinguishing these sub-critical and supercritical liquid-vapor phase transformations into the three paths rather artificial. Moreover, in this context, ultrafast-laser heated supercritical fluids exhibit much more pronounced explosive evolution into a vapor-droplet mixture than that during "phase explosion" [10-13] of superheated sub-critical liquids, while in a case of molecular liquids this explosive evolution (fragmentation [6]) of their supercritical fluids should be clearly separated from thermal fragmentation of their molecular species.

Originally, the argotic term "phase explosion", introduced at first in laser-related literature in the 90's [12-13] as a reminiscence of fast discharge-driven boiling phenomena [10-11] in relation to thermally-induced explosive-like, soudible expulsion of laser ablation products leaving behind a violently disturbed re-solidified surface [14], mimics the more relevant term "vapor explosion" (also, rapid phase transformations, thermal explosions, explosive boiling, fuel-coolant interactions) [9]. Historic meaning of the latter term and its other names was a sudden, catastrophic rise of internal pressure inside tankers, filled by liquefied propane gas, via explosive spontaneous liquid-vapor transformation, as a result of uncontrolled increasing temperature or depressurization in the gas containers, sudden and explosive vaporization of a volatile liquid in paper and aluminum production, reactive metal processing and nuclear power generation resulting in industrial hazards via injury to personnel and structural damage [9]. Furthermore, there is a whole set of other diverse thermo-physical phenomena (the most familiar one is the hot milk suddenly running out of its container during boiling), such as cavitation in pressurized liquid jets or around ship propellers [15], boiling crisis on a hot solid wall [16], soudible and visible sudden rupture of liquids under tension [15,17-18], and explosive boiling in pulsed laser-irradiated surface molten layers [19], which exhibit similar visible signatures, but came in different argotic technical names.

The common line in all these different appearances is rather fast nucleation of vapor bubbles in a liquid as its relaxation response to non-equilibrium ambient conditions, making the liquid stretched by the internal tensile stress or superheated at the fixed temperature due to the sub-equilibrium external compressive pressure [9-18]. The fundamental kinetic processes providing such phase relaxation in the non-equilibrium liquid, are the homogeneous (bulk) nucleation and spinodal decomposition, representing in P,T-coordinates (Figure 1), respectively, liquid-vapor phase transformations of metastable (arbitrarily unstable) and labile (absolutely unstable) superheated or stretched liquids in the region between the liquid-vapor equilibrium curve (binode, B) and liquid stability limit curve (spinode, S_1), with the latter extending along both positive and negative pressure scales [18]. The basic physical origin of the absolute liquid stability limit (spinode curve)...

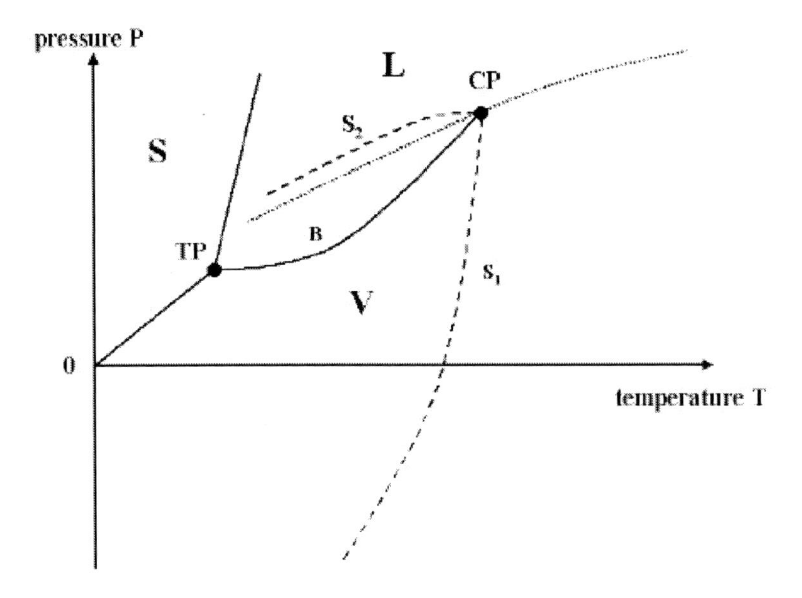

Figure 1. Schematic of phase diagram in P,T-coordinates with B, S_1, S_2, TP and CP denoting, respectively, the liquid-vapor binode, liquid-vapor and vapor-liquid spinodes, triple and the critical points. Symbols S,L,V show the regions of equilibrium solid, liquid and vapor phases, respectively, while the dotted line represents the hypothetical II-order liquid-vapor phase transition.

...is the balance of kinetic and potential energies of molecules composing the liquid, while the external factors may either raise their kinetic energies (heating), or reduce their potential energies (stretching). When approaching the liquid-vapor spinode curve through the metastability region, the radius r_{cr} of spontaneously nucleated critical vapor bubbles and their Gibbs formation energy ΔG_{cr} fall down as a function of temperature or pressure at the simultaneous exponential increase of nucleation rate $J(P,T, r_{cr})$ [9,18], yielding at the spinode in strongly spatially and temporally correlated corrugation of the unstable liquid accompanied by roughening and Ostwald ripening of the resulting pressure, temperature and density fluctuations. [20] (For review of previous studies of Cahn, Hilliard etc. see the review [21] and the recent book [9]). Typically, the liquid-vapor spinode curve is approached at increasing temperature or reducing pressure from their ambient positive magnitudes, i.e., by passing the metastability region (Figure 1), and, thus, the transient homogeneous nucleation process may provide initial stochastic thermodynamic (pressure, temperature, density) perturbations high and fast enough to trap the system in this regime. Importantly, the closer the spinode, the higher the isobaric heat capacity C_p, thermal expansion coefficient β_T, and isothermal compressibility K_T of the metastable liquid [9-10,18], exhibiting at the spinode their singularities. This makes the spinode, together with the extremely increasing nucleation rate, practically unattainable as the absolute stability limit, by the common means to provide the necessary high heating rate and to follow the drastically changing volume [22]. However, very close proximity of the liquid-vapor spinode can be attained, when the latter two conditions are satisfied on open liquid surfaces heated at rates of 10^{10}-10^{12} K/s by high-power short-pulse lasers [23], though the thermalization issue for laser energy density deposited in a liquid may become more and more crucial at higher nucleation rates when approaching the curve. Interestingly, according to the results of the recent MD simulations [1-6], near-spinodal homogeneous nucleation of vapor bubbles at negative and positive pressures, resulting, in the former case, in rupture (the so-called "thermo-mechanical spallation" [1-6]) of liquids under tensile stress due to space-sensitive "quasi-homogeneous" bubble nucleation driven by a propagating rarefaction wave [24-25], and, in the latter case, in hydrodynamic expansion (expulsion) of a vapor-droplet mixture [24-30] accompanying propagation of a liquid-vapor phase separation wave front in the superheated molten material [31,32], are rather hardly distinguishable in their microscopic dynamics prior to ablative removal, though they are very different in resulting ablation topologies – rather flat and shallow versus more deep and highly violent craters, respectively.

Likewise, liquid-vapor phase separation in a supercritical fluid, occurring continuously under *equilibrium* thermodynamic conditions via nano-heterogeneous mixture of coexisting vapor and liquid phases represented by stochastic, short-living density fluctuations [33], may proceed very differently under highly *transient* phase transformation conditions, provided by ultrafast (typically, picosecond) isochoric heating of liquids (or initial solid targets) by ultrashort (femtosecond or short picosecond) laser pulses [34-41]. Currently there are no experimental means to track microscopic dynamics of transient phase transformations in supercritical fluids (the only time-resolved x-ray scattering study has been recently reported to visualize picosecond nucleation and cavitation dynamics in femtosecond-laser superheated molten bismuth [41]), MD simulations remain the only way to shed light on these phenomena, revealing formation of vapor-droplet mixture as a result of sub-critical "phase explosion" and supercritical "fragmentation" of laser-heated liquids [1-6]. However, one should note that the current MD models presumably employing binary "6/12", Morse or these modified potentials for intermolecular interactions [1-6], are not strictly applicable in the near-critical region (far asymptotic, large-distance region of such potentials), where not the "exhausted" binary, but multi-particle interactions, presented by corresponding density fluctuations and rather high correlation radii, are important [9,33]. Typically, such binary potentials derived from basic spectral characteristics of low-excited vibrational states of diatomic molecules are not very accurate at large inter-particle distances, which is the case in the strongly correlated near-critical fluids. Hence, quantitative and even qualitative results of MD simulations, employing binary intermolecular potentials to visualize internal structure and dynamics in the near-critical fluids (including, apparently, also the liquid-vapor near-spinodal region), are yet to be confirmed experimentally, using the time-resolved x-ray scattering [41] or other advanced techniques.

The abovementioned sub-critical and supercritical liquid-vapor phase transformations induced on surfaces and in bulk materials by short- and ultrashort-pulse lasers will be the main focus of this chapter. We intentionally exclude cold rupture (spallation) phenomena in liquids far from their corresponding critical points, which are still the subject of ongoing experimental studies [36,41], and rather "simple" surface vaporization (sublimation-like) phenomena, which are just believed to be well understood (see the recent observations of pronounced differences of laser heating and vaporization on graphite surfaces in vacuum and ambient atmosphere [42]). Thus, we limited the scope of this chapter by laser ablation processes, occurring in the near-critical region and resulting, as a rule, in expulsion of vapor-droplet mixtures.

B. Laser-Induced Near-Critical Liquid-Vapor Phase Transformations: Surface and Bulk Cases

Taking expulsion of a near-critical vapor-droplet mixture as a basic criterion of laser-induced phase transformations of interest, one can easily find other experimentally observable attributes distinguishing them, e.g., from a preceding – in terms of incident laser intensity or fluence – regime of the common surface vaporization [42-45]. The latter could be either quantitative – typically, fluence- or intensity-dependent magnitudes of removal rate (crater depth Z_{abl} per laser pulse) [31-32,42-48] and recoil pressure P_{rec} [24-25,31-32,48-50], or qualitative – shapes of ultrasonic waveforms $P(t)$ [24-25,31-32,48-50], phase composition of ablated [24-32,34-39,43-45,50-52] and re-deposited species [43-45,53-54], respectively, crater topologies [47] and appearance of near-crater debris [52-54]. Most of these attributes sharply vary (or appear) upon onset of the near-critical phase transformations enabling rather unambiguous identification of their thresholds on laser fluence (intensity) scales – for opaque targets, at typical threshold fluences of a few J/cm^2 (sub-GW/cm^2 intensities) for nanosecond laser pulses [24-25,31-32,42-50] against sub-J/cm^2 values (TW/cm^2-level intensities) for femtosecond laser pulses [34-41] (for broader bibliography, see the reviews [40,55]).

Historically, material removal (drilling) rates defined as an average crater depth per one laser pulse, and related quantities (removed mass per unit square per pulse) were employed in the 90's as the simplest and the most obvious attributes of short-pulse laser-induced sub-critical phase transformations [31-32,45-48]. A drastic – an order of magnitude – threshold-like increase of a removal rate near the phase transformation threshold fluence (Figure 2) after the preceding slow surface vaporization can be associated with the rather low specific removal energy density ε_{abl} for a vapor-droplet mixture, which is lying somewhere between the enthalpies of the material of interest in its melting and critical points (figure 3) – $\Delta H(l, T_{melt}, P)$ and $\Delta H(l, T_{crit}, P_{crit})$, respectively [the latter value is usually unknown, but can be reasonably evaluated as equal to the evaporation heat (enthalpy) $\Delta_{ev}H(l, T \geq T_{melt}, P)$ of the material, considering the enthalpy as an elementary binding (potential) energy preserving the continuous structure of the liquid against the counteracting kinetic energy of thermal molecular motion]. This low removal energy density contrasts to that of surface vaporization, which includes both the $\Delta H(l, T_{melt}, P)$ and $\Delta_{ev}H(l, T, P)$ additives, typically yielding, in theory, in two or three times higher consumption of laser energy, while in ablation experiments, performed in ambient atmosphere or in other lower- or higher-pressure gas environments, the actual difference could be even higher (Figure 3), potentially, because of re-deposition of evaporated species back to the surface at their non-zero sticking coefficients [44,56]. Likewise, a similar drastic drop of the specific removal energy density can be observed near threshold fluence, representing the transition between two characteristic ablation mechanisms for femtosecond laser heated mono-crystalline graphite (Figure 4) [57], which can be related to the rupture (spallation) and liquid-vapor expulsion processes [35,58].

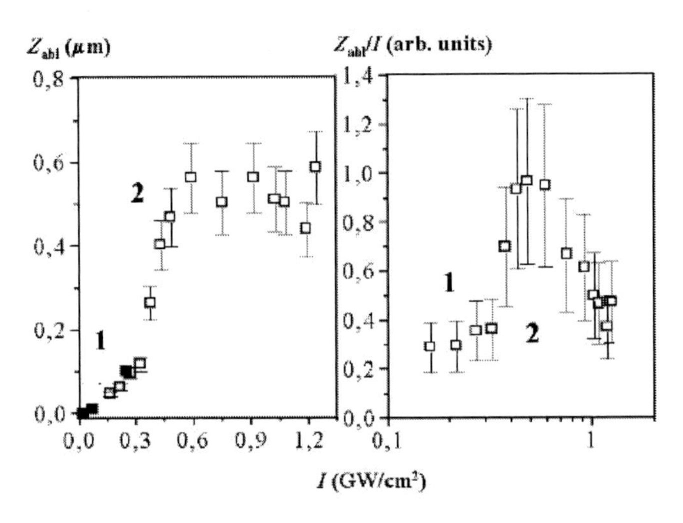

Figure 2. (left) Removal rate (crater depth per pulse) Z_{abl} versus laser intensity I on polycrystalline graphite surfaces ablated by 532-nm, 25-ns laser pulses (adapted from Ref. 32). The drastic threshold-like increase of the removal rate shows the sub-critical liquid-vapor phase transformation (2) succeeding the surface vaporization regime (1). (right) Normalized rate Z_{abl}/I is shown for comparison.

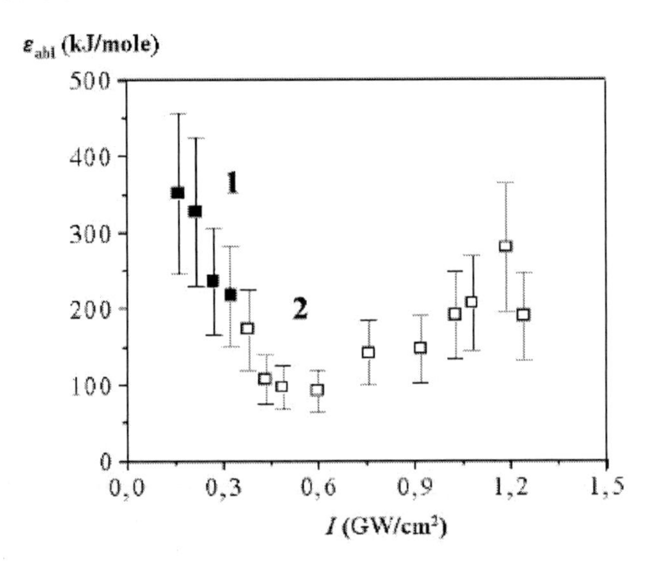

Figure 3. Specific removal energy density ε_{abl} versus laser intensity I for the polycrystalline graphite ablated by 532-nm, 25-ns laser pulses (adapted from Ref. 32).

In many previous experimental studies the prominent transitions between the surface vaporization and sub-critical liquid-vapor phase transformation, or between the rupture and liquid-vapor expulsion, were not observed in measurements of removal rates (crater depths, removed masses), varying in these regimes from a few nanometers to a few micrometers per laser shot (Figures 2,4), when these ablation parameters were measured by common experimental means (weighting [46], mechanical profilometry and through-drilling [48] measurements), for the reason of negligible surface vaporization removal, additionally effectively damped by debris re-deposition on walls inside measurable multi-shot craters. As a result, more sensitive, accurate and informative (though less available) optical profilometry [47] and interferometry [57], scanning near-field optical or atomic force microscopy [59], mass-spectroscopy [32,34,60-62] and electric probe [27] techniques, as well as rear-side ultrasonic measurements of single-shot recoil pressure [24-25,28,32,49-50,63] or multi-shot crater depths [32] were required to detect threshold-like onsets of the abovementioned near-critical liquid-vapor phase transformations in the form, respectively, of corresponding deeper craters, appearance in a gas phase of multiply-charged heavy (charged droplets) and light (multiply-charged atomic ions) fractions of ablation products, leveling off a difference between thermoacoustic and ablative components of ultrasonic signals during ablative thermal expansion of vapor-droplet mixtures. Moreover, an appearance of extensive lateral debris and violent crater topology could also be a good qualitative indication of such near-critical liquid-vapor phase transformations, resulting from considerable backward and lateral mass fluxes in laser plumes with initial hemispherical and later spherical expansion geometries [64]. Importantly, following the rapid and extensive near-critical mass removal, which is usually accompanied by ignition of above-surface hot and opaque ablative laser plasma via optical breakdown in the vapor-droplet mixture of laser plumes [32], complete optical screening of the corresponding target surface was achieved for laser fluences about two times higher than the threshold for the sub-critical liquid-vapor phase transformation (Figure 5) with the simultaneous saturation of the reflected laser fluence (Figure 6) [63]. Hence, this set of characteristic signatures – surface plasma ignition and reflectance saturation – may also work well for qualitative identification of sub-critical phase transformations driven by short laser pulses.

Surprisingly, at even higher laser fluences – tens of J/cm^2 (multi-GW/cm^2 intensities) for nanosecond laser pulses [46-48,65-66] and hundreds J/cm^2 (sub-PW/cm^2 intensities) for femtosecond laser pulses [67-68] – there is another regime of enhanced, multi-micrometer ablative removal from various transparent and opaque – metallic, semiconductor or dielectric – targets. The other key signatures of this ablation regime are the presence of the near-surface highly emissive ablative plasma [46-48,65-66], strongly delayed (up to several microseconds) and highly directed expulsion of the target material in the form of multi-micron droplets [47,65] without considerable debris deposition [47] around the fabricated craters with high-aspect ratios [47]. Unfortunately, accurate rear-side ultrasonic measurements of internal pressures within the cylindrical molten channels in the plasma-heated targets are challenging in this ablation geometry because of the high reverberation background in employed ultrasonic detectors from the much higher preceding plasma pressure pulse [48]. However, estimates have shown that GPa-level peak pressure and high deposited energy density (ε » 10 kJ/cm^3) magnitudes achieved at rather moderate laser intensities ($I > 10$ GW/cm^2) inside the multi-micron deep channels, filled by the molten material [69], far exceed the corresponding critical values for the materials of interest [70]. Therefore, one may consider this ablation regime as a variety of bulk near-critical phase transformations, induced by a bulk heat source provided by the highly penetrating short-wavelength bremsstrahlung and line radiation of the short-living near-surface ablative plasma [48].

To sum up, this provided brief overview of the most characteristic signatures of laser-induced near-critical liquid-vapor phase transformations at moderate and high laser fluences (intensities) which enables their straightforward classification as surface and bulk ones, which we will follow further in this chapter.

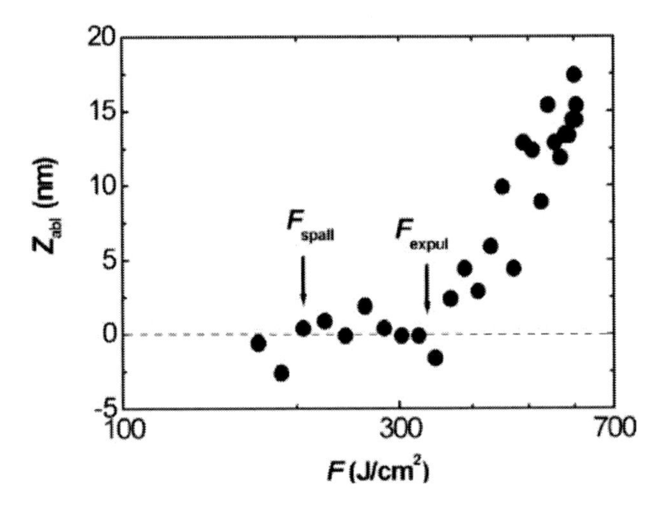

Figure 4. Removal rate Z_{abl} versus laser fluence F for mono-crystalline graphite ablated by 744-nm, 100-fs laser pulses (adapted from Ref. 57). The characteristic ablation regimes are the spallation ($F \geq F_{spall}$) and liquid-vapor expulsion ($F \geq F_{expul}$).

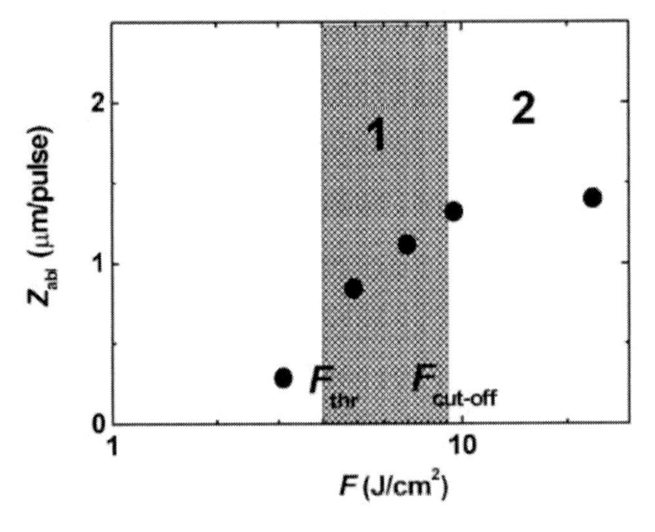

Figure 5. Removal rate Z_{abl} versus laser fluence F for polycrystalline graphite ablated by 248-nm, 25-ns laser pulses (adapted from [63]). The arrows show the thresholds for the sub-critical liquid-vapor phase transformation (1, F_{thr}) and saturation of Z_{abl} (2, $F_{cutt-off}$) as a result of surface screening.

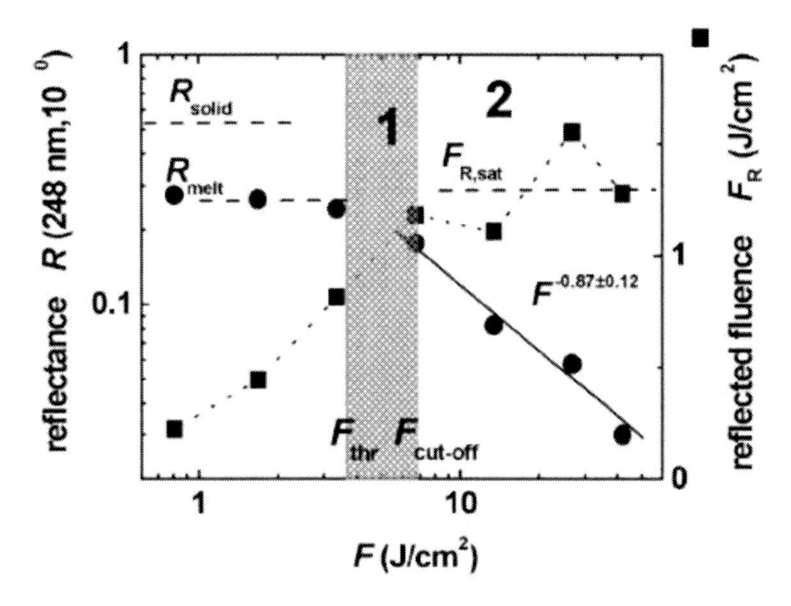

Figure 6. Reflectance R (dark circles) and reflected fluence F_R (dark squares, dotted line) of the pump pulse versus incident laser fluence F for mono-crystalline graphite ablated by 248-nm, 25-ns laser pulses (adapted from [63]). The shaded region shows the reflectance cut-off fluence $F_{cut\text{-}off}$ in relation to the thresholds for the sub-critical liquid-vapor phase transformation (1, F_{thr}) and saturation of F_R (plateau, 2). The dashed horizontal lines show the reflectivity levels of solid (R_{solid}) and molten (R_{melt}) graphite phases, and the saturated reflectivity magnitude $F_{R,sat}$. The solid line represents the linear fit of $R(F)$ curve.

C. Theoretical Concepts Underlying Modeling of Laser-Induced Near-Critical Liquid-Vapor Transformations

In this section we overview main theoretical grounds underlying surface or bulk deposition of laser energy inside condensed matter, which is accompanied by sub-surface superheating of corresponding surface or bulk liquid phase, as a pre-requisite for its sub-critical or supercritical liquid-vapor phase transformations.

I) Modeling of Laser-Induced Surface and Bulk Energy Deposition: Looking for Sub-Surface Superheating?

Historically – since the 1970's until recently – laser-induced near-critical surface and bulk liquid-vapor phase transformations were not explicitly distinguished as such instantaneous (within the heating laser pulse) surface and strongly delayed (on a microsecond timescale) bulk liquid-vapor expulsion could be hardly resolved in a time domain during typical high-power nanosecond and longer laser pulses, and standard experimental characterization techniques were not sensitive enough to measure minute material removal events in the former regime. As a result, bulk phase transformation phenomena with naked-eye visible, very pronounced material removal and plasma sparking were more thoroughly studied and theoretically modeled in their various phenomenological aspects [43-48], even though sometimes rather unexpected suggestions – e.g., the so-called "optical transparency" of near-critical molten metals and other materials [71] (as a brief review of previous studies, see introduction in Ref. [72]), or the multi-microsecond long heat conduction over multi-micron distances within solid targets after their intense nanosecond laser excitation [73] –

were invoked to interpret corresponding very high removal rates of a few kilometers/s (on a order of solid-state longitudinal sound velocity) within common surface removal models [43-45]. Moreover, since a "generalized" near-critical liquid-vapor phase transformation – either surface, or bulk one – can be realized via the abovementioned two main fundamental pathways, which, depending on degree of superheating, are thermo-mechanical spallation at moderate superheating and moderate tensile stress provided by rarefaction wave propagation [24-25,32], or spontaneous explosive expansion/expulsion at positive (compressive) external pressures of a strongly superheated (metastable or unstable) molten phase [24-32] or supercritical fluid [34-41]. These ablation pathways have determined two basic theoretical approaches developed to describe laser-induced near-critical liquid-vapor phase transformations in condensed matter.

In the first approach dealing with the spontaneous explosive expulsion of a strongly superheated molten phase, a preceding sub-surface superheating until near-critical temperatures (sometimes, up to two times higher [46,73]) is suggested as its main driving force [10-14,24-25,45-47,75] (see also the review [76]). According to a solution of a corresponding heat conduction equation in the following form… [77]

$$C_p \frac{\partial T}{\partial t} = \kappa \frac{\partial^2 T}{\partial x^2} + C_p u \frac{\partial T}{\partial x} + \alpha \left[1 - R(I)\right] I e^{-\alpha x}, \qquad (1)$$

…such temperature maximum appears under a recessing surface of an absorbing and thermally conducting melt (the absorption coefficient α, the thermal conductivity κ and diffusivity χ) as a result of its surface vaporization at a vaporization front velocity u and evaporation enthalpy $\Delta_{ev}H \equiv \Delta_{ev}H(l, T, P)$, supported by heat conduction from a surface layer, which is heated by a laser flux with an incident intensity I {an absorbed intensity $[1 - R(I)]I$ for the instantaneous reflectivity of the melt $R(I)$} [77]. The sub-surface superheating may be characterized either by the intensity parameter $B = (\alpha\chi\Delta_{ev}H)/\{[1 - R(I)]I\}$ [74], or by the similar parameter $b = \alpha\chi/u$ [75] (assuming $[1-R(I)]I = u\Delta_{ev}H$ [75]), that shows a balance between the thermal flux $\alpha\kappa\Delta_{ev}H/C_p$ from hot bulk melt to its outer surface resulting from its intense surface vaporization, and the incoming laser flux, $[1 - R(I)]I$. Then, there arise two important cases: 1) $B \ll 1$ {the so-called "bulk" absorption regime, which is relevant for weakly-absorbing, thermally insulating low-boiling organic or some inorganic (e.g., water) liquids or dielectric molten phases of solid insulators and conductors}, when heat conduction from the absorbing bulk is insufficient (slow) to affect the bulk heat source $\alpha(1 - R)I$ and the sub-surface heat accumulation evolves in the corresponding rather smooth sub-surface temperature maximum at the depth $\sim 1/\alpha$ [76], approaching characteristic near-spinodal (near-critical) temperatures and providing intense homogeneous nucleation of vapor bubbles and explosive expansion/expulsion of the superheated liquid phase [56]. In contrast, in the case of $B \approx 1$ corresponding to the "surface" absorption regime, which is relevant for all strongly absorbing molten materials (mostly, metals and semiconductors), the fast heat conduction completely exhausts the near-surface heat source and surface vaporization of the melt pool occurs at the nearly steady-state, monotonously decreasing temperature profile in the bulk [77].

Obviously, since the key "tuning" parameter $B = (\alpha\chi\Delta_{ev}H)/\{[1 - R(I)]I\}$ for the "bulk" and "surface" absorption regimes is intensity-dependent, it appears that for any material with its intrinsic set of characteristics $\{\alpha,\chi,\Delta_{ev}H\}$, the bulk absorption regime with $B \ll 1$ must be achieved at sufficiently high laser intensities. However, as many opaque materials – metals, semiconductors and semimetals – have rather narrow ranges of temperature-dependent variation of these basic optical and thermal characteristics, there are doubts about realization of sub-surface superheating in such opaque materials at the laser wavelengths. Nevertheless, it was recently experimentally demonstrated and theoretically proved that such effect could be readily realized in a number of both opaque and transparent materials (crystalline silicon and amorphous silica) at high (multi-GW/cm^2) nanosecond laser intensities, when presumable bulk heating ($\alpha < 10^3$ cm^{-1}) is provided not directly by laser irradiation, but indirectly by short-wavelength, highly penetrating thermal (bremsstrahlung) and line radiations of near-surface hot ablative plasmas [48].

According to previous numerical calculations [46,74,77], the sub-surface superheating, preceding near-critical liquid-vapor phase transformations, exhibits a difference of the maximum (sub-surface) and surface temperatures in the broad range from a few K up to several thousands K. The corresponding maximum temperatures sometimes considerably – by a few tens of thousands K – exceed typical critical temperatures of the most refractive substances [46,74], being unphysical for the reasons discussed below. Particularly, this discrepancy apparently results from the omission of "internal" vaporization losses in bulk superheated liquids, i.e., of near-critical homogeneous nucleation of both sub-critical and supercritical vapor bubbles [9,18], which, as an important thermal sink, will significantly reduce the sub-surface superheating at the expense of the liquid corrugation by the growing vapor bubbles. The fundamental appearance of this bubble nucleation effect is related to the fact of singularities of the isobaric heat capacity C_p and thermal expansion coefficient β_T in the proximity of the liquid-vapor spinode [9-10,18-20], similarly to exponential rise of the same thermodynamic quantities near other phase transition points – e.g., of C_p in the so-called pre-melting regime [78], when intense generation and nucleation of point defects – precursors of the emerging liquid phase – occurs across the solid phase. Moreover, such homogeneous nucleation of vapor bubbles impacts acoustic and hydrodynamic characteristics of near-critical fluids by 1) rising up longitudinal sound velocities in vapor-droplet mixture with increasing liquid phase and dropping down those in cavitating liquid (the higher vapor content, the lower sound velocity, comparing to its value in non-cavitating liquid) [79,80]; 2) exhausting of transient tensile stresses (e.g., thermoacoustic ones) to support bubble nucleation and growth, while, vice-versa, increasing their compressive components [24,25]. However, to date there were just a few theoretical phenomenological attempts (at the almost absent experimental measurements), accounting for cavitation effects in modeling and calculations of sub-surface superheating, which were focused mostly on volumetric and bubble nucleation (phase transformation) phenomena – ultrasonic pressure generation and its influence on dynamics of a sub-surface gas cavity formed from separate vapor bubbles [24,72]. More thoroughly, the issue of the near-critical rises (singularities) of C_p, β_T and K_T quantities will be discussed below in section II.A.

II) Modeling of Near-Critical Liquid-Vapor Phase Transformations: Equations of State Are Precious!

So far, there were a limited number of theoretical studies modeling near-critical liquid-vapor phase transformations under realistic conditions of surface laser ablation. First, one of such realistic conditions is the presence in a computation cell of a free liquid surface, which exempts vaporization with the resulting recoil vapor pressure, justifying applicability of one common (e.g., van der Waals) equation of state with its two – liquid-vapor and vapor-liquid – spinode curves (Figure 1), while, in the opposite case, in a computation cell with periodic boundary conditions, but without such open liquid surface, the liquid has no external (vapor) pressure applied, which is required to transform its near-critical second-order phase transition curve into the two abovementioned spinodes and the critical point (see the Landau model of II-order phase transitions [81]). The second realistic condition, as was mentioned above, is a proper choice of thermodynamic equation of state (EOS), since long-scale spatial correlations require accounting for higher terms in the virial EOS [15], which are intrinsically absent in various versions of generalized van der Waals-like EOS [82,83] with its underlying binary interactions. Since strength of binary interactions drops down at longer distances, where higher-order interactions become important, [9,15,33] (note that realistic – nanometer or longer – spatial scales of laser-heated regions in liquids or solids are typically comparable to characteristic magnitudes of correlation radius in the near-critical region) the presumable use of binary potentials or simplified van der Waals-like EOS may give a crude picture of corresponding near-critical effects missing their coherent, strongly-correlated counterpart.

With these two conditions in mind, one can separate the relevant previous theoretical works into phenomenological studies [84-86], proceeding via numerical calculations in the framework of a set of hydrodynamic equations, supplied by some EOS, and pure or combined MD simulations [1-6,87]. In the first group, the main results are observations of near-critical liquid-vapor phase separation in terms of mass density perturbations [84-85], and estimates of ejected liquid fraction and ablation threshold fluences [84]. In the second, larger group different MD simulations have given similar results, predicting, basically, two main – spallation and liquid-vapor expulsion – regimes, while at high, near-critical superheating these regimes become rather undistinguishable [87], exhibiting intense cavitation and ejection of larger, or smaller droplets in the former and latter regimes, respectively. This qualitative picture is shown to be consistent with the present experimental findings reported below.

II. Laser-Induced Surface Near-Critical Liquid-Vapor Phase Transformations

A. Surface Liquid-Vapor Phase Separation Driven by Short Laser Pulses

As was mentioned above, subsurface superheating of molten materials until near-spinodal temperatures is an important pre-requisite of their following surface phase transformations induced by short (long picosecond, nanosecond or longer) laser pulses [10-14,19,23-32,42-50,63,74-77,86] (in different words, "surface heating" [88] or "thermal confinement" [87] conditions), while ultrashort (short picosecond or shorter) pulsed laser irradiation enables not only picosecond or sub-nanosecond liquid rupture [1-6,35-36,89] under internal stress confinement ("bulk heating" [90]) conditions [89], but also slow delayed – nanosecond or sub-microsecond – expulsion of vapor-droplet mixture [37-38,90], occurring just under deposition of the proper energy density in the ablated surface layer. Below, we present a consistent physical picture of the most important and characteristic experimental signatures of the near-critical liquid-vapor phase transformation regime under short- and ultrashort-pulse laser irradiation, including a drastic rise of removal rate and recoil pressure of a vapor-droplet mixture of ablated products, with ignition of optical breakdown plasma in laser plumes at almost complete screening the ablated surfaces in the former case of short laser pulses.

I) Ablation of Graphite and Other Opaque Materials

Though various experimental indications of short-pulse laser-induced near-critical liquid-vapor phase transformations on surfaces of opaque materials were previously reported in numerous works [10-14,19,23-32,42-50,63,74-77,86,91], so far the most unambiguous and enlightening insights into this phenomenon were provided only via simultaneous use of a number of diverse complementary – optical, ultrasonic, mass-spectroscopic, surface microscopic characterization and other – experimental techniques. Among a few such comprehensive studies we will distinguish those two [32,63] which were focused on nanosecond laser ablation of graphite in lieu of numerous fundamental and application-oriented goals, e.g., studies of high-temperature, high-pressure physics of refractory materials near their critical thermodynamic points [70], generation of carbon nanostructures (fullerenes [60-61], nanotubes [92]), cleaning and healing of graphite wall surfaces of nuclear and thermonuclear reactors (e.g., those of ITER one).

In the first study, nanosecond laser surface ablation of graphite samples (both mono-crystalline – UPV-1TMO trademark – and polycrystalline ones) was investigated by means of optical reflectance, ultrasonic and surface profiling measurements [63]. First, energies of single specularly reflected 248-nm, 25-ns (FWHM) laser pulses on fresh surface spots of the optical-quality mono-crystalline graphite film were measured at a small incidence angle of 10^0 and variable incident laser energy (fluence) values using two similar thermocouple energy meters VChD-2 (OKB FIAN). At moderate incident laser fluences $F < 5$ J/cm^2 the measured energies of specularly reflected single laser pulses increase sub-linearly versus F showing gradually decreasing spatially and temporally averaged optical reflectance $R(248\text{ nm}, 10^0) = R_{248} < R(\text{solid}, 248\text{ nm}, 10^0) \approx 0.51$ [93] (Figure 6) as an expected thermo-optical UV response of solid graphite [94] or molten carbon [95] (the latter phase is known to show the 400-nm normal incidence reflectance about 0.25 under complete acoustic relaxation conditions [96] at the absence of its other reliable UV reflectance values). This observation is in agreement with the melting threshold of graphite, $F_{\text{melt}} \approx \Delta H(l, T_{\text{melt}}, P)(\chi\tau_{\text{las}})^{1/2}/(1 - R_{248}) \approx$ 1.2 J/cm^2, evaluated accounting for the "thermally thin" skin depth in graphite at 248 nm ($\delta_{248} \approx 10$ nm « $[\chi\tau_{\text{las}}]^{1/2} \approx 18$ nm for $\chi \approx 1.4\times10^{-2}$ cm^2/s [97]) and the enthalpy of graphite in its melting point, $\Delta H(l, T_{\text{melt}}, P) \approx 2\times10^5$ J/mole [70,98]. Surprisingly, at higher fluences $F > 5$ J/cm^2 the reflected laser energies saturate in a threshold-like manner at the cut-off fluence $F_{\text{cut-off}} \approx 7$ J/cm^2, indicating drastically enhanced, *complete* scattering and/or absorption on the molten graphite surface for the trailing part of the laser pulse which begins at the instant during the pulse when the accumulated fluence exceeds $F_{\text{cut-off}}$. The saturated laser fluence $F_{\text{R,sat}} \approx 1.5$ J/cm^2 corresponds to the characteristic reciprocal dependence of graphite reflectance $R_{248} \propto 1/F$ (Figure 6), observed earlier on graphite surfaces at high-power 1064-nm laser irradiation and related to formation of ablative carbon plasma [99].

Second, rear-side ultrasonic measurements were carried out in a single-shot mode at variable fluence F in the acoustic near-field using a piezoelectric transducer SHAPR-13 ($LiNbO_3$, bandwidth > 100 MHz, sensitivity of 1mV/atm) manufactured by UC VINFIN [23], with the supported mono-crystalline graphite film sliding on the 5-mm thick protective front window of the transducer via a thin lubricating vacuum grease layer. The 1.5-mm wide active front electrode of the transducer was centered relative to the laser focal point on the front surface of the sample. The voltage transients from the transducer were recorded using a 50-Ω input of a digital storage oscilloscope Tektronix TDS-2024, which was triggered via another 50-Ω input by an electric pulse from a fast photodiode DET-210 (Thorlabs) fed by a weak split laser beam. In these measurements, the waveforms of the acquired acoustic transients were found to change suddenly near a similar threshold fluence $F_{thr} \approx 4$ J/cm^2, showing a transition from the almost symmetrical bipolar pulses (compressive and rarefaction amplitudes $P_{comp} \geq P_{rare}$) of the presumably thermoelastic character [25,28,32,100], including a small compressive unipolar contribution from recoil pressure P_{rec} provided by surface vaporization, to the purely unipolar compressive pulses ($P_{comp} \gg P_{rare} \approx 0$) of the ablative character [32,63] (Figure 7). Moreover, the recoil pressure P_{rec} of ablation products (Figure 8), taken equal to P_{comp} for $F \geq F_{thr}$ and $(P_{comp} - P_{rare})/2$ for $F < F_{thr}$ [28,32], rises sharply in figure 8 for $F \geq F_{thr}$ together with the P_{comp} magnitude at the almost disappearing P_{rare}. This sharp increase of P_{rec} coincides with the abovementioned saturation of single-shot reflected laser fluence at $F_{cut\text{-}off}$ in figure 6 and thus indicates more efficient pressure generation in the superheated interfacial molten carbon layer representing a possible change of ablation mechanism, while another potential explanation – stronger pressure generation in the layer due to its enhanced absorbance (see the decreasing reflectance in figure 6) appears to be much less probable due to the rather moderate absorbance variation.

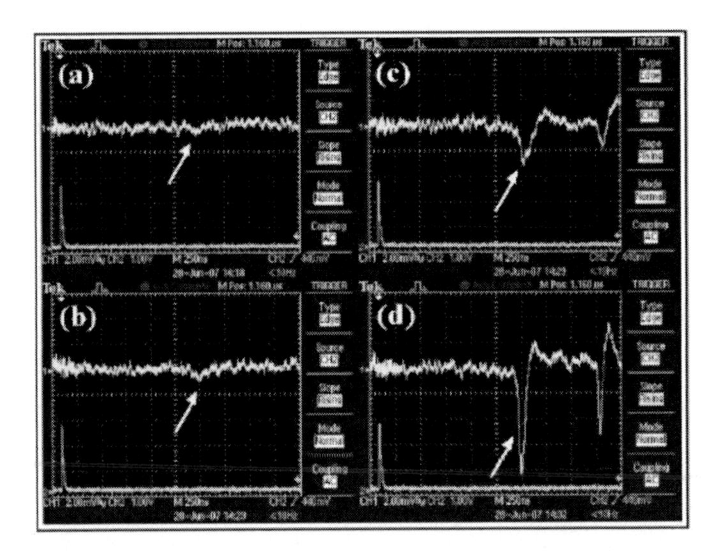

Figure 7. Ultrasonic transients recorded at different laser fluences during ablation of mono-crystalline graphite by 248-nm, 25-ns laser pulses (adapted from Ref. 63): 0.4 (a), 1.5 (b), 4 (c) and 17 (d) J/cm^2. The arrows show the main compressive (ablative recoil pressure) pulses.

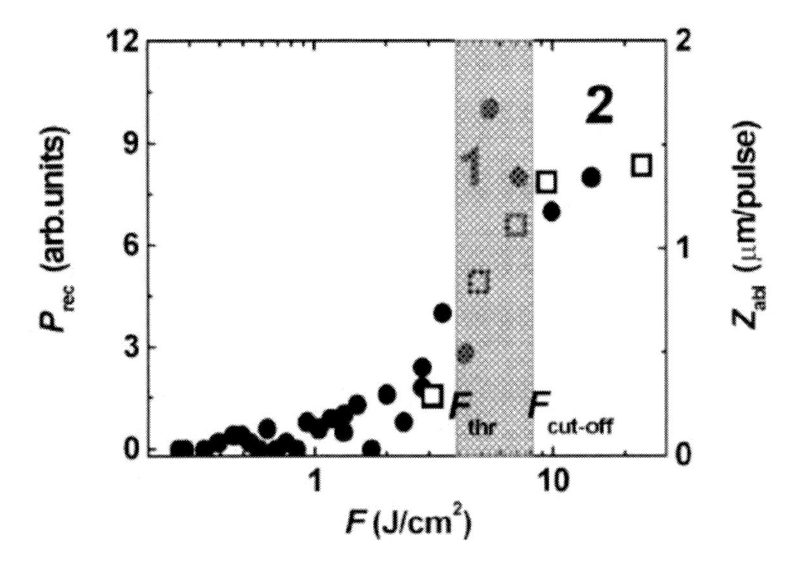

Figure 8. Ultrasonic recoil pressure amplitude P_{rec} versus incident laser fluence for mono-crystalline graphite ablated by 248-nm, 25-ns laser pulses (adapted from Ref. 63); the removal rate dependence $Z_{abl}(F)$ is shown for comparison. The shaded region shows the thresholds for the sub-critical liquid-vapor phase transformation (left, 1, F_{thr}) and saturation of ablation in terms of P_{rec} and Z_{abl} (right, 2, $F_{cutt-off}$) as a result of surface screening.

Finally, a number of millimeter-deep multi-shot trenches were fabricated at different laser fluences on a surface of polycrystalline graphite plate placed onto a PC-driven motorized three-axis stage, with the corresponding maximum trench depths optically measured at the plate edge under an optical microscope. These multi-shot removal measurements enable to resolve the ambiguity of the compressive (recoil) pressure rise in figure 8 demonstrating for $F \geq F_{thr}$ a strong increase of average ablated depth per pulse Z_{abl} and its subsequent saturation at a level Z_{plat} (Figure 5), thus closely correlating with the $P_{rec}(F)$ trend in figure 8. Similar crater depth or ablated mass dependences on nanosecond (or even microsecond) laser fluences – with sharp rise of removal rate and its following saturation – were earlier observed at different laser wavelengths for different materials and were assigned to their surface spinodal decomposition [31-32] or phase explosion [46], respectively. Moreover, previous 532-nm, 25-ns laser ablation studies of graphite, performed by means of the same ultrasonic technique to measure recoil pressure amplitudes and average crater depths per pulse (by recording a variation of an ultrasonic signal arrival time on the sample rear side in each laser pulse during multi-shot laser irradiation increasing the crater depth, see Refs. 31,32 for details), demonstrated the same close correlation of P_{rec} and Z_{abl} curves versus incident laser fluence [32]. Such correlation between the ablated depth and ablation recoil pressure magnitudes measured, respectively, in multi- and single-shot modes, demonstrates the absence of potential significant memory (reduced mass removal due to a limited crater aperture, variation of the target or crater absorbance due to its surface modification) effects during the fabrication of the deep trenches on the graphite surface. Second, this close correlation both at the rise and saturation stages represents the concerted character of mass removal and pressure generation during laser ablation for $F \geq F_{thr}$, as compared, e.g., to delayed pressure generation in an absorbing laser plume above the target surface via a sudden optical breakdown [99] or inside the superheated surface molten layer via a sudden phase explosion [10-14,19,46,56,72-

73]. In contrast, this correlation indicates that both the removal rate and recoil pressure instantaneously track all main details of the laser energy deposition during the incident laser pulse, i.e., the mass removal and pressure generation occur continuously during the pulse, rather than as a single explosive event. Hence, this correlation, together with the complete absence of the rarefaction phase in the acquired ultrasonic transients at $F \geq F_{thr}$, may indicate continuous near-critical liquid-vapor transformation in the superheated molten surface layer of graphite, possibly driven by the rarefaction pulse of the propagating thermoacoustic wave [24-25,32,101].

Very important qualitative information on the phase explosion threshold fluence on graphite surfaces was provided by electric probe and time-of-flight mass spectroscopic studies in vacuum, using a 532-nm, 25-ns laser source (see Refs. 27,32 for more details). These studies have revealed a drastic change of gas-phase composition of charged ablation products – both positively and negatively charged species – near the threshold of surface phase explosion, resulting from a threshold-like appearance of a continuous high-mass fraction including nano-cluster ions of the fullerene family (Figure 9) and larger clusters sized up to a few millions of atoms per cluster (Figure 10) [27,32]. Relating the sharp appearance of high-mass species in the laser plume to near-spinodal expulsion of vapor-droplet mixture, one can find reasonable agreement in the phase explosion threshold fluences between these vacuum studies of ablated products and the abovementioned ultrasonic and mass removal measurements at ambient atmosphere conditions (note that in a vacuum laser ablation starts at different – about twice as high – thresholds, possibly, because of re-condensation and counter-pressure effects in the Knudsen interfacial layer [44,56,64]). Moreover, an enhanced yield of multiply-charged positive atomic carbon ions from the emerging visible luminous hot and dense plasma plume was detected at laser fluences [32], just slightly exceeding the appearance threshold fluence for the high-mass fraction, being well consistent with the previous observations of complete screening of the ablated graphite surface by ablation products in figure 2 and the following entire deposition of the rest laser pulse energy (fluence) inside the laser plume.

II) Optical Visualization of Bubble Dynamics and Phase Separation in Weakly Absorbing Liquids

A separate highly informative set of experimental studies of surface phase explosions induced by short-pulse laser radiation is associated with optical and ultrasonic imaging of pre- and explosion boiling phenomena, their spatial and temporal scales and dynamics in weakly-absorbing liquids [23-26,28-30,49-50,72,102]. Among these studies, we selected those focused on liquids exhibiting true homogeneous absorption of nanosecond IR laser radiation (e.g., water [23,28,49-50,72,102] and glycerol [28-30]), since colored liquids – inks containing carbon dust particles [26], K_2CrO_4 and other inorganic salt solutions [24-25] – demonstrate presumably heterogeneous (or impurity-enhanced) boiling occurring on the absorbing molecular or particulate centers, i.e., "heterogeneous" absorption.

Water, as the most common and widespread liquid, has a long pre-history of its ablation studies starting in the 60's and peaking in the late 70's – early 80's, which were performed, mostly, by means of ultrasonic (photoacoustic) spectroscopy and heating TEA CO_2 lasers [23,28,49-50,72]. Significant progress has been achieved in the previous studies in terms of comprehensive description of basic ultrasonic (thermoacoustic, surface vaporization, phase explosion/explosive boiling and optical breakdown) generation regimes, and ultrasonic measurements of their thresholds, as well as diverse optical, thermal and acoustic parameters of water [49-50,103]. However, microscopic bubble dynamics characterized by bubble dimensions, oscillation frequencies, damping times, amplitudes of pressures and intensities/spectra of radiation emitted during bubble collapse phases [50], during water boiling in the phase explosion regime were not accurately acquired at that time. Just recently, with the big progress in fast data acquisition electronics (multi-GHz oscilloscopes, pre-amplifiers) and technical design of ultrasonic transducers, broadband (up to sub-GHz frequencies) ultrasonic measurements have become usable to study microscopic details of vapor bubble dynamics in homogeneously superheated liquids [23,102].

In the recent series of contact far field ultrasonic measurements, both short-term (nanosecond) near-spinodal homogeneous nucleation of multiple nanoscale steam bubbles of sub- and near-critical dimensions in a μm-thick layer on a free water surface superheated by a focused radiation of a 10.6-μm TEA CO_2 laser above its explosive boiling threshold at heating rates $\sim 10^{10}$ K/s, and their subsequent long-term (multi-microsecond) collective interactions (coalescence and percolation) were for the first time revealed at variable incident laser fluences F = 0.8-11 J/cm^2 using an experimental setup described elsewhere as a fast, state-of-the art ultrasonic transducer SHAPR-13 ($LiNbO_3$ piezoelement, flat response in the 1-100 MHz range, manufactured by UC VINFIN) [23,102]. The corresponding weak random oscillations of short-living nanometer-sized near-critical bubbles and stimulated synchronous oscillations of μm-sized supercritical steam bubbles and their coalescence products were recorded in the form of a broad weak high-frequency spectral band and a number of strong low-frequency spectral lines described in details below.

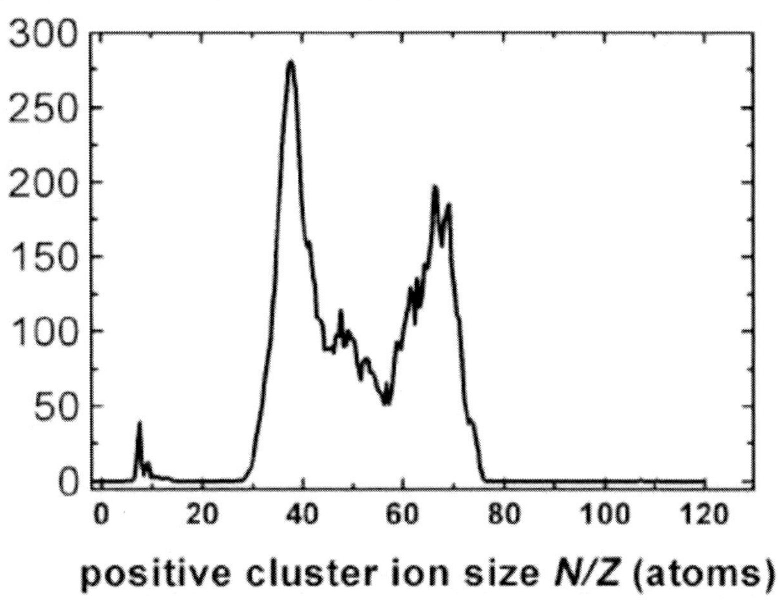

Figure 9. TOF mass spectrum of positive carbon cluster ions yielding from polycrystalline graphite surface ablated by 532-nm, 25-ns laser pulses at laser intensity 0.4 GW/cm^2 slightly exceeding the threshold value of 0.3 GW/cm^2 for the near-critical phase transformation (adapted from Ref. 32).

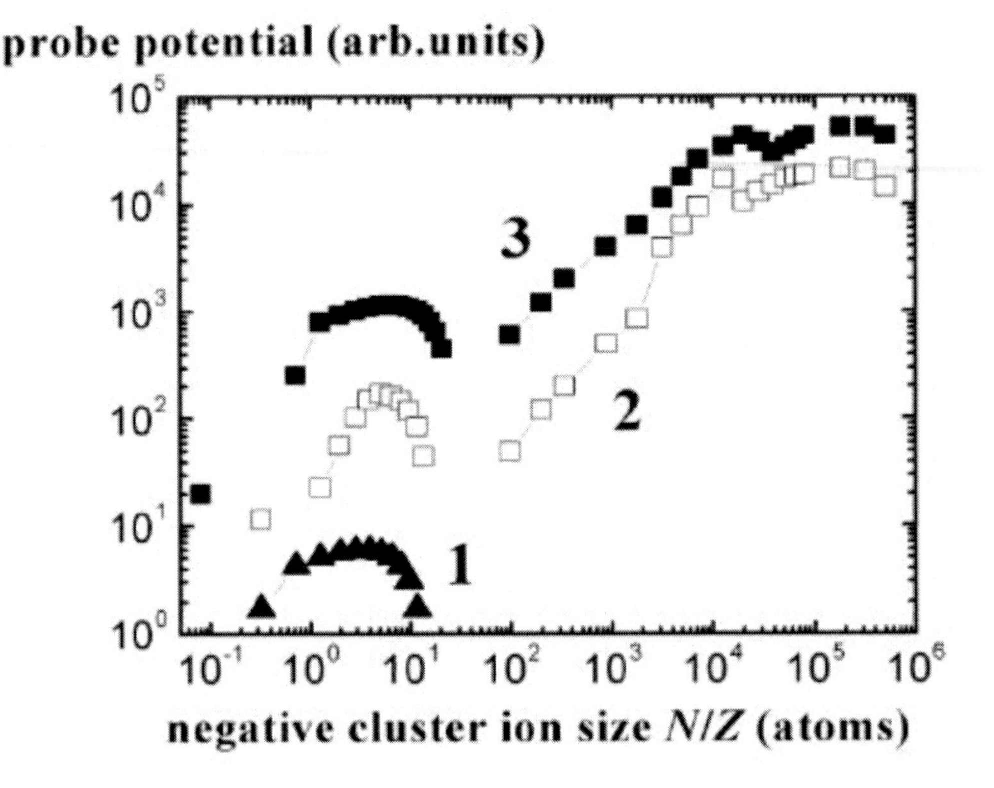

Figure 10. Electric probe signal of negative high-mass cluster ions (charged carbon nanodroplets) yielding from polycrystalline graphite surface ablated by 532-nm, 25-ns laser pulses at laser intensities 0.21 (1), 0.33 (2) and 1.2 (3) GW/cm^2 (adapted from Ref. 27,32). The near-critical liquid-vapor transformation threshold is 0.3 GW/cm^2.

In particular, these measurements showed characteristic waveforms $P(t)$ consisting typically of a main pulse (t = 0-0.15 μs) and an oscillatory tail at $t > 0.15$ μs (figure 11). The main pulse increased slowly for sub-threshold fluences $F \leq F_1 = 1.7\pm0.3$ J/cm^2, but then rose rapidly at fluences exceeding this liquid-vapor phase transformation threshold value (Figure 12) (see similar thresholds for water and for similar TEA CO_2-laser temporal pulse shapes in [50]). Moreover, near the threshold the main acoustic pulse transformed from tripolar to bipolar (see transients 1-3 in figure 11), where both the tripolar and bipolar waveforms were first time derivatives of the actual bipolar and unipolar waveforms generated during the preceding 70-ns laser spike (their FWHM equal that of the laser spike) via the thermoacoustic or explosive liquid-vapor phase transformation mechanisms [32,100], respectively, and slightly perturbed by surface vaporization [50] and cavitation in the superheated water. The differential effect resulted from diffraction of the acoustic transients in the far field [100] where data acquisition was performed, and explained the slower increase of the compressive pressure amplitude P_{comp} (positive phases in figure 11) at higher $F \approx$ 4-5 J/cm^2 after its initial rapid rise at $F \geq F_1$ by large increases of the lateral size for the explosively boiling region in the superheated interfacial water layer versus F. Visible formation of mm-sized surface bubble and expulsion of a water jet accompanying laser ablation for $F > F_1$ strongly supported the explosive character of the threshold F_1.

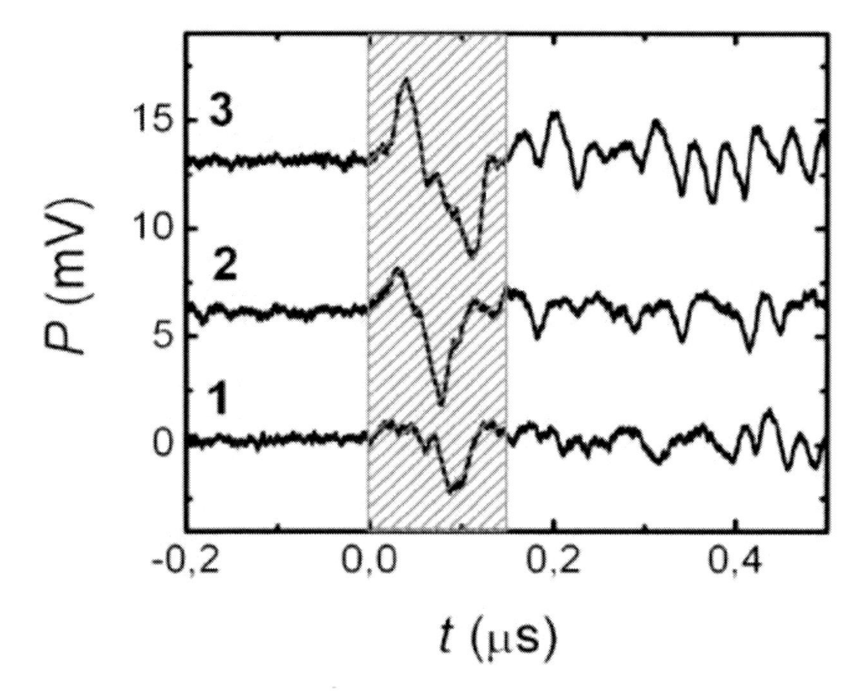

Figure 11. Ultrasonic waveforms recorded in water at $F \approx 1.4$ J/cm^2 < F_1 (1), 1.7 J/cm^2 $\approx F_1$ (2) and 2.4 J/cm^2 > F_1 (3) (adapted from Ref. 23). The patterned region shows the transformation of the main acoustic signal near the explosive liquid-vapor transformation threshold.

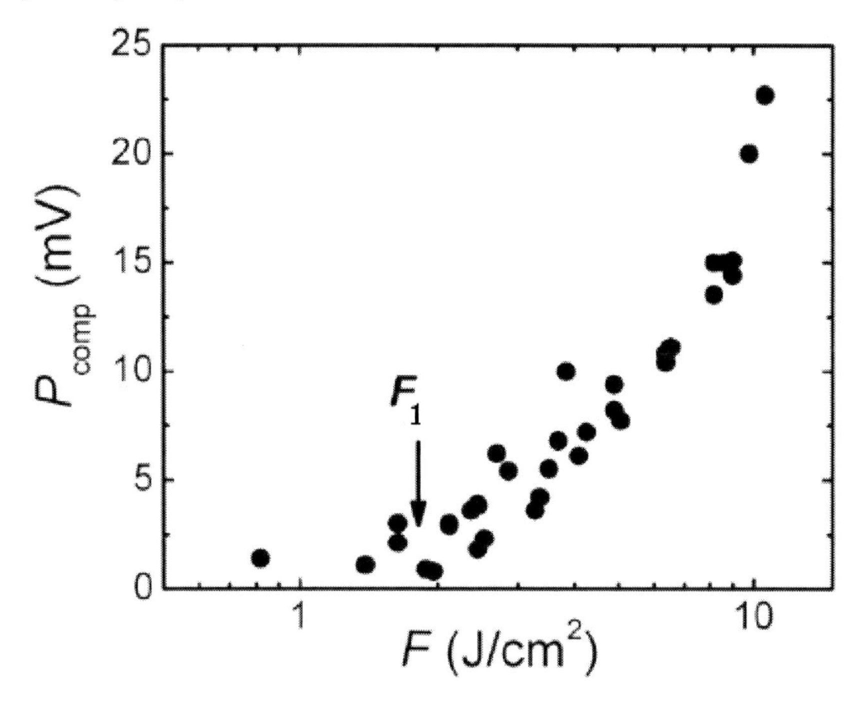

Figure 12. Compressive pressure P_{comp} in water versus laser fluence F. The arrow shows the explosive liquid-vapor transformation threshold F_1 in water. Adapted from Ref. 23.

This conclusion was also confirmed by a simple energy balance analysis. Near-critical liquid-vapor phase transformation of water was expected to occur during the laser pulse near its liquid-vapor spinode at positive pressures P (P_0 = 1atm < P < P_{cr}) and temperatures $T \approx (0.9\text{-}1)T_{cr} \approx (5.9\text{-}6.5)\times10^2$ K [18,70] (the critical pressure and temperature of water are $P_{cr} \approx$ 22.4 MPa and $T_{cr} \approx$ 647 K [70], respectively) in the "thermal confinement" regime [87], when the threshold volume energy density $\varepsilon_{thr} \approx (1 - R)\times F_2/\delta \approx 1\times10^3$ J/cm^3 [18,103] was supplied by the incident laser fluence $F_2 \approx$ 0.9 J/cm^2 for $R \leq 0.01$ [93] and $\delta \approx$ 9 μm [103] (the reflectivity and penetration depth for a flat surface of bulk water at normal incidence and 10.6-μm laser wavelength). For the increasing total laser fluence F, the explosive transformation onset in water at the instantaneous fluence $F(t) \approx F_2$ was achieved at an earlier time t during the CO_2-laser heating pulse and was indeed resolved using the nanosecond acoustic transducer. For example, for the energy content ratio $\gamma \approx$ 50:50 of the CO_2-laser spike and its tail there were two reference total fluence thresholds $F_2 \approx$ 0.9 J/cm^2 and $F_1 \approx$ 1.8 J/cm^2 ($F_1 \approx F_2/\gamma \approx 2F_2$), which should provide such explosive liquid-vapor phase transformation of water at the end of the tail and right after the spike, respectively. Therefore, at $F_2 \leq F \leq F_1$ one can see the main bipolar thermoacoustic pulse accompanied by oscillatory explosive transformation (cavitation) signal during the laser pulse tail, while at $F \geq F_1$ the explosive transformation effect and, to a minor extent, surface vaporization builds up the main unipolar pulse and the accompanying cavitation signal [50]. Note that, according to such energy balance analysis and the experimental data of other studies, surface vaporization of water started at lower $F \approx$ 0.3 J/cm^2 [50], corresponding to the normal boiling temperature $T_{boil} \approx$ 373 K at $P = P_0$ [70].

In accordance with the reference thresholds F_1 and F_2, at $F \approx$ 0.8 J/cm$^2 \leq F_2$ only the main symmetric tripolar thermoacoustic pulse was recorded followed for $t >$ 0.2 μs by a few low-amplitude oscillations at a background level (transient 1 in figure 13). In contrast, at $F \approx$ 1.4 J/cm^2 > F_2 accompanying the main asymmetric tripolar pulse is a pronounced oscillatory tail which sets up after the laser spike at $t >$ 0.4 μs (transient 2 in figure 13) and represents characteristic cavitation dynamics of steam bubbles with resonant frequencies $f \approx$ 10-40 MHz (Figure 14b), which μs-scale dynamics has been studied in our previous work [102]. The amplitude FFT spectra in figure 14 were obtained for the first 0.7-μs (t = 0.2-0.9 μs) slices of this and other acoustic transients for $F > F_2$. At higher $F \approx$ 2.1 J/cm$^2 \geq F_1$ an oscillatory tail starts at the end of the laser spike at $t \geq$ 0.1 μs (transient 3 in figure 13). Similar pressure-tension cycles in the range of 15-30 MHz were earlier detected in water at similar CO_2-laser fluences (typical f = 16, 20 or 25-30 MHz [50,104-105]), if point acoustic detectors were used to avoid the destructive interference of acoustic waves emitted by multiple bubbles at different positions in the near-surface laser-superheated water layer, but were interpreted as oscillations of bubbles – bubble growth ($P(t) > 0$), shrinkage and collapse ($P(t) < 0$), and rebound (next positive stage of $P(t)$) [9,15,18] – produced around micron-sized solid dust species (e.g., soot) suspended in water of industrial districts. However, since in this work no visible oscillations were observed in the acoustic transients on sub-μs or μs timescales at $F < F_2$ (Figure 13), such multi-MHz bubbles can only be related to sub-μs explosive homogeneous nucleation in superheated interfacial water layers at $F > F_2$. Surprisingly, the spectral amplitudes of bubble modes in figure 14 do not change significantly at higher $F <$ 6 J/cm^2 in spite of increasing axial and radial dimensions of the superheated surface water layer, while in agreement with the abovementioned F-independent onset of explosive liquid-vapor phase transformation at $F(t) \approx F_1$ and the virtual diffraction-limited increase of P_{comp} in figure 12 at $F = 1.7\text{-}6$ J/cm^2 at the slowly increasing actual P_{comp} amplitude. Altogether these facts demonstrate that under our experimental conditions the explosive liquid-vapor transformation is driven by thermodynamic (e.g., degree of superheating) rather than kinetic (e.g., heating rate) factors and exhibits a *single* threshold F_{thr} for different laser fluences $F > F_1$, indicating such transformation at some superheating limit which could be the liquid-vapor spinode (see below). Also, the main oscillation modes in figure 14 show their damping during the laser tail with characteristic times of about 1 μs consistent with our previous measurements [102].

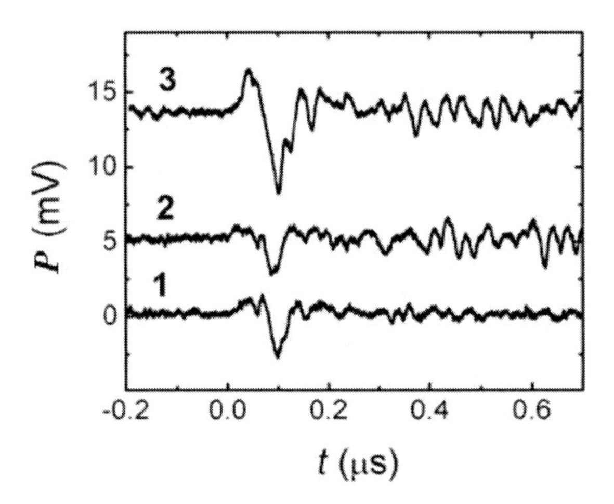

Figure 13. Oscillatory tails ($t > 0.1$ μs) of acoustic waveforms at $F \approx 0.8$ J/cm^2 $\leq F_2$ (1), $F_2 \leq 1.4 \leq F_1$ (2) and $2.1 \geq F_1$ (3) J/cm^2. Adapted from Ref. 23.

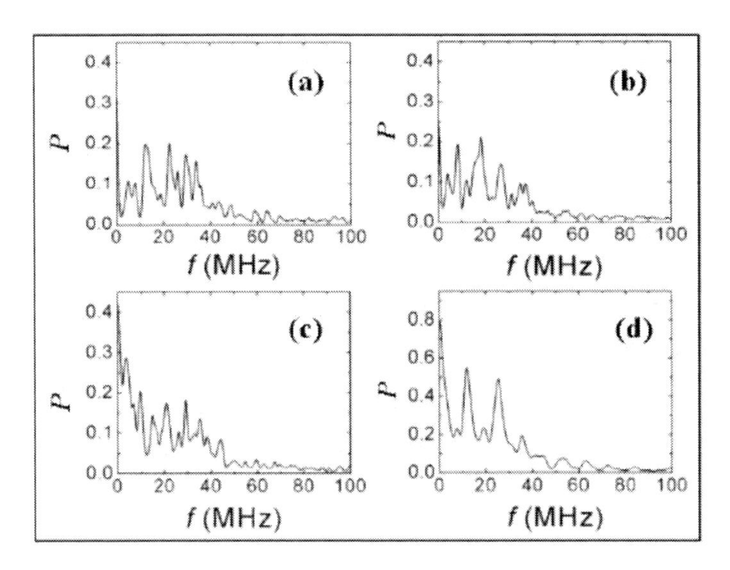

Figure 14. FFT spectra $P(f)$ taken in the time interval $t = 0.2\text{-}0.9$ μs at different fluences of 1.4 (a), 1.7 (b), 3.5 (c) and 6.5 (d) J/cm^2. Adapted from Ref. 23.

More complex, but noisy acoustic transients exhibiting higher-resolution FFT spectra more rich in higher f were obtained at $F \approx 2.1$ J/cm^2 (Figure 15) using thinner water layers ($H \approx$ 1-1.5 mm), since, in this case, the diffraction effect for the multi-MHz acoustic waves is considerably weaker. In particular, one finds in figure 15 that the bubble oscillation mode at $f \approx 32$ MHz has much shorter lifetime ($\approx$ 150 ns according to its FWHM parameter $\Gamma \approx 6$ MHz) than the other modes in figure 14, while another mode at $f \approx 63$ MHz exhibits slightly longer lifetime $\approx$ 250 ns ($\Gamma \approx 4$ MHz in figure 15). Potentially, the latter mode could be the second harmonic of the former one with its lifetime longer because of higher thermal agitation. Supercritical steam bubbles representing both of these modes may be the precursors of the larger bubbles coalesced on a ns timescale [106] from the smaller ones and oscillating at lower $f < 30$ MHz as shown in figures 14,15. Furthermore, FFT spectra in figure 15 show also a weak diffuse band at $f \approx$ 70-250 MHz with the central frequency $f_C \approx 1.6\times10^2$ MHz and FWHM $\Gamma \approx 1\times10^2$ MHz, which has the average amplitude at the level of RF noise from the CO_2-laser gas discharge.

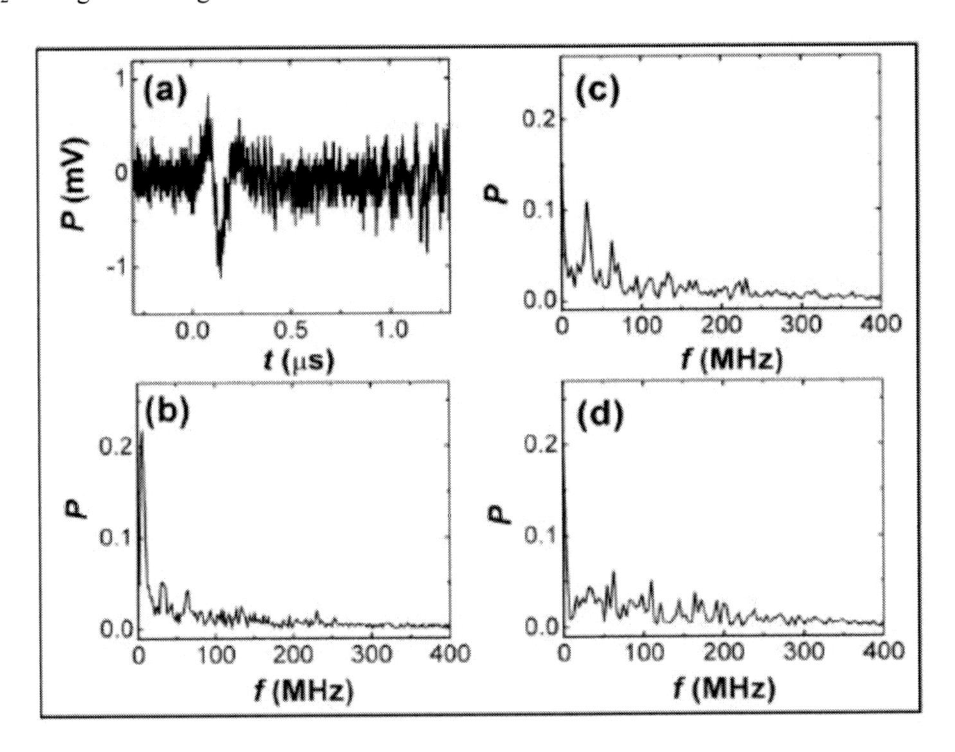

Figure 15. $P(t)$ transient at 2.1 J/cm^2 (a) and FFT spectra $P(f)$ taken in the intervals t = 0-0.3 μs (b), 0.3-0.5 μs (c) and 0.5-0.7 μs (d) showing transient abundances of different steam bubbles. Adapted from Ref. 23.

In our opinion, the observed band may represent a multitude of transient near-critical steam bubbles (rather than non-oscillating periodic spinodal structures) which are precursors of the 32- and 63-MHz supercritical bubbles and grow very slowly and randomly for $(dr/dt) = 0$ and $(d^2r/dt^2) = 0$ at $r = r_{cr}$ [9,18-19] driven, presumably, by a random force $\Psi(T,P,r,t)$ resulting from thermal fluctuations. Indeed, the Gibbs free energy $\Delta G(T,P,r)$ for homogeneous nucleation of bubbles has a maximum at $r \approx r_{cr}$ and the thermodynamic driving force for bubble growth $(\partial\Delta G/\partial r)_{T,P,r_{cr}} \approx 0$, defining the critical bubble radius as r_{cr} =

$2\sigma(P,T)/[P_S(T)-P_0]$ [9,18-19], where $P_S(T)$ is the saturated vapor pressure at T. Such critical and near-critical bubbles have a chance to make a few random oscillations near $r \approx r_{cr}$ under the influence of $\Psi(T,P,r,t)$ during their slow evolution in size and frequency domains in the near-critical region, emitting a number of short and low-amplitude acoustic wave packets with random frequencies and phases at strongly broadened linewidths, as seen in figure 15 from the close correspondence of their f_C and Γ. The number of near-critical bubbles per unit volume and periods of their near-critical oscillations (random acts of growth and shrinkage) are strong (exponential) functions of T [9,18-19]. Therefore, one can expect to observe such random oscillations of tiny near-critical bubbles in FFT spectra in the form of a very weak and broad band with amplitudes and central frequencies rapidly changing versus F, which seems to be the case in this work. Since critical bubbles are "bottleneck species" in the homogeneous nucleation kinetics exhibiting the minimal possible growth rates [9,18-19], their oscillation frequencies $f \sim f_C$ characterize a bubble nucleation frequency $f_{nucl}(P,T,r_{cr})$, the parameter used to define at a given P,T a steady-state homogeneous nucleation rate…

$$J(P,T) = N_0(P,T) \times B \times \exp[-\Delta G(P,T,r_{cr})/kT], \qquad (2)$$

…where $N_0(P,T) \sim 10^{22}$ cm^{-3} is the density of molecules in a liquid and $B \sim 10^{10\text{-}12}$ Hz is the kinetic pre-factor [9,18-19]. The product $B \times \exp[-\Delta G(P,T,r_{cr})/kT] = f_{nucl}(P,T,r_{cr})$ and, as a result, for the explosive liquid-vapor transformation conditions assumed here – $T^* \approx 0.9T_{cr} \approx 5.9\times10^2$ K, $P_S(T^*) \approx 8.6\times10^6$ Pa [70] and $f_{nucl}(P_S(T^*),T^*,r_{cr}) \approx f_C \approx 1.6\times10^2$ MHz – one finds $\Delta G(P_S(T^*),T^*,r_{cr}) \approx (4\text{-}8)\times10^{-20}$ J. Simultaneously, one can write for $\Delta G(P,T,r_{cr}) = 1/3 \times (4\pi r_{cr}^2) \times \sigma(P,T) = (2\pi r_{cr}^3)/3 \times [P_S(T) - P_0]$ [18] which gives for the known $\Delta G(P_S(T^*),T^*,r_{cr})$ and $[P_S(T^*)-P_0]$ the estimate $r_{cr} \approx 1.5\text{-}2$ nm consistent with $r_{cr} \sim 1\text{-}3$ nm in Ref. [5,9]. Moreover, for the known r_{cr} and $\Delta G(P_S(T^*),T^*,r_{cr})$ one can estimate $\sigma(P_S(T^*),T^*) = 3 \times \Delta G(P_S(T^*),T^*,r_{cr})/(4\pi r_{cr}^2) \approx (4\text{-}5)\times10^{-3}$ N/m, which corresponds, according to data for $\sigma(P_S,T_S)$ on the water binode [70], to the near-spinodal temperature $T^* \approx 6.2\times10^2$ K $\approx 0.95T_{cr}$ in good agreement with the energy balance analysis above and typical spinodal temperatures $T_{spin}(P > 0) > 0.92T_{cr}$ for real liquids [18]. The extracted bubble nucleation parameters – their dimensions, nucleation frequency and work of formation, as well as temperature and surface tension of the superheated water, and energy balance analysis strongly support the near-critical phase transformation in the laser-superheated water near its liquid-vapor spinode in the thermal confinement regime.

When the abovementioned critical steam bubbles grow to supercritical dimensions blown up by $P_S(0.95T_{cr})$, their maximum diameter D_{max} for maximal $f \approx 32$ MHz can approach $D_{max}(32\text{ MHz}) \approx 6$ μm $\leq \delta \approx 9$ μm [103] calculated using the Rayleigh's formula [19] written as $D_{max} \approx 1.1 \times [P_S(0.9T_{cr}) \times 1/\rho(0.9T_{cr})]^{1/2}/f$ and values of the water density $\rho(6.2\times10^2\text{ K}) \approx 0.5\times10^3$ kg/m^3 [18] and $P_S(6.2\times10^2\text{ K}) \approx 1.5\times10^7$ Pa [70]. This result shows single-layer packaging of the supercritical bubbles in the superheated interfacial water layers and explains why acoustic waves emitted by the bubbles can be readily recorded in the normal direction in the acoustic far field without significant attenuation by other bubbles or loss of coherence of their oscillations in the recorded acoustic signals. As a result, the subsequent nanosecond coalescence of these micron-sized bubbles presumably occurs along the surface of the superheated water layers. Such coalescence is clearly observable as a temporal evolution of bubble clouds in the explosively boiling water in the time frame t = 0-12 μs (Figure 16), where at $F \geq F_{thr}$ there is a pronounced oscillatory tail beginning right after the laser spike (t > 0.2 μs) (Figure 16) not perturbed by the acoustic echoes in the water layer appearing after time instants $t \approx$ 12, 24 and 36 μs, respectively (inset *b* in figure 16). Particularly, low-amplitude pressure oscillations occur at different frequencies f decreasing as a function of time: $f \approx$ 15-40 MHz at t = 0.2-1.4 μs, $f \approx$ 5-15 MHz at t = 1.5-3 μs, while for t > 3 μs there are very pronounced 1-MHz and 2-MHz oscillations.

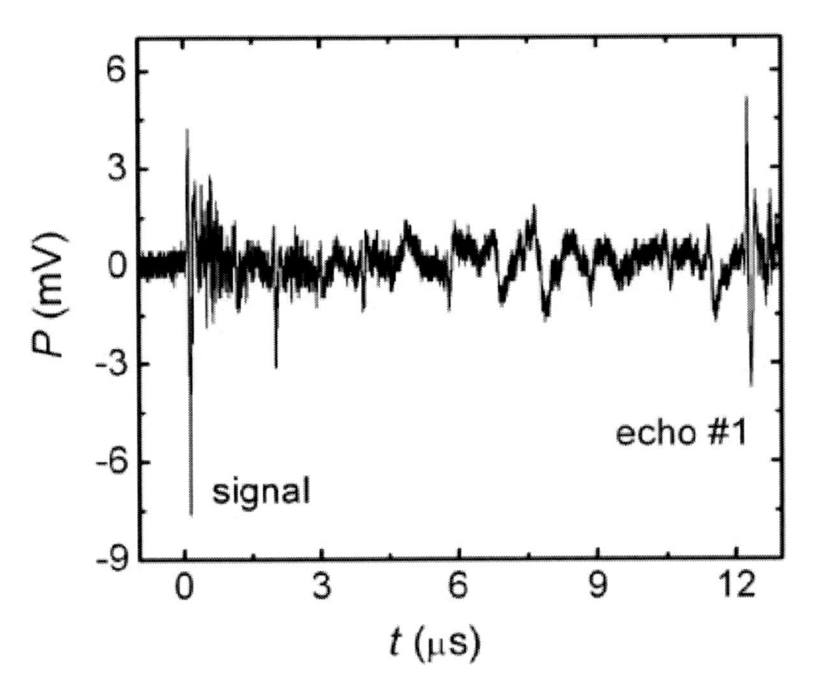

Figure 16. Long-term oscillatory ultrasonic transient $P(t)$ recorded in water at laser fluences about 3.5 J/cm^2, including the first echo of the main signal. Adapted from Ref. 102.

This dynamic of various frequency components is more evident in amplitude FFT spectra of the waveform (Figure 17) taken in the interval t = 0-12 μs using a 1-μs time window. These spectra show rapid population of the lower-frequency (1-5 MHz) oscillation modes at the expense of the higher-frequency (20-40 MHz) modes. However, when plotted vs. time (Figure 18a,b), the amplitudes $P(f)$ of *all* these modes demonstrate a damping effect during t = 2-4 μs with lifetimes $\tau(f)$ about 1 μs ($\tau(f)$ values extracted from half-widths of corresponding spectral modes in figure 17 are 0.7-0.9 μs) which is consistent with complete permanent decrease of the total acoustic power $\Pi(t) \sim |P(t)|^2$ for $t \leq$ 1.5-2 μs (see inset in figure 18). Importantly, at the end of this process the amplitude P(1 MHz) of the characteristic oscillation at the lowest frequency $f \approx$ 1MHz (Figure 17) exhibits a gradual increase until $t \approx$ 8 μs (Figure 18), when the final population decay of the mode occurs at a rate of 0.7 MHz (τ(1 MHz) $\approx$ 1.5 μs). Other modes with intermediate frequencies $f \approx$ 18.4 and 5.9 MHz shown in figure 18 are similar, but with smaller peaks at earlier $t \approx$ 3 and 4 μs, respectively.

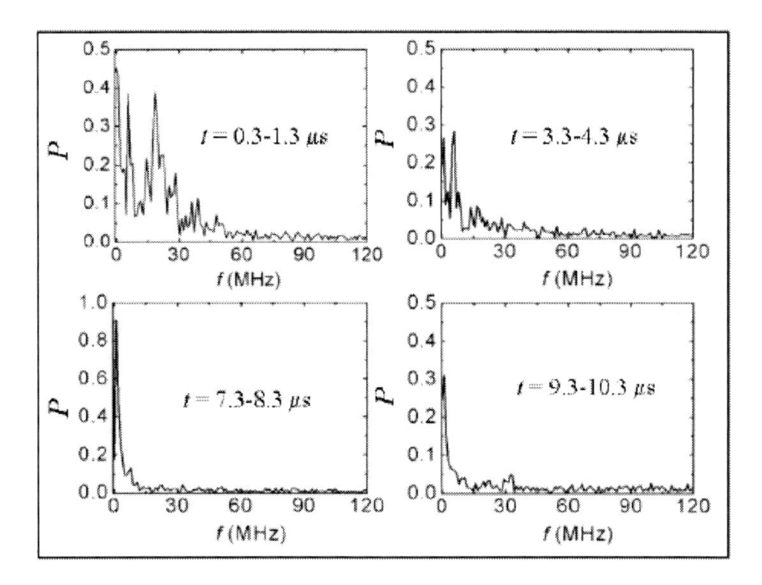

Figure 17. FFT spectra $P(f)$ for water at laser fluence about 3.5 J/cm^2, taken using a 1-μs time window in the time interval t = 0.3-10.3 μs. Adapted from Ref. 102.

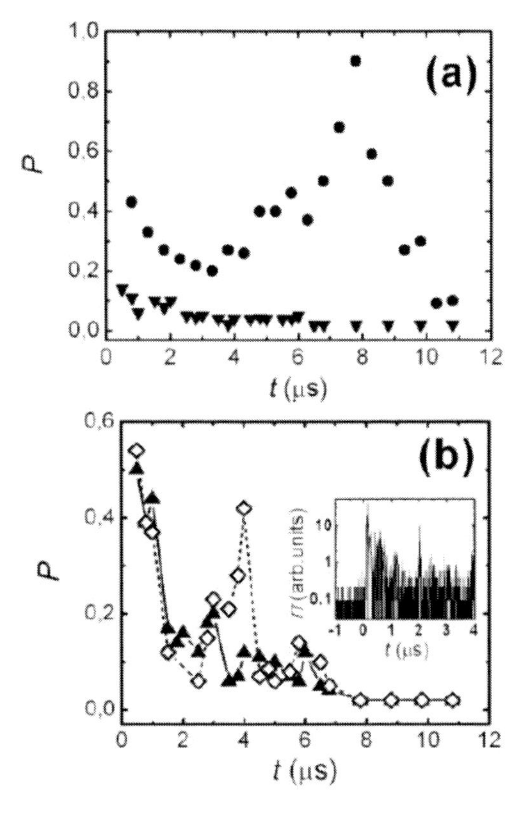

Figure 18. Time-dependent FFT amplitudes $P(f)$ for the main spectral modes at 1 MHz (a, circles), 28.4 MHz (a, squares) and 36.5 MHz (a, down triangles), 5.9 MHz (b, rhombs) and 18.6 MHz (b, up triangles). Inset: relaxation of the acoustic power Π over the time interval t = 0-1.5 μs. Adapted from Ref. 102.

Surprisingly, these experimental results, including characteristic dynamics of vapor bubbles in *bulk* water, their oscillation frequencies and damping times, are in good agreement with similar results for *surface* micro-bubbles in liquid nitrogen [106]. The microsecond damping of steam bubble oscillations and the consequent "red" shift of their amplitude FFT spectra accompanied by the final damping of bubble oscillations at the lowest frequency of 1 MHz can be interpreted in terms of separate physical processes experimentally demonstrated for individual or twin bubbles [107] and illustrated in MD simulations [5]. In particular, one can consider, respectively: 1) dissipative losses to acoustic emission, heat diffusion and viscous effects during $t \leq 3$ μs for *all* initial steam bubbles (oscillation frequencies of 1-40 MHz) nucleated via homogeneous boiling mechanism at temperatures $T \approx 590$-600 K and pressures $P^* \sim 10^7$ Pa (the saturated vapor pressure of water $P_w^S(593\ K) \approx 10^7$ Pa [18]) during the 100-ns laser spike; 2) the resulting reduction of size and energy of these bubbles leading to their enhanced coalescence (in the collapse phase [107]) during $t \approx 3$-8 μs and giving larger bubbles oscillating more slowly (the "red" spectral shift); 3) coalescence and coarsening of the intermediate bubbles for $t \leq 8$ μs to the maximum possible size (diameter) $D_{max} \leq \delta(10.6\ \mu m) \approx 9\ \mu m$ followed during $t \approx 8$-11 μs by collapse of the final bubbles resulting in expulsion of micro-jets (micron-sized water droplets) [24-25,50,107] or by further growth of the bubbles which is confined within the superheated cavitating water layer (percolation) and may ended up with spallation of the partially disintegrated top liquid layer or its expulsion in the form of vapor-droplet mixture [5].

In contrast, during recent similar broadband contact ultrasonic studies of explosive liquid-vapor phase transformations in sub-micron thick layers on a free water surface strongly superheated by a 2.94-μm Q-switched Er:YAG laser (the FWHM pulse length of 200 ns), no indications of any such supercritical vapor bubble oscillations were observed [72]. Furthermore, stochastic separate or double sharp and intense sub-nanosecond pressure spikes superimposed on less intense, slow thermoacoustic and surface vaporization pressure signals were detected (Figure 19) and interpreted as attributes of sub-surface phase explosion and expulsion of the entire underlying cooler surface layer (see another in Chapter 4), since their sub-nanosecond duration is consistent with previous recent experimental measurements and theoretical estimates of characteristic phase explosion onset times [41,109-111]. Though these experimental results are yet to be verified and understood, one can assume that the very high laser fluences and high volume energy density deposited in water (ranging from three to eight times the threshold energy density for explosive liquid-vapor phase transformations evaluated above) in these experiments may give rise to optical breakdown in water in hot spots of the laser beam, or other unidentified surface ablative laser plume or plasma effects. Alternatively, such surface vaporization and explosive liquid-vapor transformation of ultrasonic signals without any oscillatory signals from supercritical bubbles may indicate that near-critical vapor bubbles can't grow to supercritical sizes, which are "ultrasonically visible" through their multi-MHz oscillations [23,50,104-105], in the too thin superheated surface layer, but can move to the outer surface and break there, resulting in the observed fast pressure spikes [72].

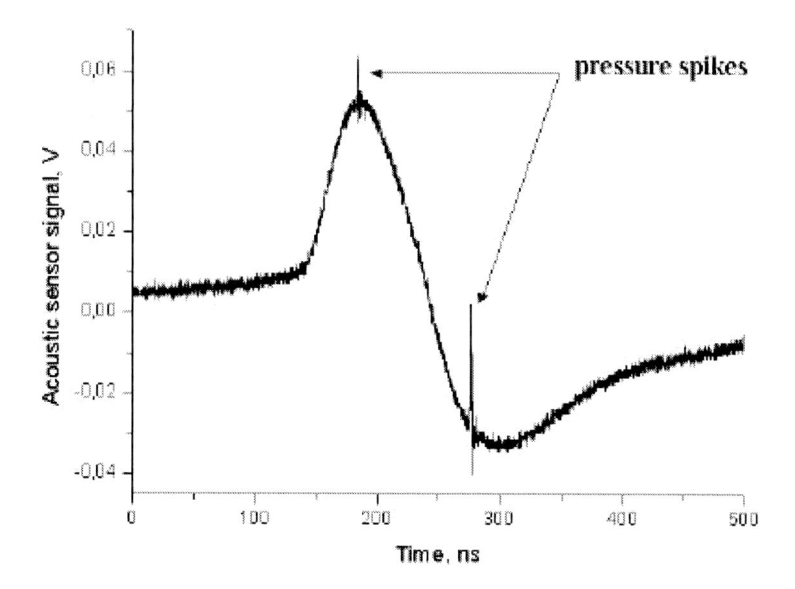

Figure 19. Ultrasonic transient in water heated by Er:YAG laser at laser fluence about 0.6 J/cm^2 (courtesy of A.A. Samokhin and N.N. Il'ichev).

Another set of informative experimental studies of near-critical surface liquid-vapor phase transformations in homogeneously laser-superheated pure liquids was related to glycerol heated by nanosecond Q-switched Er:YAG and OPO lasers in a typical matrix-assisted laser desorption ionization (MALDI) environment [28-30]. In the first work [28], a MALDI setup was equipped with a rather slow ultrasonic transducer to measure ablation rate of the glycerol matrix and to relate it to molecular ion yield. The authors separated the acquired ultrasonic signals into thermoacoustic and surface vaporization (recoil pressure) components, showing that the tensile (rarefaction) pressure phase of the former one is completely cancelled by the significantly enhanced compressive pressure of the latter one at the explosive phase transformation threshold (nevertheless, referring to previous studies [24], this fact was interpreted in that work as a strong consumption of the corresponding tensile stress by the growing vapor bubbles). Importantly, the drastic rise of the recoil pressure near the explosive liquid-vapor transformation threshold was found to correlate with a strongly nonlinear (ninth power) yield of molecular glycerol ions detected using a TOF MS setup [28], rather unexpected for the used mid-IR laser pulses and, potentially, indicating that laser intensity and density of gas phase ablation products were sufficient for optical breakdown in the bulk liquid glycerol or its expelled droplets. Further MALDI studies on glycerol employing similar IR OPO laser sources [29-30] demonstrated that both sub- and micrometer-sized glycerol particles (droplets) appear as a result of laser-induced near-critical surface liquid-vapor transformation, (the authors erroneously identified the removal process as photomechanical spallation in the stress confinement regime [5]) with their concentrations scaling proportionally to IR absorption coefficients at the particular laser wavelengths and their mean sizes (and corresponding PE thresholds) scaling inversely proportional to the coefficients [29]. Moreover, it was observed that the average size of large, micron-sized droplets decreased continuously with increasing laser fluences representing either reduction of characteristic spatial dimensions in vapor-droplet mixtures produced at higher deposited energy densities [30], resembling eventual spontaneous or laser-driven vaporization of expelled droplets in the laser plume [32] (see, e.g., delayed emission of C_2 fragments from gas phase, laser pre-heated C_{60} fullerene molecules in vacuum [112]).

A piece of important information on bubble dynamics underlying near-critical surface liquid-vapor phase transformations in superheated liquids, complementary to the abovementioned ultrasonic and mass-spectroscopic works [23,26,28-30,49-50,101-105], was brought by optical imaging and ultrasonic studies performed in colored liquids [24-26]. These studies performed on multi-micron (10 - 10^2 μm) thick surface liquid layers of water colored by variable concentrations of ink and $CuCl_2$ [26], orange G dye [24] and K_2CrO_4 [25] to tune its absorption coefficient with respect to the heating 1064-nm Q-switched Nd:YAG, 532-nm Q-switched Nd:YAG and 248-nm KrF excimer laser radiations, respectively, have revealed several stages of mass removal during the surface cavitation phenomena. The first, fast removal stage occurs during the heating laser pulse in the form of the pronounced cavitation of micrometer-sized vapor bubbles underlying the near-critical surface liquid-vapor transformation, being accompanied by intense acoustic emission both in ambient atmosphere (shock wave) and in water (thermoacoustic and recoil pressure waves), as well as by expulsion of an opaque vapor-droplet mixture [24,25]. However, at the later – microsecond – times the eventual vapor-droplet expulsion pressurizes the liquid producing a surface transient surface depression (cavity) [24], and results in onset of a quasi-continuous hydrodynamic liquid flow at the lateral boundary of the ablated water region, while the entire process resembles the key stage during deep laser drilling of materials [43-45,48,113]. Finally, a release of the transient recoil pressure ends up with another stage of much deeper water cavitation and, at higher deposited energy densities, with collapse of the surface cavity, expelling the next portion of vapor and droplets [24]. As a result, multi-modal distributions of liquid droplets were observed at different deposited energy densities [26], tracking these different mass removal stages. Surprisingly, no ultrasonic emissions were recorded at the later removal stages, [24-26] indicating their late relaxation character; in contrast, the main ultrasonic signals, which appeared during the heating laser pulse, were interpreted as tracking either sub-surface micro-scale rupture of the weakly superheated low-absorbing liquids via their cavitation driven by the corresponding thermoacoustic tensile stress components [24-25], and their following expulsion, as well as in the case of explosive expulsion of the strongly superheated high-absorbing liquids [87].

III) Derivation of Mechanisms and Modeling of Surface Phase Explosion

The unique set of experimental findings, presented in section II.A(i), regarding the KrF-laser ablation of graphite [63], including observation of the cut-off fluence $F_{\text{cut-off}} \approx 7$ J/cm^2 in optical reflection of the laser-heated graphite, the correlating drastic rise of its removal rate and recoil pressure near the similar threshold fluence $F_{\text{thr}} \approx 4$ J/cm^2 and their subsequent saturation enable us to draw a few important conclusions and suggestions. The most obvious one is that the intense mass removal starts during the heating laser pulse at the onset instant t_{rem}, when the accumulated instantaneous fluence $F(t) \approx F_{\text{thr}}$, and drops down the good specular reflection of the pulse by the graphite sample because of significant absorption/scattering of the trailing part of the pulse in the laser plume. The intense mass removal proceeds between F_{thr} and $F_{\text{cut-off}}$, as the instant t_{rem} shifts at increasing $F > F_{\text{thr}}$ to the leading part of the laser pulse, until complete surface screening is achieved by the ablated products, resulting in saturation of the reflected fluence at $F_{\text{cut-off}}$. This removal regime ($F >$

F_{thr}) in figure 5 with $Z_{plat} \approx 1$ µm » $\delta_{248} \approx 0.01$ µm, more probably corresponds to the abovementioned surface near-critical liquid-vapor phase transformation [87], rather than to surface vaporization, since the maximum recoil pressure $P_{rec} \sim \rho_{gr}V_{ev}V_{vap} \sim 10^3$ bar estimated using the mass and momentum conservation rules [44] to derive P_{rec} as a function of the graphite density $\rho_{gr} \sim 10^3$ kg/m^3, removal rate $V_{ev} \sim Z_{plat}/\tau_{las} < 10^2$ m/s for $Z_{plat} \sim 1$ µm and $\tau_{las} \sim 10^{-8}$ s, and the velocity of vapor species in the Knudsen layer $V_{vap} \sim (RT_{crit}/M)^{1/2} \sim 10^3$ m/s for the carbon molar mass $M = 1.2\times10^{-2}$ kg/mole and the estimated critical temperature of carbon $T_{crit} \approx 7\times10^3$ K [70,98], approaches the characteristic sub-critical equilibrium vapor pressures $\sim 10^3$ bar of carbon, where the predicted critical pressure of carbon equals to 2240 bars [70,98].

Furthermore, the strong correlation between Z_{abl} and P_{rec} magnitudes demonstrates no significant plasma contribution and rather continuous mass removal until its saturation at Z_{plat}. This correlation and the F-dependent onset of the explosive liquid-vapor phase transformation (see also the same observation for water in figure 11 [23]) may indicate that this transformation occurs not as an instantaneous event due to sub-surface superheating [74-77] in the molten carbon layer, but via quasi-continuous removal of a carbon vapor-droplet mixture from the superheated surface molten layer destabilized by a thermoelastic rarefaction (tensile) wave, which represents its preceeding compressive counterpart reflected at the free melt surface [24-25,32,100]. This is consistent with the fact that under such experimental conditions removal rate (front propagation speed for the near-critical liquid-vapor transformation) $V_{ev} < 10^2$ m/s is lower, than typical longitudinal sound velocities in a vapor phase with some droplet content, ($\sim 10^2$-10^3 m/s depending on the liquid content [79]) driving the removal, i.e., the correlation between Z_{abl} and P_{rec} for $F > F_{thr}$ shows the nearly constant – fluence- or intensity-independent – expansion velocity of the resulting vapor-droplet mixture, corresponding to the same initial (apparently, near-spinodal) thermodynamic state and the same removal dynamics of the superheated carbon.

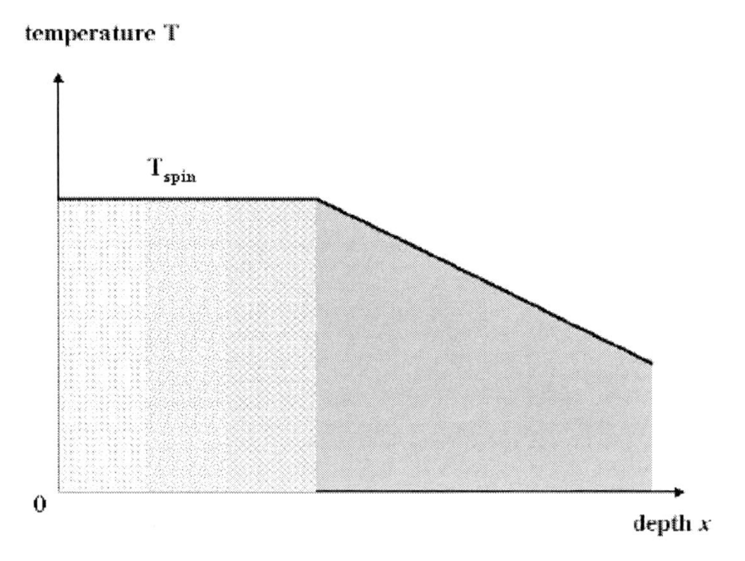

Figure 20. Schematic of spatial in-depth temperature profile during laser-induced surface near-critical liquid-vapor transformation. The patterned regions show different corrugation of the liquid via homogeneous bubble nucleation, given by lighter grey color.

Moreover, in the lieu of the revealed quasi-continuous explosive removal during the heating laser pulse one can assume a temporally steady-state and spatially monotonous temperature profile in the superheated liquid carbon layer prior to the propagating liquid-vapor transformation front (Figure 20). Presence of a subsurface temperature maximum (subsurface superheating) is usually related to evaporative surface cooling of bulk absorbing liquids with the surface vapor energy flux dominating over the heat flux to the surface from the heated bulk, thus satisfying the criterion $\alpha_{liq}\chi_{liq}/V_{ev} \ll 1$ [56,74-77], where the surface vaporization rate V_{ev} is known to be an *intensity-dependent* quantity. In contrast, the *fluence-dependent* onset of such explosive liquid-vapor phase transformation was observed in Refs. [23,63], pointing out the potential absence of such sub-surface superheating. In our opinion, in the proximity of the liquid-vapor spinode curve a potential laser-induced subsurface temperature maximum can be cleared out by exponentially increasing bulk homogeneous nucleation of vapor bubbles [9,18], thus consuming the excessive deposited laser energy in the superheated molten carbon layer due to the sharply rising (diverging) temperature dependences of isobaric heat capacity C_p and thermal expansion coefficient β_T [9,18], with the nucleated subsurface bubbles strongly driven by the tensile wave, [24-25,101] providing decomposition of the molten carbon layer in the form of a vapor/droplet mixture. Indeed, the singularity of C_p and β_T magnitudes is favorable for appearance of a monotonously decreasing spatial temperature profile inside bulk liquid with a flattened near-surface part. This can be demonstrated by including the divergence of these quantities in the common one-dimensional heat conduction equation in the form, adapted from [56,114,115] and supplied by a laser heating source and bulk nucleation sink (in Eq.(3a) this sink is contained in the $C_p^*(T)$ dependence)…

$$T\frac{\partial S}{\partial t} = C_p^*\frac{\partial T}{\partial t} + \beta_T V\frac{\partial P}{\partial t} = \kappa\frac{\partial^2 T}{\partial x^2} + C_p^* u\frac{\partial T}{\partial x} + \alpha[1 - R(I)]Ie^{-\alpha x}, \quad (3a)$$

$$C_p\frac{\partial T}{\partial t} + \beta_T V\frac{\partial P}{\partial t} = \kappa\frac{\partial^2 T}{\partial x^2} + C_p u\frac{\partial T}{\partial x} + \alpha[1 - R(I)]Ie^{-\alpha x} - \int_0^{r_{crit}} \Delta G(P,T,r)\, j(P,T,r)\, dr, \quad (3b)$$

…where S and V are the molar entropy and volume of the liquid phase, the quantity $(\partial P/\partial t)$, requiring another equation for pressure dynamics [56,72], describes evolution of the internal liquid pressure during the initial stage of the near-critical liquid-vapor phase transformation, ΔG and j are the Gibbs energy and specific nucleation rate for the vapor bubble of the radius r (r_{crit} is the critical bubble radius). Accounting for the abovementioned singularities near the corresponding spinode curve, one finds very slow variation of the liquid temperature T as a function of the deposited laser energy, corresponding to $(dT/dx)_{surf} \approx 0$ and $T \leq T_{spin}(P)$ (Figure 20), canceling the convective and diffusive terms on the right side of the Eqs.(3). Additionally, one can note for the diverging β_T very slow pressure dynamics in the near-spinodal liquid, which is consistent with its rather slow – on the nanosecond timescale (see microsecond removal timescale in Ref. 24) – acoustic relaxation at high internal pressures, but very low longitudinal sound velocities in cavitating liquids [79-80]. This observation provides a unique opportunity for additional heating and ongoing cavitation in the near-spinodal liquid further to supercritical states, while this fact was indeed experimentally observed as continuous dissociation of liquid droplets, expelled in the form of a vapor-droplet mixture near the explosive liquid-vapor transformation threshold, in the laser plume at higher (above-threshold) laser fluences, which may finally end up with its optical breakdown and ignition of ablative plasma [32]. However, when the heating laser pulse is gone, the explosive expansion becomes the dominating factor in the vapor-droplet mixture dynamics, as shown in the case of ultrashort (femtosecond or short picosecond) heating laser pulses [1-6, 37-38].

Finally, the abovementioned experimental results and their theoretical analysis enable us to bring a few enlightening comments in the evergreen discussion on the existence of sub-surface superheating and related explosive surface liquid-vapor transformation phenomena. First, these multiple experimental facts indicate that during a heating laser pulse such propagating transformation front is characterized by intensity- and fluence-independent, near-spinodal thermodynamic conditions, while the increasing nanosecond laser intensity just faster pre-heats the liquid until some characteristic, generally pressure-dependent, near-spinodal temperature T_{spin}, with the subsequent almost instantaneous (sub-nanosecond) phase transformation process. It is this characteristic temperature – not normal boiling or other irrelevant temperature values [12,32],– that can be apparently chosen to set up a boundary condition; in contrast, the C_p singularity makes the common boundary conditions, considering interface vapor and thermal fluxes in the form [12,32]

$$u\Delta_{ev}H + \kappa\left(\frac{\partial T}{\partial x}\right)_{surf} = \left[1 - R(I)\right]I, \qquad (4)$$

non-applicable, as heat conduction transport to the outer surface becomes negligible for the flattened spatial temperature profiles (Figure 20), comparing to energy fluxes related to the high near-spinodal removal rates, while the surface vaporization front is not well-defined in the strongly corrugated absorbing near-surface layer of the cavitating liquid or vapor-droplet mixture, corresponding to internal vaporization (boiling) within the layer. However, the simple boundary condition in the form (similar to that used in Ref. 32)

$$V_{ev}\Delta H^{*} = A\left[1 - R(I)\right]I, \qquad (5)$$

with the pre-factor A representing the fraction of the laser energy, which goes to the pre-heating of the liquid beyond the liquid-vapor transformation front to T_{spin}, the removal energy density (enthalpy) $\Delta H^{*} \approx \Delta_{ev}H$, can be used for easy evaluation of removal rates in the surface ablation regime. Third, it appears that sub-surface superheating, though not ever experimentally observed yet, may take place in absorbing liquids rather far from their corresponding liquid-vapor spinode curves, but then result in sub-surface cavitation [24], but not in the *explosive* liquid-vapor transformations. Vice-versa, such explosive liquid-vapor transformations may occur on liquid surfaces superheated until their near-spinodal temperatures, but the considerable sub-surface superheating, as calculated in Refs. 12,74-77, ignoring the crucial divergences of C_p and β_T in the near-spinodal region, hardly appears to be possible.

B. Surface Near-Critical Liquid-Vapor Phase Transformations Driven by Ultrashort Laser Pulses

Compared to the abovementioned surface near-critical liquid-vapor transformations induced by short (sub-nanosecond and longer) laser pulses, in the case of ultrashort (femtosecond and short picosecond) laser pulses such explosive phase transformations still occur with the considerable temporal delays required for incident ultrashort laser pulses to photo-excite the material of interest; then, depending on the type of the material – metal, semiconductor or dielectric, for the deposited energy to convert partially from the resulting electronic excitation (direct laser excitation of ionic subsystem is extremely rare) into quasi-continuum of phonon modes [116], or separate specific hyper-populated "hot" coherent phonon modes [117] and, subsequently, into long-living point defects [118]. The latter process, directly competing with common heating of materials via picosecond electron-lattice thermalization, is especially significant for those of them which exhibit strong electro-lattice coupling – primarily, dielectrics [119], since generation of point defects via self-trapping of free carriers or excitons, having all signatures of self-accelerated sub-picosecond Bose condensation-like lattice deformation process [120], may precede or occur simultaneously with lattice heating via incoherent carrier emission of various relevant phonons (phonon quasi-continuum), which satisfy two energy and quasi-momentum conservation principles, and selection rules for electron-phonon interactions [119]. Surprisingly, there were a few observations of similar point defect generation process in femtosecond-laser excited materials with much weaker electron-lattice coupling – semiconductors and even semimetals, taking the form of catastrophic ultrafast (sub-picosecond) disordering [121-124] at the incompletely equilibrated phonon modes (it is also called "non-thermal melting", though its detailed physical mechanism – non-thermal, thermal or combined one – is still under study [125]). In any case, at somewhat (in some cases, approximately 1.5 times [126]) lower photo-excitation levels melting may occur on a picosecond timescale via a thermal path, though at significantly higher deposited energy densities, comparing to the corresponding equilibrium phase transformation [127].

Finally, when enough energy in the thermal form is deposited by ultrashort laser pulses in the material of interest to initiate its thermal explosive removal under quasi-static (short-pulse) laser irradiation conditions (typically, one can talk of near-critical enthalpies of corresponding superheated fluids [33]), a few other ultrafast processes come into play to compete for the energy with the thermal surface near-critical liquid-vapor transformation. First, rapid compressive thermoelastic stress relaxation via expansion of the superheated melt cools it down considerably [1-6,115,128], thus requiring initial supercritical enthalpies to be deposited to guarantee, at least, its sub-critical superheated liquid state upon the hydrodynamic expansion (spallation of the resulting molten surface layers in the form of blown up single bubbles with nanometer-thick outer walls [1-6,36,129] may happen under these ultrashort laser irradiation conditions, but it is not in the scope of this chapter). On the other hand, once the energy density deposited by ultrashort laser pulses exceeds some critical value at the early electronic excitation stage, a purely non-thermal Coulomb explosion may take place in the form of a plasma double-layer lift-off [129-132], when hot electrons escaping the photo-excited solid target via the thermionic effect [130], drag off the cold ions of the positively charged surface layer. The ultrafast (sub-picosecond) Coulomb explosion, whose signatures were observed not only in dielectrics [130], but also in semiconductors and even in metals [129,131-132], suppresses not only fast energy transport into bulk [129], but also its *in situ* conversion into thermal lattice excitations, thus, dynamically competing with all thermal processes. Hence, there is a rather narrow range of ultrashort laser irradiation conditions and corresponding deposited energy densities that provide the necessary pre-conditions for such surface near-critical phase transformations. Nevertheless, generally, the relaxation character of ultrashort pulsed laser ablation makes ablative signatures of all transient – supercritical, near-critical and sub-critical – thermodynamic states almost equally visible, thus, requiring use of temporally- and spatially-resolved experimental techniques to resolve and to track separate specific ablation stages. Below, we will give an overview of some techniques, enabling monitoring of surface explosive liquid-vapor transformations in materials excited by ultrashort laser pulses.

I) Time-Resolved Optical and X-Ray Imaging

Time-resolved optical imaging of ultrafast laser ablation is a well-developed experimental technique, combining a "pump/probe" principle with common optical microscopy with corresponding output reflectivity [126,133] or interference [118] signals. While the latter variant requires good transient surface quality and reflectance of ablated matter, the former one indicates just variations of such transient surface reflectance and, thus, is more suitable in studying removal of vapor and droplet components resulting from surface explosive liquid-vapor transformations. Typically, such time-resolved reflective optical microscopy (TRROM) provides snapshots of spatially resolved surface or near-surface reflectance variations across the inhomogeneously (for the given real Gaussian or other beam shapes) irradiated sample surface with its characteristic clear transient features – drops or rises, Newton rings, scattering or emissive regions – representing on different timescales the corresponding photo-excitation, non-thermal or thermal melting, thermo-mechanical spallation of an optically thin, reflective molten shell (surface bubble), expulsion of a vapor-droplet mixture, near-surface plasma formation, while the final resulting images show the permanent post-pulse surface relief and structural (optical) changes (Figure 21). Since spallation (rupture) of the superheated liquid layer, occurring at lower incident laser fluences [126,133] (deposited energy densities), exhibits the prominent high-contrast Newton rings (Figure 21), surface near-critical explosive liquid-vapor transformation in the hot, pressurized melt can be identified in the TRROM images, as a succeeding ablation regime, in the form of corresponding scattering – inhomogeneous on the micrometer scale – regions, appearing closer to the laser beam center at higher deposited energies. When represented versus the pimp/probe delay time, the reflectance curves show explosive-like picosecond expansion of superheated molten materials (Figure 22), becoming faster at increasing laser fluences.

Recently, time-resolved low- and high-angle X-ray scattering was employed to monitor an initial homogeneous nucleation in nanometer-thick surface superheated molten layers of InSb semiconductor irradiated by ultrashort laser pulses [41]. These studies have revealed the characteristic nanosecond nucleation scale of nanometer-sized density fluctuations (vapor bubbles) in the metallic liquid layer, which are consistent with results of previous experimental and theoretical (MD) studies of homogeneous nucleation of sub-critical and critical bubbles in homogeneously superheated liquids [23,50], but are contrasting with sub-nanosecond heterogeneous nucleation of nano-scale near-critical steam bubbles on femtosecond laser-heated gold nanoparticles [110]. As a result, one can conclude on the existence of a rather slow – nanosecond – induction period prior to surface explosive liquid-vapor transformations in superheated liquids under sub-critical thermodynamic conditions, even in the case of ultrashort heating laser pulses, though at higher deposited energy densities femtosecond laser ablation occurs as ultrafast (picosecond) hydrodynamic expulsion of vapor-droplet mixture (see figures 21 and 22 above), passing through a set of supercritical states [1-6,36].

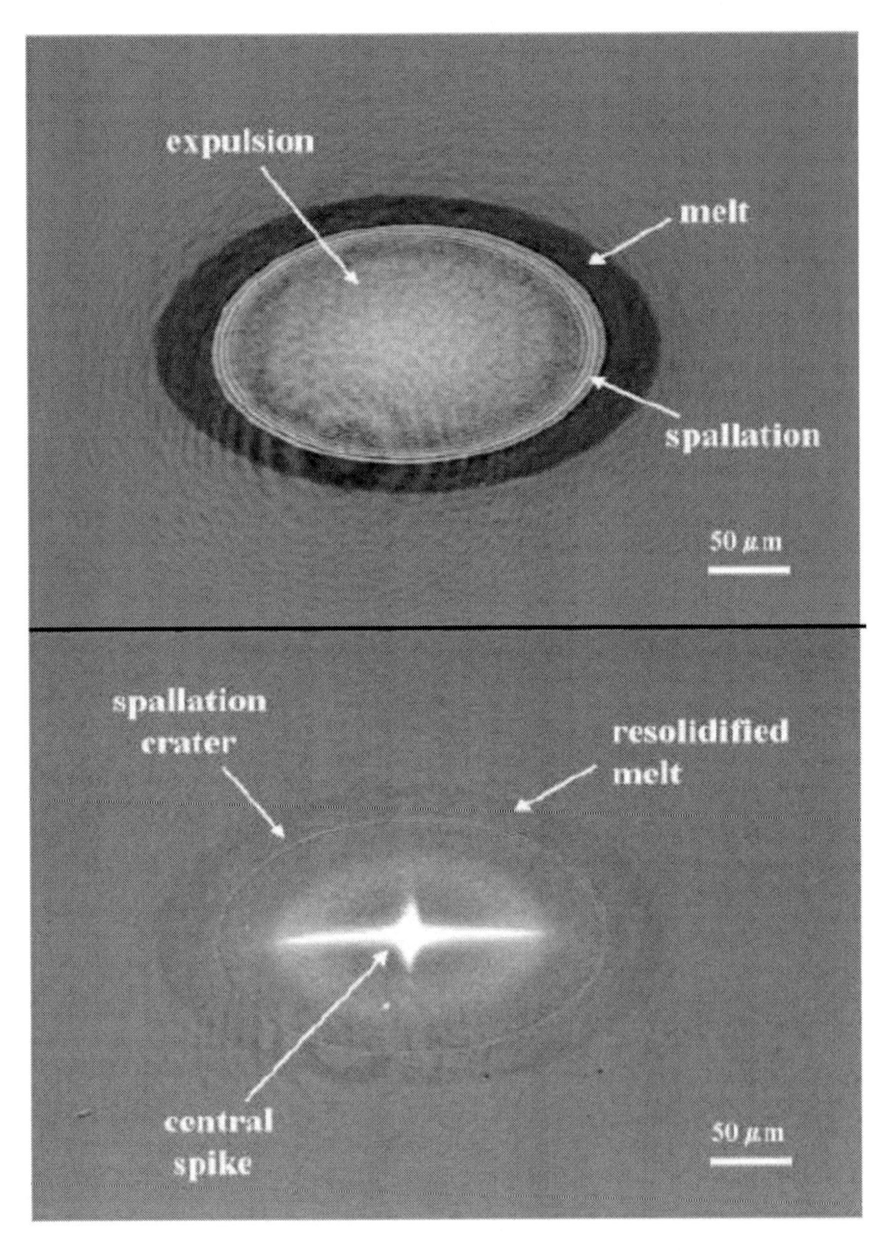

Figure 21. Characteristic TRROM images taken at 372-nm probe wavelength for silicon wafer ablated by 120-fs, 744-nm laser pulses with fluence about 1 J/cm^2. (top) Delay time equals to 300 ps; (bottom) final image (adapted from Ref. 57).

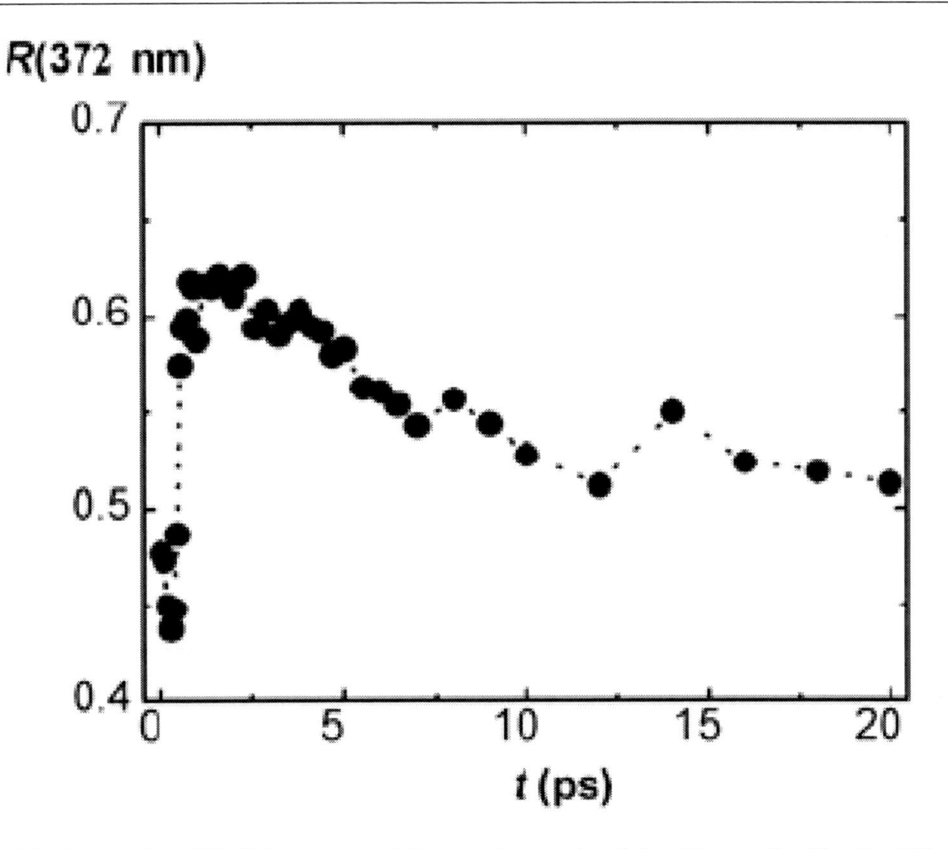

Figure 22. Characteristic TRROM transient at 372-nm probe wavelength for silicon wafer ablated by 120-fs, 744-nm laser pulses with fluence 0.75 J/cm^2 (adapted from Ref. 57).

The existence of such two, fast (picosecond) and slow (nanosecond), stages during femtosecond laser surface ablation of various material has been recently demonstrated for steel samples at different laser wavelengths, variable (sub-picosecond and short picosecond) pulse widths and rather high fluences ≤ 20 J/cm^2 (intensities of 10^{13} - 10^{14} W/cm^2) [38]. The authors have shown that under the femtosecond laser irradiation conditions steel exhibited two – short and long nanosecond – peaks of ablative plasma luminescence and similar peaks of absorption/scattering in the ablative plasma/plume revealed by time-resolved optical transmission measurements along the steel surface, and interpreted exactly in terms of the abovementioned ultrafast – sub-picosecond or picosecond – supercritical and delayed nanosecond sub-critical thermodynamic trajectories, respectively. However, in some other time-resolved shadowgraphy studies [37] three ablation stages were distinguished (Figure 23), including expansion of highly pressurized, hot Cu and Al fluids up to 200 ns after the irradiation at much lower femtosecond laser fluences ≤ 1 J/cm^2 (intensities $\leq 10^{13}$ W/cm^2). The subsequent removal stages involved explosive emission of micrometer-sized melt droplets during $\approx$ 700 ns, and massive melt ejection at laser sub-microsecond and short microsecond times [37]. Apparently, one can consider the second, intermediate removal stage as an indication of surface near-critical liquid-vapor transformation in the femtosecond laser heated metals, while the next stage may represent expulsion of the melt, struck by its recoil pressure.

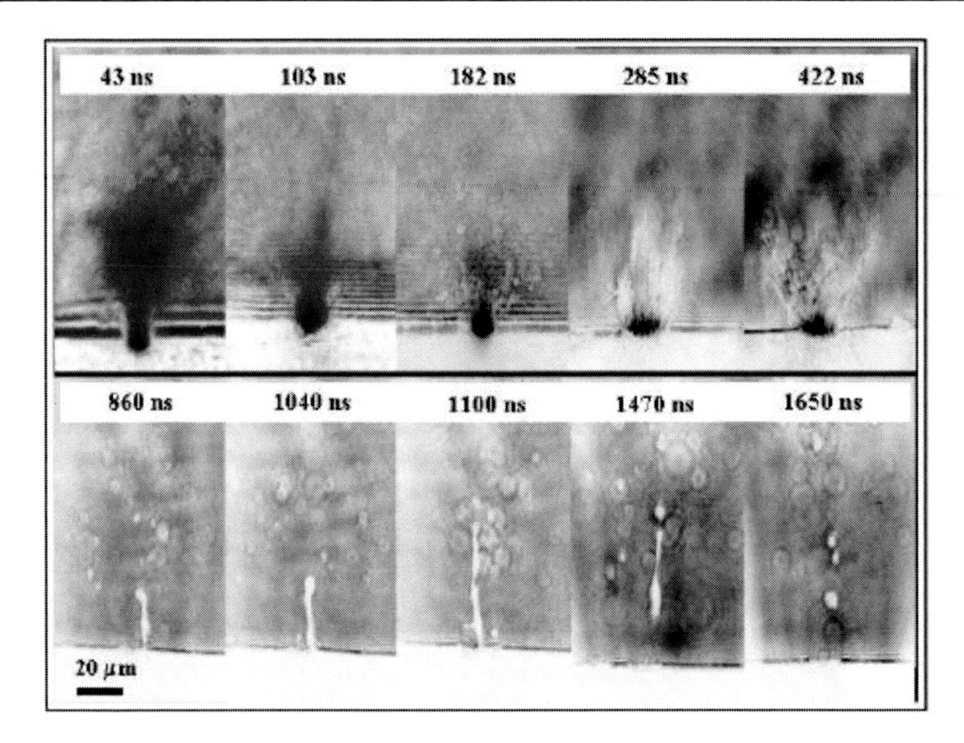

Figure 23. Time-resolved shadowgraphy images of femtosecond laser plume on aluminum surface (courtesy of I. Mingareev and A. Horn, for details see Ref. 37).

II) Debris Re-Deposition and Nano-Particle Generation

There are many experimental indications that femtosecond laser-induced surface explosive liquid-vapor transformation in superheated materials provides extensive picosecond expulsion of nanoscale ablative particulates representing original species of vapor-droplet mixture and their coalescence products [1-6,39,53-54], as well as expansion of other, much larger micrometer-sized liquid droplets [37,52] expelled, however, at much later times. In the first case, the intrinsic nanoscale character of the resulting particulates was confirmed by their appearance both in micro- and nano-scale (Figure 24) femtosecond laser ablation studies. As previously noted, such nano-particulates are more readily formed during ablation of semiconductor and dielectric materials with higher cohesive energies [70], more favorable for faster evaporative cooling and subsequent solidification of the species. These processes preserve shapes and integrity of such particulates during their re-deposition back onto the ablated target caused by ambient atmosphere effects [134-136]. However, recently, a few interesting examples of nanoscale debris deposition as a residue of nano-particulates (metal black) formed via multi-shot femtosecond laser ablation of various metals and silicon were demonstrated in Refs. 53 and 137. Moreover, when femtosecond laser ablation of metals and semiconductors is performed in a liquid environment, the yielding sols of nano-particulates [39,138] represent a practically important research and commercial product for biomedical targeting and sensing, as well as for other nano-plasmonic applications (for bibliography, see ref. 138). Importantly, variation of femtosecond laser fluence enables us to manipulate the average size of such nano-particulates [138] providing, potentially, enlightening information about characteristic spatial scales of ultrafast laser-induced surface near-critical liquid-vapor phase transformations in metals and other materials, as well as the residual violent surface topology (Figure 25).

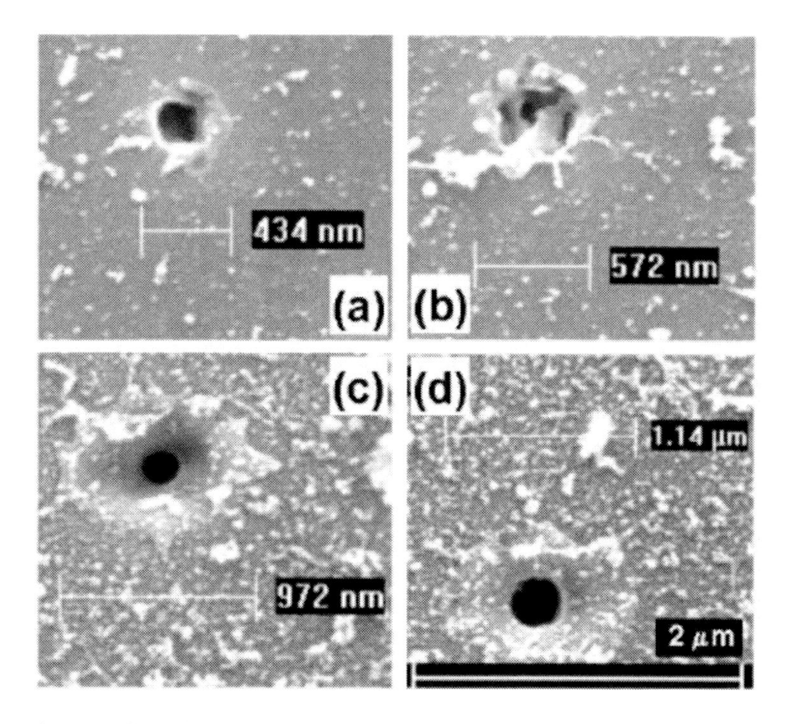

Figure 24. SEM images of sapphire surface ablated by single 1.06-µm, 600-fs laser pulses at fluences of 110 (a), 120 (b), 150 (c) and 180 (d) J/cm^2 (courtesy of A.J. Hunt).

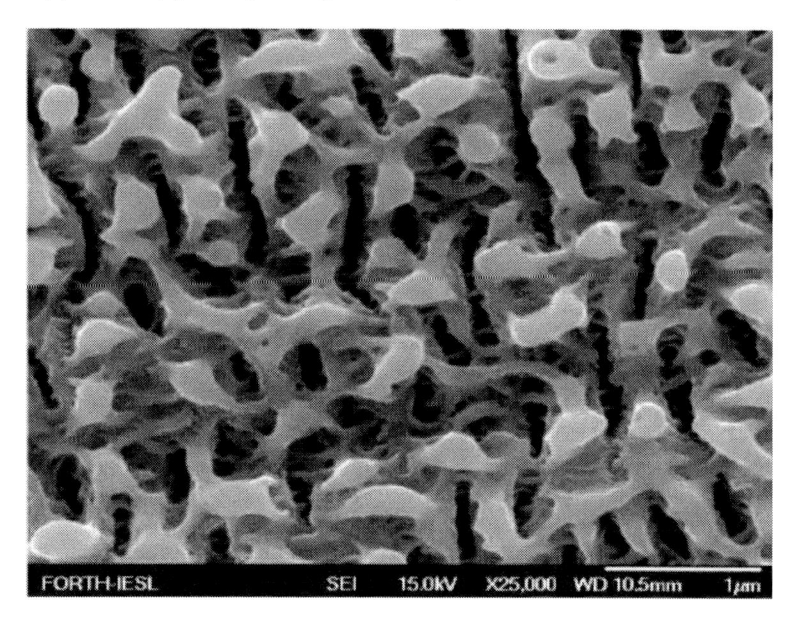

Figure 25. Field-emission SEM image of Ni surface ablated in ethanol environment by 800-nm, 50-fs laser pulses with peak fluence about 14 J/cm^2 (courtesy of E. Stratakis and G.A. Shafeev).

III) Removal Rates

Depth measurements on surface craters produced by ultrashort laser pulses give clear indications of a transition between the regimes of spallation (rupture) and surface near-critical liquid-vapor transformation. In the former regime, a nanometer-thick liquid shell is spalled from the surface molten layer above the corresponding spallation threshold laser fluence F_{spall} under action of thermoacoustic tensile stresses [1-6,35-36,57-58,139], which exceeds the tensile (rupture) strength of corresponding molten material, finally revealing the rather shallow surface crater (Figure 26). In contrast, the laser-induced surface explosive liquid-vapor transformation, occurring at higher laser fluences (but above its well-defined threshold fluence F_{trans}), exhibits more pronounced, much deeper surface craters with expelled ablation products visibly residing around the crater (Figure 27). Similar pictures can be observed not only in femtosecond-laser ablated dielectrics [54,139-140], but also in semiconductors and semimetals (Figure 3) [57]. The corresponding fluence dependences of single- or multi-shot crater depth Z_{abl} demonstrate almost flat regions between F_{spall} and F_{trans}, while at higher fluences $F > F_{trans}$ the single-shot crater depth or multi-shot removal rate (average crater depth per shot) increase drastically, enabling unambiguous identification of F_{trans}. Much higher multi-shot removal rates in the surface explosive phase transformation regime, comparing to the preceding spallation regime, represent its higher nano- and micro-machining efficiency, important for diverse applications.

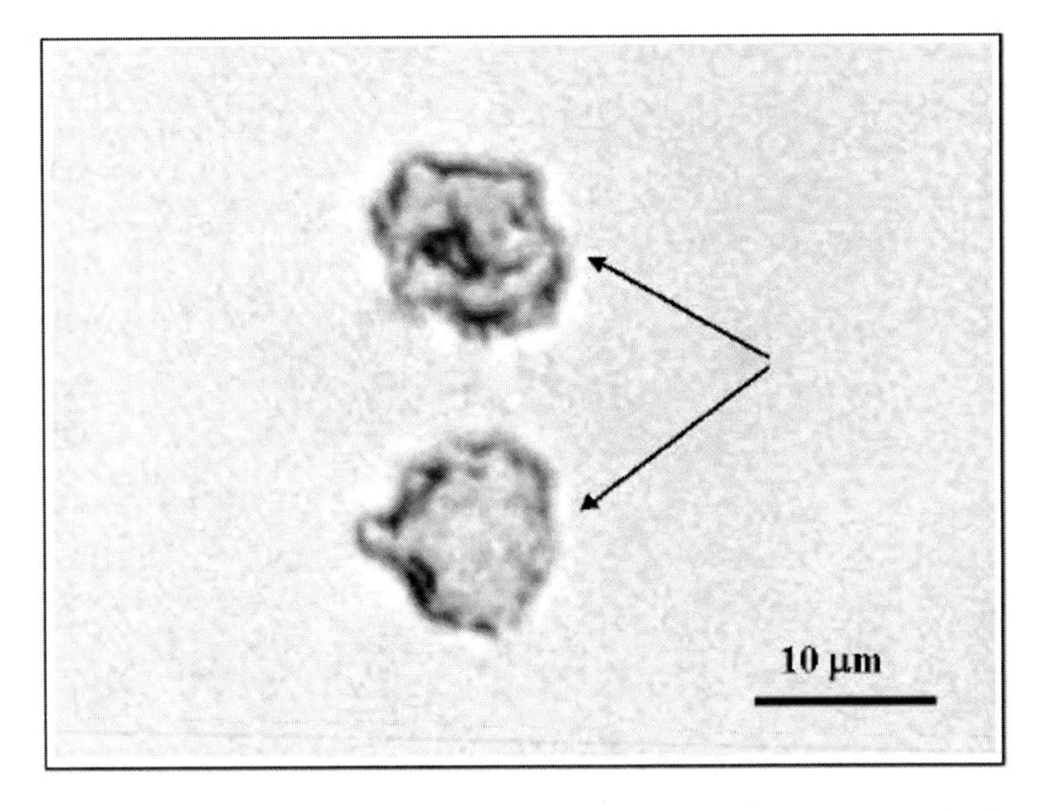

Figure 26. Optical microscopy image (bottom illumination) of spallation craters on silica surface ablated by 248-nm, 50-fs laser pulses with fluence of 7 mJ/cm^2. Adapted from Ref. 139.

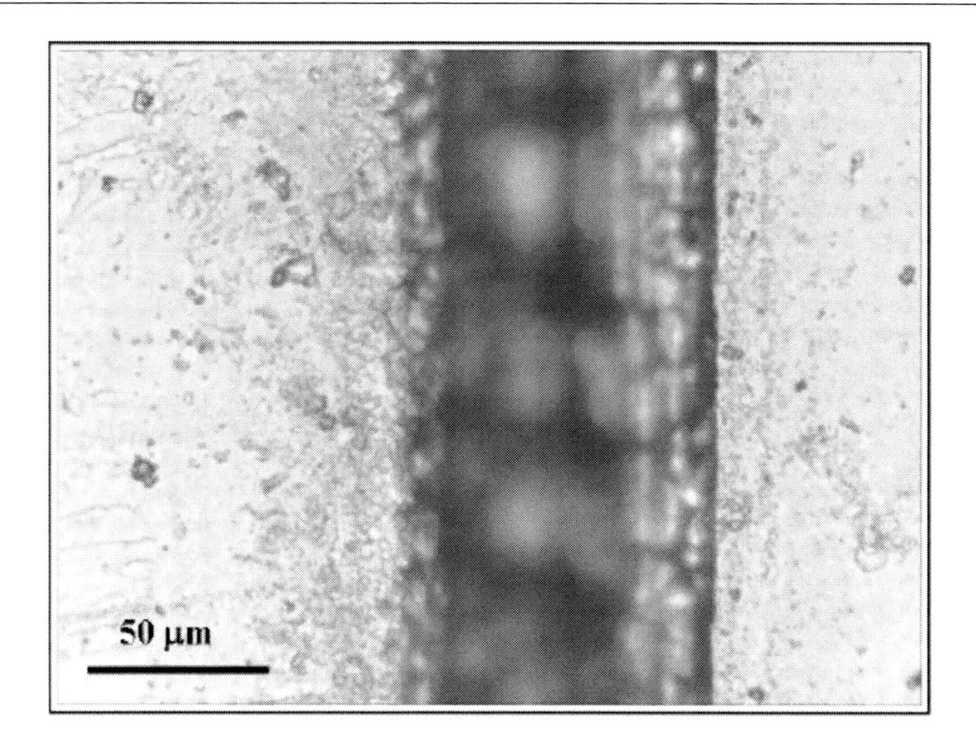

Figure 27. Optical microscopy image (bottom illumination) of expulsion craters (trench) on silica surface ablated by 248-nm, 50-fs laser pulses at fluence about 1 J/cm^2. Adapted from Ref. 139.

IV) Contact Ultrasonic Measurements of Ablative Recoil Pressure

Contact ultrasonic studies provide important information about key parameters of femtosecond laser ablation, even though this technique probes multi-MHz components of the actual multi-GHz or sub-THz hypersonic ablative responses [141].

Recently, contact broadband ultrasonic monitoring has been employed to investigate details of femtosecond UV (248 nm) laser ablation of silica glass in its spallation and surface explosive phase transformation regimes, using the experimental scheme described elsewhere [139]. First, it was observed that at low laser fluences $F \geq 8$ mJ/cm^2 the ultrasonic transients sharply transformed into more complex asymmetric waveforms [Figure 28(b)] with $P_{comp} \geq P_{rare}$ and with the succeeding weaker and asymmetrical inverted pulse, while both P_{comp} and P_{rare} amplitudes simultaneously dropped down versus F still preserving their asymmetry (Figure 29). Under these irradiation conditions, single-shot laser exposures of the glass surface exhibited for $F \geq 6$ mJ/cm^2 quite accurate (no debris), sub-micron thick elevated glass shells (Figure 26) of a spallative character [54], correlating with the appearance of the abovementioned ultrasonic features. It is noteworthy that at slightly higher fluences ($F \geq 8$ mJ/cm^2) these shells became open, (spalled) revealing very shallow, almost flat spallative craters [54,139]. As a result, one can explain these waveform shapes and the fluence trends of the ultrasonic amplitudes by a sequence of a bipolar thermoelastic pulse, representing reversible thermal expansion, and of a stress release pulse resulting from thermo-mechanical spallation of the corrugated softened surface layer. At much higher laser fluences – in the broad fluence range $F = 0.08$-2 J/cm^2 – this glass shows ultrasonic transients with the two – first short and intense, and the subsequent long and weak – compressive pulses [Figure 28(c)], correlating with a formation for $F \geq 0.09$ J/cm^2 of visible, micrometer-deep ablative craters with debris re-deposited on the surfaces outside the ablated region (Figure 27). Hitherto, one can relate the first compressive pulse to the instantaneous expulsion of the near-critical vapor-droplet mixture via thermal and hydrodynamic relaxation of the hot, compressed silica fluid via a sequence of supercritical thermodynamic states [1-6,36]. Then, the second, weaker 100-ns long compressive pulse delayed also by 100 ns, can be associated with delayed, long-term sub-critical surface liquid-vapor phase transformation in the less superheated surface melt. Such two-stage fs laser ablation with its strongly delayed and slow second stage is consistent with experimental findings of previous femtosecond [37-38] and nanosecond [24-25] laser ablation studies. Importantly, these delayed ablation processes can be more readily observed by means of the rather slow (multi-MHz) contact ultrasonic acquisition technique or other optical techniques [37-38], while much faster, but more elaborate time-resolved reflectivity or diffraction studies of fs laser ablation, using optical pump/optical probe or optical pump/x-ray (electron) probe combinations, respectively, typically do not cover the multi-scale set of ablation events in the broad time range from femtoseconds to microseconds [35-36,40-41].

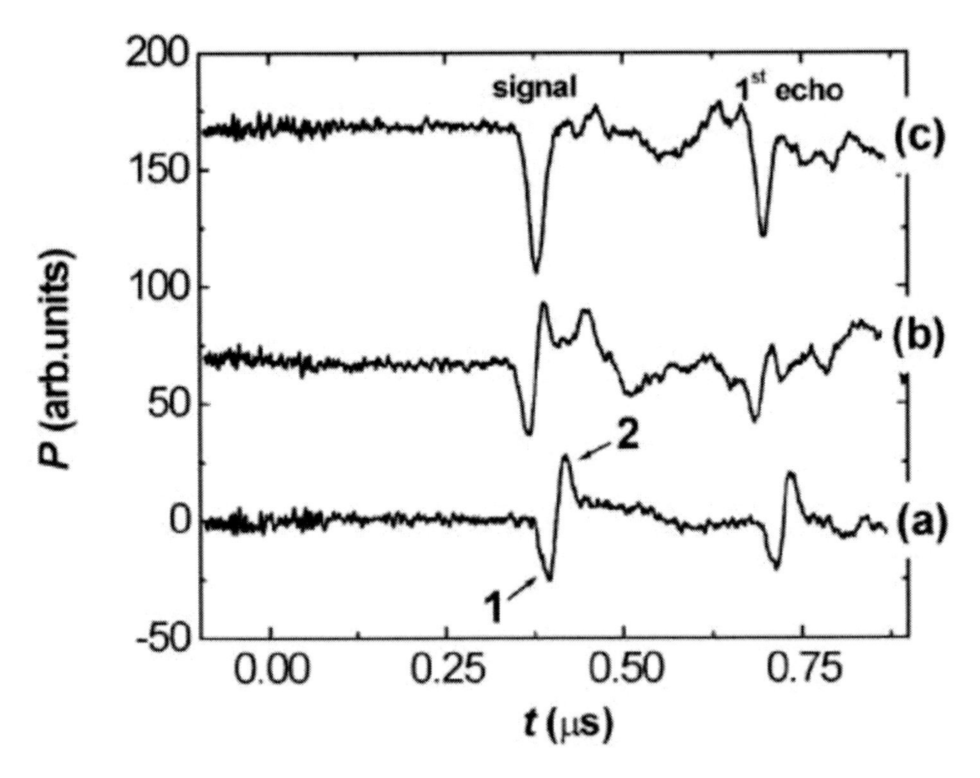

Figure 28. Ultrasonic transients $P(t)$ generated in silica glass by 248-nm, 50-fs laser pulses at $F = 2$ (*a*), 29 (*b*), and 250 (*c*) mJ/cm^2 with the arrows showing their compression (*1*) and rarefaction (*2*) phases. Adapted from Ref. 139.

Similarly, contact ultrasonic diagnostics were also very informative in studying UV femtosecond laser ablation of crystalline silicon in the spallation and surface near-critical liquid-vapor transformation regimes, using the same experimental setup and detection scheme [139]. In these studies, thermo-mechanical spallation of the nanometer-thick liquid shell from the surface molten silicon layer at $F \geq 8$ mJ/cm^2 was identified in the form of the strongly (almost doubly) broadened, non-monotonous ultrasonic pulses of the compressive character (Figure 29) (at lower fluences the ultrasonic transients exhibited their tensile character resulting from silicon densification either due to melting, or due to deformation-potential electron-acoustic phonon interactions [119], see the same figure), propagating in the silicon sample at the ordinary sonic velocities (Figure 30). However, at slightly higher laser fluences $F > 30$ mJ/cm^2 the ultrasonic transients transformed into the intense unipolar (compressive) pulses with the specific temporal shapes, exhibiting sharp fronts and exponential tails (Figure 29) characteristic of shock waves [142] and propagating in the silicon sample at supersonic velocities (Figure 30). Moreover, the acquired shockwave-like transients demonstrated a linear fluence dependence of their amplitudes (Figure 31), which is, at the first glance, consistent with the single-photon absorption of the 248-nm (5 eV) laser photons near the *X*-points in the quasi-momentum space of the energy band structure of silicon [119], providing the necessary energy deposition for its laser ablation, though, on the other hand, such linearity across the different – spallation, surface near-critical liquid-vapor transformation – ablation regimes requires an additional special study and modeling [143].

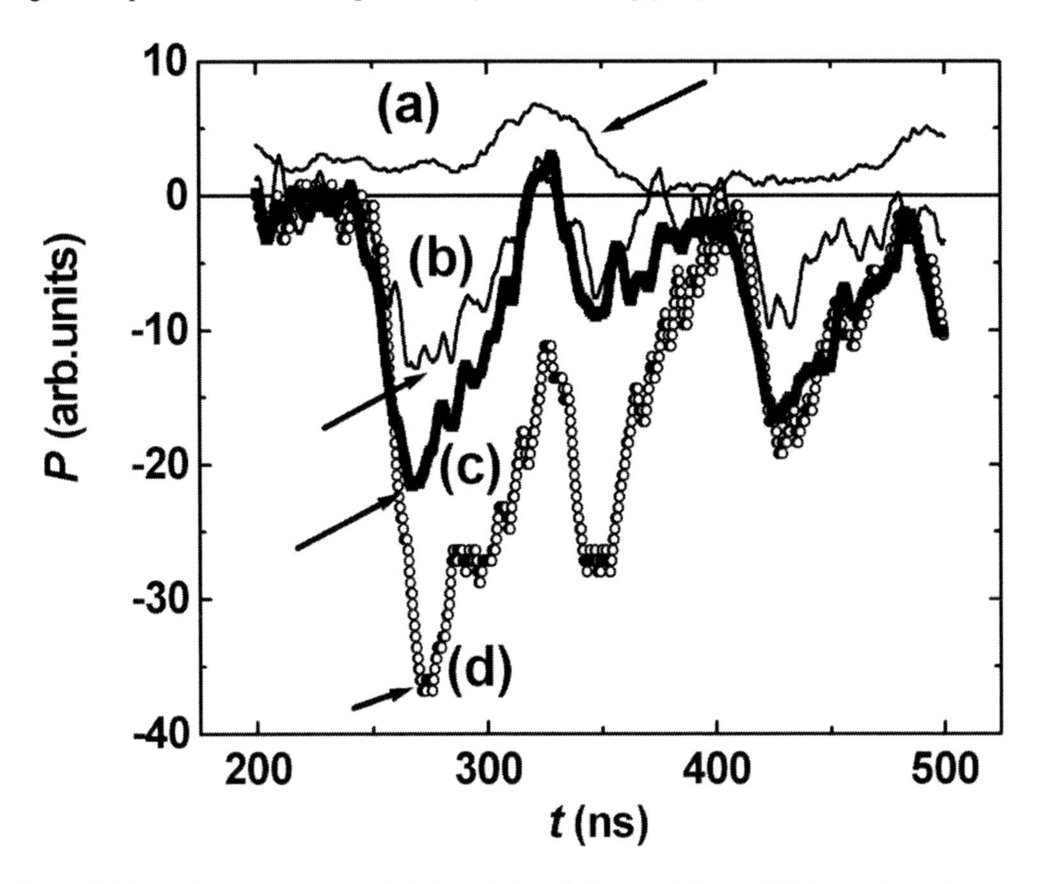

Figure 29. Ultrasonic transients recorded during ablation of silicon by 248-nm, 50-fs laser pulses at laser fluences 5 (a, upper solid curve), 15 (b, bottom solid curve), 30 (c, bottom thick dark curve) and 300 (d, bottomf thick light curve) mJ/cm^2 with the arrows showing the main tensile/compressive (a, EHP) and other pure compressive (b-d) ultrasonic pulses. Adapted from Ref. 143.

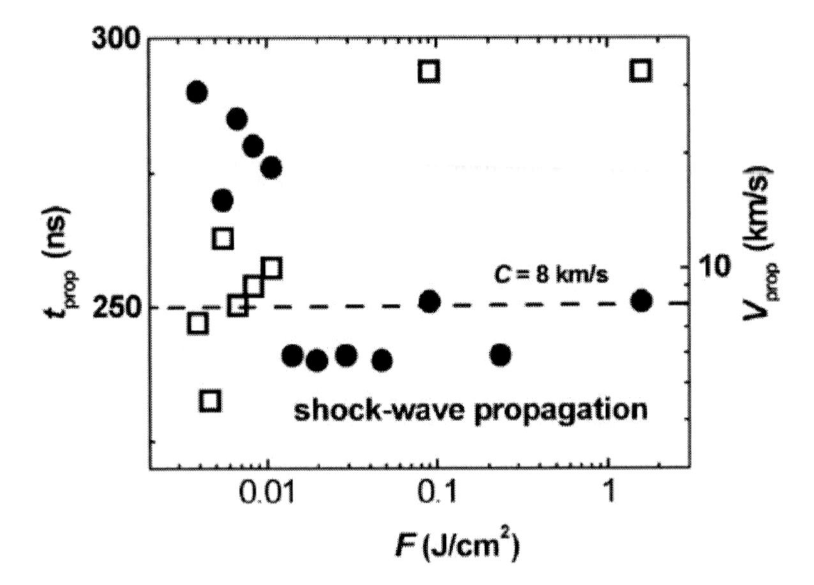

Figure 30. Fluence dependence of propagation time t_{prop} (dark circles) and velocity V_{prop} (light squares) of ultrasonic waves across the 360-μm silicon ablated by 248-nm, 50-fs laser pulses (adapted from [143]). The ordinary longitudinal sound velocity in [100] direction (dashed line), $C \approx 8$ km/s [70], is given for comparison.

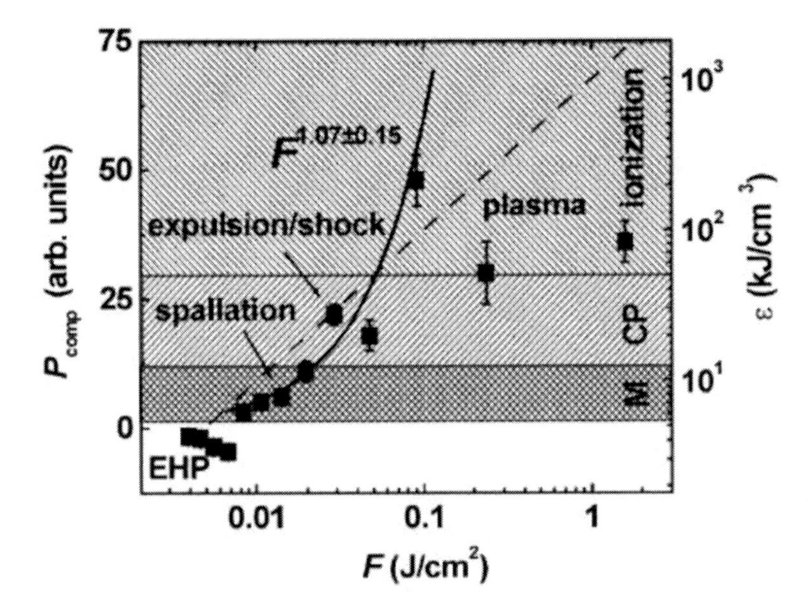

Figure 31. Fluence dependences of compressive pressure amplitude P_{comp} (dark squares) and of calculated deposited energy density ε (straight dashed line) in silicon ablated by 248-nm, 50-fs laser pulses (adapted from [143]). The arrows show the observed ultrasonic and optical effects (formation of electron-hole plasma, onset of spallation, of expulsion/shock and surface plasma), while the shaded regions represent the characteristic energy density ranges corresponding to silicon melting (M), critical point (CP) and atomic ionization [70]. The dark curve indicates the linear ($\propto F^{1.07\pm0.15}$) slope of P_{comp}(F) dependence.

V) Theoretical Modeling

The abovementioned femtosecond laser ablation paths via near-critical – supercritical and sub-critical – surface liquid-vapor phase transformations were first theoretically predicted

[36] and then thoroughly simulated over the last decade presumably in numerous illustrative molecular dynamics simulations [1-6]. Briefly summarizing their main results, the fast expansion of the femtosecond laser heated supercritical fluid results in its cooling either to supercritical states, or to sub-critical metastable states. In the first case, the supercritical fluid rapidly – on the picosecond timescale – expands further, providing its separation into vapor phase and nanoscale liquid droplets (the supercritical liquid-vapor phase separation), while on the second case explosive liquid-vapor transformation via homogeneous vapor bubble nucleation in the sub-critical superheated liquid can be considerably delayed [37-38,139]. In some works, strongly delayed heterogeneous nucleation of vapor bubbles at liquid/solid interfaces (so-called normal boiling [12,16]) was predicted [4], representing, respectively, rather extraordinary materials, which do not melt, but sublimate at normal conditions (i.e., possess very high vapor pressures in their corresponding triple points). However, keeping in mind some internal flaws of the present MD simulations – binary potentials and rather approximate EOS, there is still enough space for experimental research on femtosecond laser ablation fundamentals using more and more advanced experimental tools [41].

III. Laser-Induced Bulk Near-Critical Liquid-Vapor Phase Transformations

A. Dense Screening Surface Laser Plasma Heats the Target!

A pre-requisite condition of laser-induced bulk near-critical liquid-vapor phase transformations is multi-micron deep energy deposition inside ablated materials, which can be naturally provided by any type of highly-penetrating electromagnetic radiation, e.g., that of hot plasmas of corresponding ablative laser plumes via their short-wavelength thermal bremsstrahlung plasma emission [48]. Indeed, it was earlier noted for nanosecond laser deep drilling regimes that hot ablative plasmas existed in near proximity to target surfaces [47-48,65-66], heating the targets directly through their continuous thermal radiation [144-145]. Radiative energy transfer from laser-induced ablative plasmas to the targets is usually considered insignificant at low laser intensities compared to thermal conductivity and direct absorption of light [144], while only the undesirable plasma shielding of the targets from incident laser radiation is usually accounted for [47,146-147]. Nevertheless, in contrast, at high laser intensities such dense (critical or supercritical) hot plasma can support intense laser ablation by electronic heat conduction and short-wavelength (UV, VUV, EUV or X-ray, depending on plasma temperature T_e) radiative energy transport from the hot plasma to the target surface [144,146-148]. However, electronic heat conduction accompanied by a picosecond electron-phonon thermalization process [130,149] can provide heat penetration only on sub-micrometer scales, $(\chi\tau_{las})^{1/2} \leq 1$ μm, for the characteristic electronic diffusivity in metals $\chi \approx 1$ cm^2/s [70] and short or ultrashort laser pulses of duration $\tau_{las} < 10^{-8}$ s. In contrast, radiative energy transport is limited only by a penetration depth δ_λ and spectral exitance $M_\lambda(T_e)$ of short-wavelength (thermal and non-thermal) plasma radiation, depending for each particular wavelength λ of the plasma emission spectrum only on the plasma temperature T_e. As a result, such short-wavelength plasma radiation appears to be very useful in laser-induced plasma-assisted ablation [150] for precise micro-machining of wide-gap dielectrics with common visible nanosecond lasers, which generate transient regions of induced absorption on dielectric surfaces via their electronic excitation by a short-wavelength radiation of an extrinsic ablative laser plasma.

The characteristic, universal parameter, underlying the transition from surface absorption of electromagnetic radiation in absorbing solids (typical absorption coefficients $\alpha \sim 10^{4-6}$ cm^{-1}) to their bulk absorption ($\alpha \ll 10^4$ cm^{-1}), is the bulk plasmon wavelength $\lambda_{pl} \sim 100$ nm [93]

$$\lambda_{pl} = \frac{2\pi c}{e}\sqrt{\frac{m^*}{N_v}} = \frac{2\pi c}{\omega_{pl}}, \tag{6}$$

where c is the speed of light in vacuum, e is the charge of an electron and m^* is its effective mass, N_v is the effective valence-electron density and $\omega_{pl} \sim 10^{16}$ rad/s is the characteristic bulk plasmon frequency. As a result, both real (ε_1) and imaginary (ε_2) parts of relative permittivity for any material vary drastically near $\lambda \approx \lambda_{pl}$

$$\varepsilon_1 = n^2 - k^2 = 1 - \frac{\omega_{pl}^2\tau^2}{1+\omega^2\tau^2} \approx 1 - \frac{\lambda^2}{\lambda_{pl}^2} \approx 1, \varepsilon_2 = 2nk = \frac{\omega_{pl}^2\tau}{\omega(1+\omega^2\tau^2)} \approx \frac{\lambda^3}{(2\pi c\tau)\lambda_{pl}^2}, \tag{7}$$

where n and k are refractive index and extinction coefficient, respectively, ω is the frequency of incident radiation and τ is the momentum relaxation time for carriers. For a radiation wavelength much shorter than the bulk plasmon wavelength (i.e., $\lambda \ll \lambda_{pl} \ll c\tau$), one finds the refractive index $n \approx 1$ and the extinction coefficient $k \propto \lambda^3/\tau$. Then, the absorption coefficient $\alpha_\lambda = 4\pi k/\lambda$ exhibits a long trend $\alpha_\lambda \propto \lambda^2/\tau$ with several sharp and narrow peaks, representing characteristic X-ray lines. As an example, the extinction coefficient $k \propto \lambda^3$ at the wavelengths 10^{-3} μm $< \lambda < 10^{-2}$ μm for Al (see figure 1 in Ref. 48), while $k \propto \lambda^4$, with $\alpha_\lambda \propto \lambda^3$ and $\tau \propto \lambda^{-1}$ in the wavelength range 10^{-5} μm $< \lambda < 10^{-3}$ μm [93].

This universal behavior in the short-wavelength range of the electromagnetic spectrum is characteristic of all other materials [93]. Such highly-penetrating short-wavelength (hard UV or soft X-ray) radiation with $\lambda < \lambda_{pl}$ could be readily produced in hot near-surface, one-dimensional (1D) laser plasmas heated to the plasma temperature T_e by incident laser radiation via the *inverse Bremsstrahlung* process [43]. X-ray quanta emitted from ablative laser-induced plasmas were observed at $\lambda_{max} \sim 10^{-4}$ μm for $I \sim$ 10-100 PW/cm^2 of femtosecond laser pulses [151] and at $\lambda_{max} \sim 10^{-(3-4)}$ μm for $I \sim$ 0.1-100 TW/cm^2 of sub-nanosecond and nanosecond laser pulses [43,148]. The corresponding thermal particle energies $k_BT_e \sim 10^{2-3}$ eV in laser plasma could be estimated, using Wien displacement law in the form $\lambda_{max}T = 2.9\times10^3$ μmK [152]. These previous observations were in agreement with scaling relationships for thermal energies $k_BT_e \approx C_1I^{1/2}$ (see figure 2 in Ref. 48) and $k_BT_e \approx C_2I^{2/3}$ in sub-critical and critical laser plasmas for sub-nanosecond and nanosecond laser pulses [47,146,148,153], respectively, where k_B is the Boltzmann constant and $C_{1,2}$ are some numerical factors. For femtosecond laser pulses k_BT_e scales as $C_3I^{1/2}$ or $C_4I^{5/6}$ [48] according to calculations [68] or experimental data [151], respectively.

Thermal emission from 1D homogeneously heated plasmas into a solid angle of 2π *sr* (steradians) is characterized, according to Planck blackbody spectral radiation law [152], by spectral source exitance M_λ [W/(m^{2}μm)]

$$M_\lambda(T) = \frac{2\pi hc^2}{\lambda^5(e^{hc/\lambda kT}-1)} = \frac{3.7\times10^8}{\lambda^5[\exp(1.44\times10^4/\lambda T)-1]}, \quad (8)$$

where the temperature T and the wavelength λ are given in degrees Kelvin and μm, respectively. Integrating M_λ over the entire plasma emission spectrum at each particular T_e, one can obtain maximum intensities $I^{pl}_{max}(z = 0, T_e)$ of the short-wavelength plasma radiation on the target surface ($z = 0$)

$$I^{pl}_{max}(0,T_e) = \int_0^\infty M_\lambda(0,T_e)d\lambda \quad (9)$$

drastically increasing at increasing laser I and the corresponding plasma T_e values. Importantly, corresponding characteristic emission times $\tau^*(T_e)$ of such super-radiant plasmas are very short, rapidly decreasing as $\tau^*(T_e) \propto 1/T_e^2$ in order to satisfy the energy conservation law in the form $I\tau_{las} \approx I^{pl}_{max}(0,T_e)\tau^*(T_e)$ ($I \propto T_e^2$, $\tau_{las} \approx$ const and $I^{pl}_{max} \propto T_e^4$ in accordance with the Stefan-Boltzmann law [152], so $\tau^*(T_e) \sim 1/T_e^2$). Nevertheless, the overall effect of such radiative transfer from hot plasmas to solid targets is not negligible, providing at high plasma temperatures comparable intensities of incident laser radiation and short-wavelength plasma radiation [154].

The target depth irradiated by a short-wavelength plasma emission can be estimated using the Wien displacement law [152] to evaluate the "most probable wavelength" of such radiation, $\lambda_{max}(T_e) = 2.9\times10^3$ μmK/T_e, corresponding to maximum $M_\lambda(T_e)$, and then to take the effective skin depth of the short-wavelength plasma radiation, $\delta_{\lambda max}(T_e) = 1/\alpha_{\lambda max}(T_e)$. In so doing, one can consider plasma thermal radiation being emitted at the *single* effective wavelength, $\lambda_{max}(T_e)$, and obtain insight into the qualitative character of the laser intensity dependence of the ablative crater depth Z_{abl}, using the Beer law in the form of the following expression, $E_{abl} = E(0,T_e)exp[-\alpha_{\lambda max}(T_e)Z_{abl}]$, for the volume energy density at the target surface $E(z = 0,T_e) = \alpha_{\lambda max}(T_e)I_{max}(0,T_e)\tau^*(T_e)$. Solving this expression for Z_{abl}, one finds

$$Z_{abl}(I) = \frac{1}{\alpha_{\lambda max}(T_e(I))}\ln\left(\frac{\alpha_{\lambda max}(T_e(I))I_{max}(0,T_e(I))\tau^*(T_e(I))}{E_{abl}}\right) \propto I^\gamma \ln I, \quad (10)$$

where the threshold volume energy density, E_{abl}, represents some thermal process underlying the laser ablation, *e.g.*, melting in a case of a melt expulsion [47,65,67]. The exponent γ in this equation can be derived from the following considerations. As was discussed above, the electron temperature in the sub-critical (critical) plasma $T_e \propto I^{1/2}$ ($T_e \propto I^{2/3}$) for sub-nanosecond and nanosecond laser pulses. Then, combining a spectral dependence of absorption coefficient, $\alpha_{\lambda max} \propto \lambda_{max}(T_e)^{x(\lambda)}$ [93], with the Wien displacement law $\lambda_{max} \propto 1/T_e$, one can show that the exponent, γ, equals to $x(\lambda)/2$ or $2x(\lambda)/3$, where $x(\lambda)$ is another exponent, describing the dependence of α on λ for each particular material.

However, the introduced "most probable" absorption coefficient $\alpha_{\lambda max}(T_e)$ for the short-wavelength plasma radiation absorption in solids does not account for the spectral asymmetry of α_λ (typically, α_λ is higher in the "red" region and is much lower for its "blue" part [93]). Taking this effect into account, one can obtain the exact "mean" effective absorption coefficient $\alpha_{eff}(T_e)$ or its reciprocal "mean" skin depth $\delta_{eff}(T_e)$

$$\delta_{eff}(T_e) = \frac{\int_0^\infty M_\lambda(0,T_e)d\lambda}{\int_0^\infty \alpha_\lambda M_\lambda(0,T_e)d\lambda}, \tag{11}$$

requiring just the knowledge of optical constants for the material of interest and plasma temperature T_e. Once the corresponding $T_e(I)$ dependence is known, the calculated $\delta_{eff}(T_e)$ relationship can be converted further to $\delta_{eff}(I) = \delta_{eff}(T_e(I))$.

Furthermore, the derived above Beer-like volume energy density distributions $E(z,T_e)$ described by the exponents $\delta_{\lambda max}$ or δ_{eff}, are not quite accurate for describing bulk energy deposition inside materials and for calculating corresponding Z_{abl} magnitudes (using the expression $E_{abl} = E(0,T_e)exp[-Z_{abl}/\{\delta_{\lambda max}(T_e), \delta_{eff}(T_e)\}]$, as these phenomena are mediated by polychromatic emission of ablative surface laser plasma. Therefore, more accurate spatial distributions of $E(z,T_e)$ can be obtained considering Beer's absorption in the form $M_\lambda(z,T_e) = M_\lambda(0,T_e)\exp(-\alpha_\lambda z)$ for each particular wavelength in the plasma emission spectrum, and calculating the integral

$$E(z,T_e) = \tau^*(T_e)\int_0^\infty \alpha_\lambda M_\lambda(0,T_e)e^{-\alpha_\lambda z}d\lambda, \tag{12}$$

where z is the depth inside the target and the effective emission time of the 1D plasma $\tau^*(T_e) \sim 1/T_e^2$ is determined directly for each set of experimental conditions from the boundary condition

$$(1 - R_{av}(F))\beta(F)F = 2\tau^*(T_e)\int_0^\infty M_\lambda(0,T_e)d\lambda, \tag{13}$$

where F is the total laser fluence incident on the target producing the ablative plasma at the very beginning of the laser pulse, $R_{av}(F)$ is the corresponding reflectivity of the ablative plasma averaged over τ_{las}, $\beta(F)$ is a fraction of the total absorbed plasma energy re-emitted towards the target, the numerical factor 2 accounts for the emission of the plasma into a solid angle of 4π *sr*, while all temporal changes of T_e are taken into account in $\tau^*(T_e)$. Hence, Eq.(12) can be re-written in the final form

$$E(z,T_e) = \frac{(1 - R_{av}(F))\beta(F)F\int_0^\infty \alpha_\lambda M_\lambda(0,T_e)e^{-\alpha_\lambda z}d\lambda}{\int_0^\infty M_\lambda(0,T_e)d\lambda}, \qquad (14)$$

that allows for estimation of $E(z,T_e)$ for different laser fluences F and the known $R_{av}(F)$, $\beta(F)$ and $T_e(I)$ dependencies. However, because of its transcendent form, this expression can be used for numerical or graphical calculations of the resulting Z_{abl} magnitudes.

Finally, with the energy deposition necessary for bulk near-critical liquid-vapor phase transformations provided by the short-wavelength emission of surface laser plasma, there appears another experimental challenge in how to probe the corresponding buried processes. Below, we will present a very limited number of experimental techniques, enabling indirect identification or even direct monitoring of such bulk explosive phase transformations.

B. Bulk Explosive Liquid-Vapor Transformations Driven by Short Laser Pulses

I) Optical Imaging

Time-resolved optical imaging has so far provided numerous visualizations of melt expulsion from laser-ablated materials, which was, however, interpreted in different ways – in terms of deep drilling via continuous lateral melt expulsion from a keyhole by high recoil vapor pressures operative presumably during short heating laser pulses [144], by hydrodynamic reaction of a deep surface molten layer on such high and short impact of recoil vapor pressure [24], or by different surface instabilities in the molten layer [45]. However, there were also experimentally observed a few highly specific features of material removal, which enabled to assign them, alternatively, to a separate plasma-mediated deep laser-drilling (ablation) regime, proceeding in the form of bulk near-critical liquid-vapor transformations.

Particularly in this regime, expulsive ablative removal of molten phase were observed to occur in a threshold-like manner via lift-off of multi-micrometer sized liquid droplets [47,65,67], whose size was close to characteristic radial dimensions $D \approx$ 30-50 μm not only for preceding near-surface ablative laser plasmas, but also for the resulting craters [47,65,67], providing their high aspect ratio (per laser shot) $Z_{abl}/D \sim 1$, as compared to $Z_{abl}/D \ll 1$ below the corresponding threshold [47,65]. Moreover, melt expulsion typically took place at a right angle to the ablated surfaces with rather high expulsion velocities $V \sim 10^2$ m/s [65,67].

II) Debris Re-Deposition

The highly directed melt expulsion in the plasma-mediated deep laser-drilling regime was found to prevent its lateral re-deposition and re-solidification as a near-edge rim [47]. Surprisingly, experimental measurements of a melt volume V_{dep} re-deposited on a target surface to form a rim around the craters, and of the total volume V_{tot} of the crater formed, have demonstrated that a threshold-like rise of Z_{abl} coincided with a simultaneous sharp drop of the ratio $\psi = V_{dep}/V_{tot}$, which changes from $\psi \approx 1$ at $Z_{abl}/D \ll 1$ to $\psi \approx 0.25$ at $Z_{abl}/D \sim 1$ near the threshold for the regime [47].

III) Removal Rates

The expulsive nature of material removal in liquid form in the plasma-mediated, deep laser drilling regime provides its high drilling efficiencies at drastically increasing single- and multi-shot cavity (crater) depths Z_{abl} [46-48,65]. Moreover, the experimentally measured removal thresholds and removal rates exhibit a number of universal scaling relationships regarding different laser parameters, including laser wavelength, energy and pulse width, while the relationships strongly resemble those characteristic of sub-critical laser plasma [48,155].

As an example, in the case of plasma-mediated deep laser drilling of commercial, atomically smooth silicon (Si) wafers [48], there were experimentally obtained two curves of average drilling rate versus laser intensity, $Z_{abl}(I)$ (Figure 32), which were qualitatively consistent with the curve $Z_{abl}(I)$ now representing single-shot multi-micron ablation crater depth Z_{abl} in Si versus I (Figure 33) obtained in study [47] in the same intensity range and at the same laser wavelength. The measured threshold intensity $I_{abl} \approx 10$ GW/cm^2 for the onset of this plasma-mediated laser drilling regime by means of the 10-ns laser pulses [48] was nearly two times lower than $I_{abl} \approx 20$ GW/cm^2 presented in Ref. 47 for their 3-ns laser pulses, possibly, reflecting the scaling relationship between sub-critical plasma temperature T_e and laser pulsewidth τ_{las}, $T_e \propto [I\tau_{las}^{1/2}]^{1/2}$ [146-147,153], which gives in the former case exactly 1.7 times lower magnitude I sufficient to heat the sub-critical ablative surface laser plasma to a similar T_e as compared to the in Ref. 47. Furthermore, linear fits were obtained for two $Z_{abl}(I)$ curves for 400-μm and 520-μm thick Si samples, respectively, represented in double logarithmic coordinates $\lg[Z_{abl}/\ln I] \sim \lg I$ in figure 34. From the slope of these curves, the experimentally measured exponents γ were found to be $\gamma_{Si}^{exp} = 0.8 \pm 0.1$ ($D = 400$ μm) and $\gamma_{Si}^{exp} = 0.7 \pm 0.1$ ($D = 520$ μm), which arc reasonably consistent with $\gamma_{Si}^{theor} \approx 0.5$ (2/3) predicted from Eq.(10) for $x(\lambda) \approx 1$ in the range 10^{-3} μm $< \lambda < 10^{-1}$ μm [93] and for $T_e \propto I^{1/2}$ ($T_e \propto I^{2/3}$) in sub-critical (critical) plasma. Similar good agreement has been obtained between our model and the experimental results for Si given in Ref. 46 (Figure 37). In the latter case, the experimental data fitted in the double logarithmic coordinates $\lg Z_{abl} - \nu \lg[I^{1/2}\ln I]$ demonstrate $\nu = 0.92 \pm 0.07$ indicating that this curve $Z_{abl}(I)$ is well described by the theoretical dependence $I^{1/2}\ln I$ predicted using Eq.(10). For the glass sample, used in those experiments (see inset in figure 38), the exponent γ was found in the double logarithmic coordinates $\lg Z_{abl} \sim \lg[I^{\gamma}\ln I]$ to equal $\gamma_{glass}^{exp} = 0.5 \pm 0.1$ ($D = 220$ μm) as compared to $\gamma_{FS}^{theor} \approx$ 0.5 predicted from Eq.(10) for sub-critical laser plasma and $x(\lambda) \approx 1$ in the range 10^{-3} μm $< \lambda$ $< 10^{-1}$ μm [93]. Finally, single-shot multi-micron ablation crater depth Z_{abl} in Al vs. I from Ref. 65 for the laser wavelengths 511/578 nm and slightly lower laser intensities, which still result in strong melt expulsion (Figure 35), demonstrated good agreement with the plasma-assisted ablation model. The double logarithmic linear fit for the data $\lg Z_{abl} \propto \lg[I^{\gamma}\ln I]$ gives the value of the fitting parameter $\gamma_{Al}^{exp} = 1.2 \pm 0.1$ consistent with $\gamma_{Al}^{theor} \approx 1$ for $x(\lambda) \approx 2$ at 10^{-3} μm $< \lambda < 10^{-2}$ μm predicted from Eq.(10).

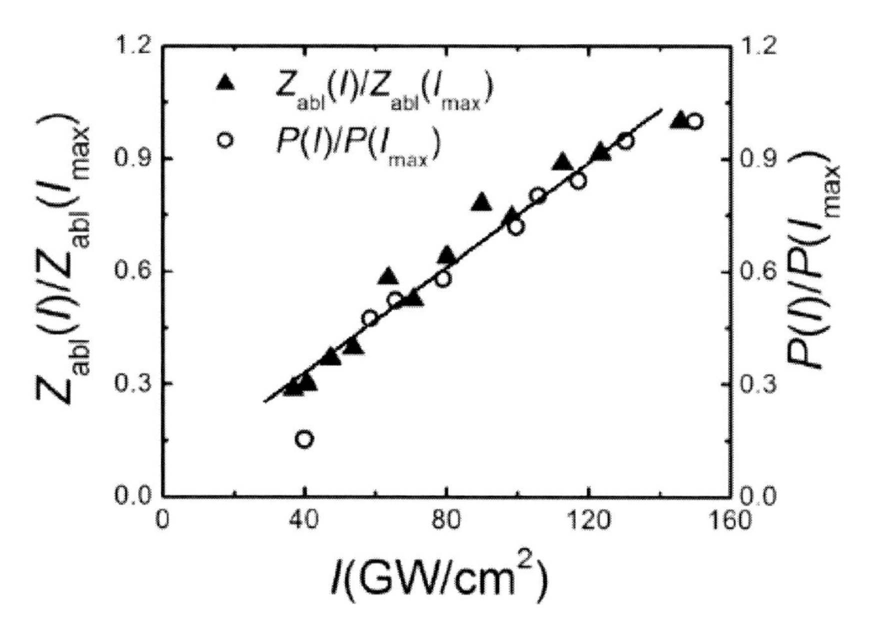

Figure 32. Normalized $Z_{abl}(I)$ and $P_{comp}(I)$ dependences for silicon surfaces ablated by 10-ns, 248-nm laser pulses; the maximum removal rate $Z_{abl}(I_{max}) \approx 12$ μm (adapted from Ref. 48).

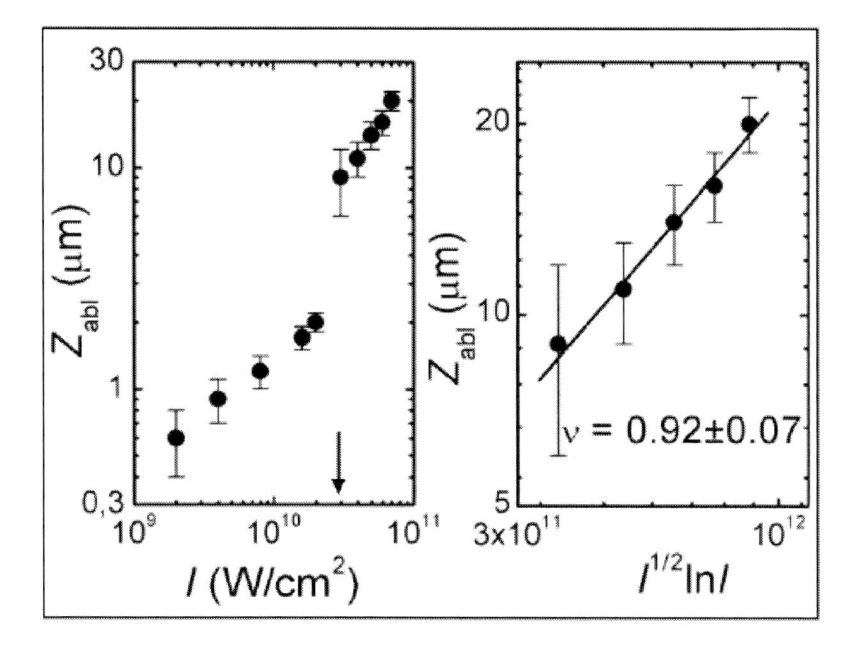

Figure 33. Dependence $Z_{abl}(I)$ (left) and its fit (right) for silicon ablated by 3-ns, 248-nm laser pulses (adapted from Ref. 47). The arrow shows the bulk liquid-vapor transformation threshold.

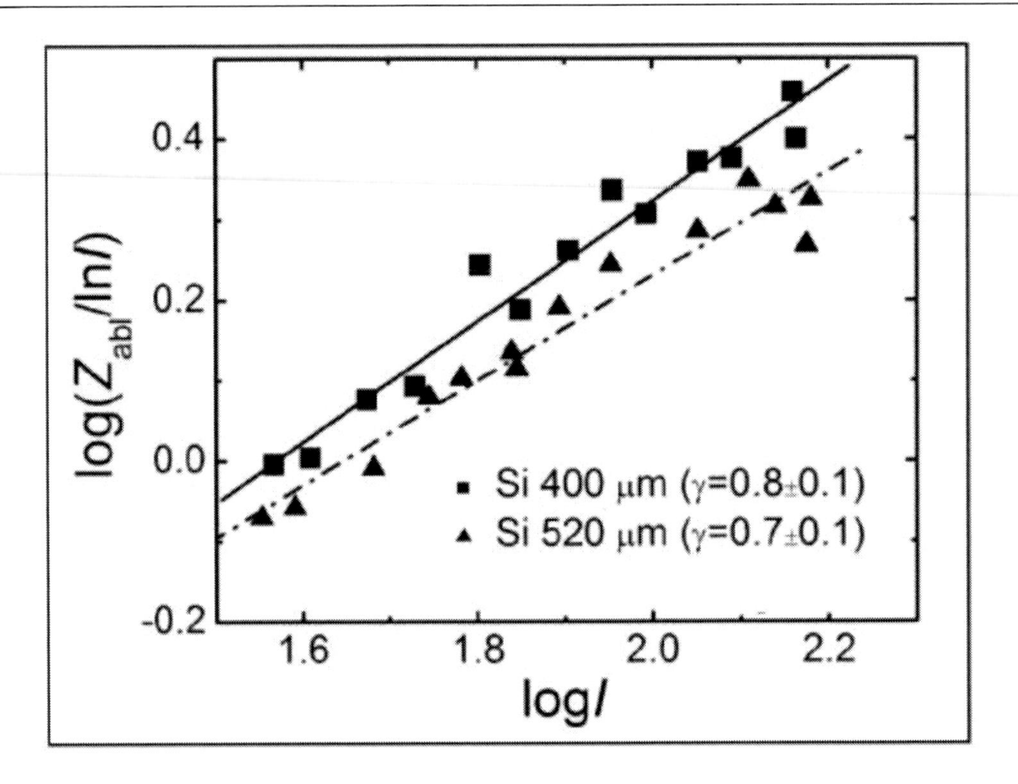

Figure 34. $Z_{abl}(I)$ curves for 400-μm and 520-μm thick Si samples represented in double logarithmic coordinates $\lg[Z_{abl}/\ln I] \sim \lg I$ (see details in figure 32 and the text).

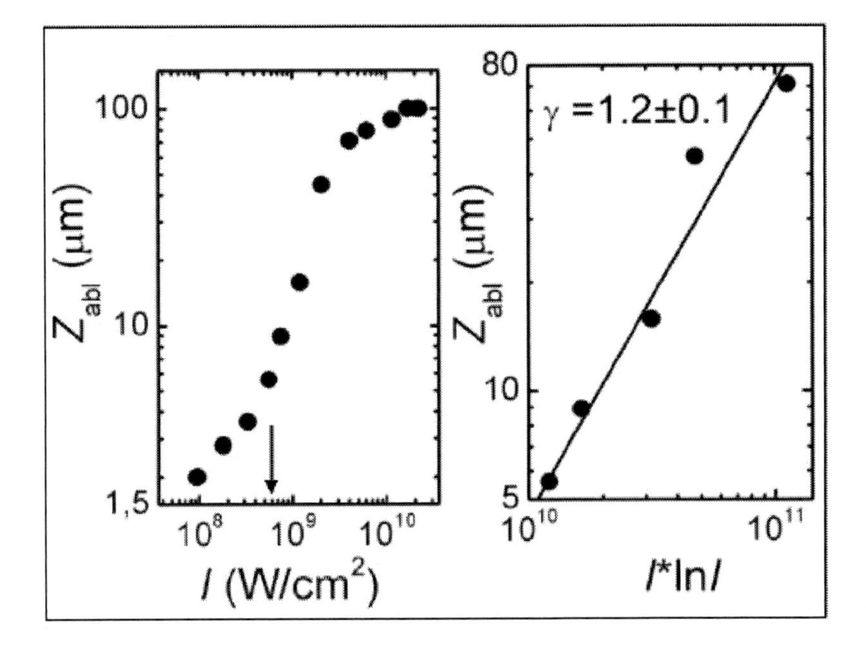

Figure 35. Dependence $Z_{abl}(I)$ (left) and its fit (right) for aluminum ablated by 30-ns, 511-nm laser pulses (adapted from Ref. 65). The arrow shows the bulk liquid-vapor transformation threshold.

Likewise, recent nanosecond UV laser ablation studies on graphite surfaces have revealed a threshold-like, second rise of removal rate Z_{abl} at high laser fluences $F \geq F_{bulk}$ (Figure 36) [156], which is consistent both in crater depths and characteristic laser intensity values with previous high-fluence studies of graphite ablation [46]. Again, the Z_{abl} magnitude in the fluence range $F \geq F_{bulk}$ exhibits the characteristic sub-linear increase with a slope $\gamma_C^{exp} \approx 0.5$ (0.46±0.36 because of the poor data set) extracted, using a double logarithmic linear fit for the data $\lg Z_{abl} \propto \lg[I^{\gamma}\ln I]$, which is in agreement with the theoretical prediction $\gamma_C^{theor} \approx 0.5$ for sub-critical laser plasma based on short-wavelength optical constants of graphite [93].

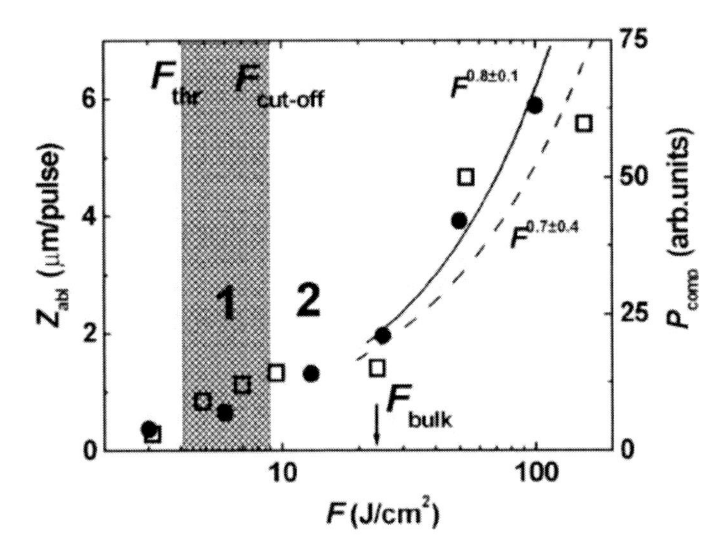

Figure 36. $Z_{abl}(I)$ and $P_{comp}(I)$ dependences for polycrystalline graphite ablated by 25-ns, 248-nm laser pulses (adapted from Ref. 156; see other details in figures 5,6,8 and in the text), with their linear fits above the bulk explosive phase transformation threshold F_{bulk} shown by the arrow.

To sum up, the existing experimental data on single- and multi-shot crater depths (removal rates) obtained at different laser fluences in previous studies [46-48,65,156] are described well by the theoretical model proposed earlier [48,145] to explain ultra-high multi-shot drilling rates and ultra-deep single-shot laser ablation craters in optically opaque and transparent solids (metals, semiconductors and dielectrics) irradiated by high-power nanosecond laser pulses in terms of ablative surface plasma mediated radiative transfer of laser pulse energy to the target via short-wavelength thermal plasma radiation.

III) Ultrasonic Imaging

In previous study [48], in order to prove that the average Z_{abl} magnitudes measured by the optical transmission technique are meaningful (i.e., the removal rate per shot is nearly uniform versus the number of laser shots N), compressive pressure amplitudes P_{comp} were measured for 400-μm thick Si wafers in the acoustic near-field region, using the contact ultrasonic technique described elsewhere [109], as a function of N (N shots on one spot, N: 1) at constant I values. These data are presented in figure 37 for different I values as normalized pressures $\pi = P_{comp}(N{:}1)/P_{comp}(N_{max}{:}1)$, where N_{max} corresponds to the laser pulse number during each pulse train, exhibiting the maximum acoustic pressure P_{comp} for the given I.

These curves showed the very small (±10%) variations of π, indeed indicating the nearly uniform average removal rates per shot, Z_{abl}, versus N, though differing for different I.

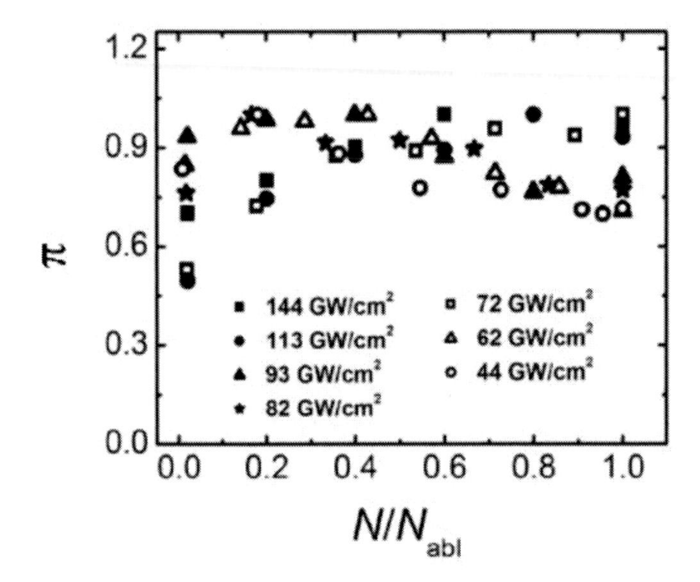

Figure 37. Normalized recoil pressure π on silicon surface versus the number N of ablating 10-ns, 248-nm laser pulses at different laser intensities (adapted from Ref. 48; N_{abl} is the intensity-dependent through-drilling pulse number).

According to their pronounced unipolar pulse shapes (Figure 38), the ultrasonic transients acquired in the ablation regime represented the near-surface pressure of ablated products and laser plasma and was proportional to the true drilling rate in each pulse [157]. Moreover, the 10-ns FWHM parameter of the recorded transients was exactly the same as the 10-ns FWHM of the Gaussian laser pulse profiles, while the laser and pressure pulses nearly coincide in time.

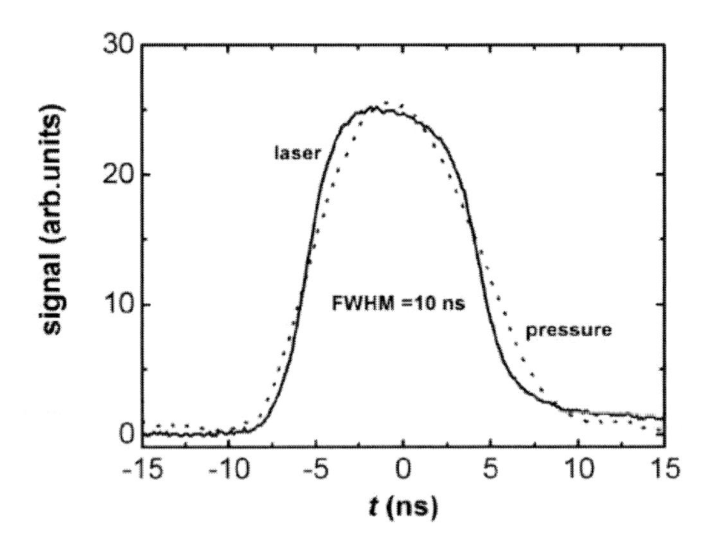

Figure 38. Ultrasonic transients acquired for silicon ablated by 10-ns, 248-nm laser pulses at the laser intensity of 62 GW/cm^2.

The amplitudes P_{comp}, measured during single-shot ablation of fresh spots on the silicon surface at different I magnitudes as an additional check of drilling rate uniformity, upon normalization to the maximum pressure amplitude $P_{comp}(I_{max})$ at the maximum laser intensity $I_{max} \approx 150$ GW/cm^2, correlated well with the average removal rate dependence $Z_{abl}(I)$ (Figure 32), normalized to the corresponding $Z_{abl}(I_{max})$ magnitude. This good agreement showed that at each value I the average removal rate was similar to that of the first laser shot, supporting the conclusion, made on the basis of experimental data in figure 37, that the average removal rate is nearly uniform. Furthermore, both the correlating $P_{comp}(I)$ and $Z_{abl}(I)$ curves demonstrate in figure 32 the same linear slope ($\propto I^1$).

To explain this coincidence of the $P_{comp}(I)$ and $Z_{abl}(I)$ curves and their linear dependence on laser intensity in this high-intensity plasma-assisted ablation regime, the intensity-dependent variations of the apparent ultrasonic source dimensions were considered. In particular, for the rather tight focusing conditions one can expect to see high-contrast spherical (point-like) optical breakdown plasma. Then, due to diffraction the acquired ultrasonic transients will have the far-field amplitude [100]

$$P_{comp}(I) = P_0(I)\frac{R_{plas}(I)}{R}, \tag{15}$$

where the plasma radius $R_{plas} \propto I^{1/3}$ for the spherical (point-like) plasma and the plasma-transducer separation distance R, $P_0(I)$ is the intensity-dependent source pressure of ablation products or ablative surface laser plasma. Since the $P_{comp}(I)$ and $Z_{abl}(I)$ curves demonstrate the same linear slope ($\propto I^1$) in figure 32, one can derive from Eq.(15) exactly $P_0(I) \propto I^{2/3}$ – the scaling relationship for critical laser plasma ($P_{plas} \sim N_{crit}kT_e \propto I^{2/3}$ for $N_{plas} = N_{crit} \equiv$ const and $kT_e \propto I^{2/3}$ [158]) – in good agreement with the derivation above, based on FWHM parameters of laser and pressure pulses in figure 38. Hence, upon proper calibration, such contact ultrasonic monitoring can be employed for the unique *in situ* tracking of the rather crucial parameters – $Z_{abl}(I)$, N_{plas} and kT_e – of buried explosive near-critical liquid-vapor transformations and optically opaque, dense *critical* ablative laser plasma .

Finally, the similar procedure was applied to derive characteristic conditions of laser-induced bulk liquid-vapor phase transformations in graphite at weak-focusing high-fluence laser ablation, represented by removal rate and compressive (recoil) pressure dependences $Z_{abl}(I)$ and $P_{comp}(I)$ in figure 36 [156]. Since the preceding laser ablation at lower fluences results in complete surface screening and saturation of removal rates (see discussion on pages 13-16 of this Chapter), one can relate the further rising $P_{comp}(I)$ curve to near-surface laser plasma pressure and the next onset of $Z_{abl}(I)$ curve at higher intensities to the plasma-mediated ablation discussed above. Then, a the weak focusing conditions one has rather large disc-like plasma and large corresponding ultrasonic source, exhibiting no diffraction and $P_{comp}(I) \approx P_0(I)$, that is $P_0(I) \propto I^{0.8}$, indicating the *sub-critical* character ($P_{plas} \propto I^{0.75\text{-}0.79}$) [146] of the ablative laser plasma on the graphite surfaces.

C. Bulk Explosive Liquid-Vapor Phase Transformations Driven by Ultrashort Laser Pulses

In the case of ultrashort laser pulses, plasma-assisted bulk heating of materials requires much higher laser intensities (see figure 2 in Refs. 48,145). For example, multi-micron deep ablative channels and corresponding high removal rates Z_{abl} (about 7 microns per shot on average) were successfully demonstrated on Cu targets at intensities $I \sim 1$ PW/cm^2 [67], when $k_B T_e \sim 10^3$ eV and $\alpha_\lambda^{-1}(\lambda \ll \lambda_{pl}) \sim 1$-$10$ μm are expected [48]. This contrasts with the common viewpoint, that higher femtosecond laser intensity results in shallower craters due to denser volume energy deposition, faster ablation and stronger plasma screening [133]. Similar high removal rates were obtained at sub-PW/cm2 laser intensities on dielectric surfaces demonstrating micron-deep single-shot craters with prominent nano-scale residue of expelled molten materials (see, e.g., figure 24). Unfortunately, not so much data was collected to date regarding such high-intensity femtosecond laser ablation regimes.

D. Modeling of Bulk Near-Critical Liquid-Vapor Phase Transformations

Despite its promising technological perspectives in fast and deep plasma-assisted laser drilling of bulk materials, to date no modeling of bulk explosive liquid-vapor phase transformations has been performed, since this effect has been just recently found [47,48,145] and its crucial parameters – spatial and temporal scales, deposited energy density distributions – were only recently evaluated [48,145].

Importantly, the abovementioned plasma-assisted laser ablation regime does provide sub-surface superheating conditions (see the corresponding spatial profiles of deposited energy density in figure 5 in Ref. 145), dictated by the criterion $B \ll 1$ [74]. It is even more surprising that highly supercritical thermodynamic states can be achieved in this short-pulse laser ablation regime at high enough – multi-micron – depths inside the target, as shown in figure 5 in Ref. 145, in agreement with the high energy density in the surface critical laser plasmas described in section III.B(iii); but this fact has yet to be experimentally verified and theoretically proved. Furthermore, the expulsion of liquid during bulk near-critical liquid-vapor transformations appears to resemble the same effect driven by recoil vapor pressure of surface vaporization in the deep drilling regime [43-45,55] and their similarity is also not well understood yet.

IV. Current Challenges in Experimental Studies and Modeling of Laser-Induced Near-Critical Liquid-Vapor Phase Transformations

First, the presented non-exhaustive overview of experimental results on short- and ultrashort-pulse laser-induced explosive near-critical liquid-vapor phase transformations unambiguously indicates that the currently existing experimental techniques are insufficient, poorly informative or even non-applicable in studies of nanoscale and fast vapor bubble nucleation processes, related near-critical singularities of basic thermodynamic quantities (isobaric heat capacity, thermal expansion coefficient, isothermal compressibility), underlying the transformations, while the new relevant experimental techniques are just emerging [41,110]. Moreover, so far there are no adequate informative experimental techniques to investigate such buried phase transformations in bulk materials. As a result, many interesting near-critical physical effects – e.g., features of sound propagation in cavitating liquids or vapor-droplet mixtures, the related potential supercritical heating of such slowly expanding cavitating liquids or vapor-droplet mixtures even by short laser pulses – and particular magnitudes of the abovementioned diverging thermodynamic quantities can not be thoroughly studied or accurately measured, respectively.

Second, theoretical modeling of such near-critical liquid-vapor phase transformations by means of numerical calculations [84-85] or MD simulations [1-6,87] is still not properly supported by adequate high-temperature, high-pressure EOS. In the former case, this is associated with ignoring the crucial near-spinodal singularities of isobaric heat capacity, thermal expansion coefficient and isothermal compressibility, governing thermal and hydrodynamic dynamics during such sub-critical liquid-vapor transformations. In the second case, the routinely used binary (two-particle) potentials do not enable for MD simulations to provide accurate representation of multi-particle correlation effects in the near-critical region. Therefore, at this time theoretical modeling may not be able to give sufficient predictive feedback to direct the related experimental research in this field.

V. CONCLUSION

In this chapter we have presented the overview of fundamentals, experimental appearances, current and potential technological implications, prospective experimental techniques and recent theoretical findings in the field of explosive liquid-vapor phase transformations in the near-critical – sub-critical or supercritical – regions of corresponding solid materials or weakly-absorbing liquids, ablated by short (sub-nanosecond, nanosecond or sub-microsecond) or ultrashort (femtosecond and short picosecond) laser pulses. We demonstrate that such laser-induced liquid-vapor transformations can be explicitly classified as surface or bulk phase transitions, depending on characteristic spatial scales of laser energy deposition inside the ablated targets, and describe the effect of surface laser plasma in providing bulk energy deposition in various materials. Based on previous and original experimental results, we emphasize the, so far, underestimated role of the thermodynamic quantities (isobaric heat capacity, thermal expansion coefficient, isothermal compressibility) diverging in the near-critical (near-spinodal) region, in mechanisms of sub-critical liquid-vapor phase transformations and their numerical modeling, while mentioning the inaccuracy of using binary potentials in molecular dynamics simulations of liquid-vapor phase separations in the near-critical region. We hope that this rather limited presentation of the previous knowledge and current advances will encourage further fruitful exploration of the multi-scale, fundamentally and technologically attractive physicochemical phenomena.

ACKNOWLEDGMENTS

The authors acknowledge the incredible contributions and technical assistance in their experimental work, invaluable discussions and friendly support from their numerous past and present co-workers – Prof. Alexander A. Karabutov and Prof. Vladimir I. Emel'yanov (both Moscow State University), Prof. Dietrich von der Linde and Dr. Klaus Sokolowski-Tinten (both – Duisburg University), Dr. Bërbel Rethfeld (University of Kaiserslautern), Prof. Susan Allen, Dr. Stanley Paul and Mr. Kevin Lyon (all – Arkansas State University), Prof. Alan J. Hunt (University of Michigan), Prof. Alexander A. Samokhin and Prof. Georgy A. Shafeev (both – A.M. Prokhorov General Physics Institute, Russian Academy of Sciences), Prof. Alexander V. Bulgakov and Prof. Nadezhda M. Bulgakova (both – Institute of Thermophysics, Siberean Branch of Russian Academy of Sciences), Prof. Andrei A. Ionin, Prof. Vladimir D. Zvorykin, Dr. Leonid V. Seleznev, Dr. Dmitriy V. Sinitsyn, Dr. Spartak I. Sagitov, Dr. Nicolai N. Ustinovskiy, Mr. Alexey O. Levchenko and Mr. Alexander A. Tikhov (all – P.N. Lebedev Physical Institute, Russian Academy of Sciences).

The authors are grateful to National Science Foundation and Russian Foundation for Basic Research for their partial financial support (Grants No. 0218024 and 08-08-00756, respectively).

REFERENCES

[1] Zhigilei, L.V.; Leveugle, E.; Garrison, B.J.; Yingling, Y.G.; Zeifman, M.I.; *Chem. Rev.* 2003, 103, 321-347.
[2] Zhakhovskii, V.V.; Nishihara, K.; Anisimov, S.I.; Inogamov, N.A.; *JETP Lett.* 2000, 71, 167-172.
[3] Cheng, C.; Xu, X.; *Appl. Phys. A* 2004, 79, 761-765.
[4] Lorazo, P.; Lewis, L.J.; Meunier, M.; *Phys. Rev. B* 2006, 73, 134108.
[5] Leveugle, E.; Ivanov, D.S.; Zhigilei, L.V.; *Appl. Phys. A* 2004, 79, 1643-1655.
[6] Upadhyay, A.K.; Inogamov, N.A.; Rethfeld, B.; Urbassek, H.M.; *Phys. Rev. B* 2008, 78, 045437.
[7] Kanel, G.I.; Razorenov, S.V.; Bogatch, A.; Utkin, A.V.; Fortov, V.E.; Grady, D.E.; *J. Appl. Phys.* 1996, 79, 8310.
[8] Moshe, E.; Eliezer, S.; Henis, Z.; Werdiger, M.; Dekel, E.; Horovitz, Y.; Maman, S.; Goldberg, I.B.; Eliezer, D.; *Appl. Phys. Lett.* 2000, 76, 1555-1557.
[9] Debenedetti, P.G. *Metastable liquids: Concepts and Principles*; Princeton University press: Princeton, NJ, 1996.
[10] Martynyuk, M.M.; *Rus. J. Phys. Chem.* 1983, 57, 494-501.
[11] aus der Wiesche, S.; Rembe, C.; Hofer, E.P.; *Heat Mass Transf.* 1999, 35, 25-31.
[12] Miotello, A.; Kelly, R.; *Appl. Phys. Lett.* 1995, 67, 3535.
[13] Bulgakova, N.M.; Bulgakov A.V.; Bourakov, I.M.; Bulgakova, N.A.; *Appl. Surf. Sci.* 2002, 197-198, 96-99.
[14] Cracuin, V.; Cracuin, D.; *Phys. Rev. B* 1998, 58, 6787-6790.
[15] Trevena, D.H. *Cavitation and tension in Liquids*; Adam Hilger: Bristol, 1987.

[16] van Stralen, S.; Cole, R. Boiling Phenomena: Physicochemical and Engineering Fundamental s and Applications; Hemisphere Publishing Corp.: Washington, DC, 1979.
[17] Blander, M.; Katz, J.L.; *J. AlChE* 1975, 21, 833-848.
[18] Skripov, V.P.; Sinitsyn, E.N.; Pavlov, P.A.; Ermakov, G.V.; Muratov, G.N.; Bulanov, N.V.; Baidakov, V.G. *Thermophysical Properties of Liquids in the Metastable State*; Gordon and Breach: NY, 1988.
[19] Xu, X.; Chen, G.; Song, K.H. *Int. J. Heat Mass Transfer* 1999, 42, 1371-1382.
[20] Skripov, V.P.; Skripov, A.V.; *Sov. Phys. Uspekhi* 1979, 22, 389-410.
[21] Abraham, F.F.; Schreiber, D.E.; Mruzik, M.R.; Pound, G.M.; *Phys. Rev. Lett.* 1976, 36, 261-264.
[22] Garrison, B.J.; Itina, T.E.; Zhigilei, L.V.; *Phys. Rev. E* 2003, 68, 041501.
[23] Kudryashov, S.I.; Lyon, K.; Allen, S.D.; *Phys. Rev. E* 2007, 75, 036313.
[24] Paltauf, G.; Schmidt-Kloiber, H.; *Appl. Phys. A* 1995, 62, 303-311.
[25] Kim, D.; Grigoropoulos, C.P.; *Appl. Surf. Sci.* 1998, 127-129, 53-58.
[26] Esenaliev, R.O.; Karabutov, A.A.; Podymova, N.B.; Letokhov, V.S.; *Appl. Phys. B* 1994, 59, 73-81.
[27] Kudryashov, S.I.; Zorov, N.B.; *Mendeleev Commun.* 1998, 5, 178.
[28] Rohlfing, A.; Menzel, C.; Kukreja, L.M.; Hillenkamp, F.; Dreiswerd, K.; *J. Phys. Chem.* 2003, 107, 12275-12286.
[29] Fan, X.; Little, M.W.; Murray, K.K.; *Appl. Surf. Sci.* doi:10.1016/j.apsusc.2008.06.033.
[30] Jackson, S.N.; Kim, J.-K.; Laboy J.L.; Murray K.K.; *Rap. Comm. Mass Spectrom.* 2006, 20, 1299-1304.
[31] Kudryashov, S.I.; Karabutov, A.A.; Emel'yanov, V.I.; Kudryashova, M.A.; Voronina, R.D.; Zorov, N.B.; *Mendeleev Commun.* 1998, 1, 25.
[32] Kudryashov, S.I. PhD thesis, Moscow State University, 1999.
[33] Arai, Y.; Sako, T.; Takebayashi, Y. (eds.) Supercritical fluids: Molecular Interactions, Physical Properties, and New Applications; Springer: Berlin, 2002.
[34] Cavalleri, A.; Sokolowski-Tinten, K.; Bialkowski, J.; Schreiner, M.; von der Linde, D.; *J. Appl. Phys.* 1999, 85, 3301-3309.
[35] Sokolowski-Tinten, K.; Kudryashov, S.; Temnov, V.; Bialkowski, J.; von der Linde, D.; Cavalleri, A.; Jeschke, H.O.; Garcia, M.E.; Bennemann, K.H.; *Springer Series in Chemical Physics* 2000, 66, 425-427.
[36] Sokolowski-Tinten, K.; Bialkowski, J.; Cavalleri, A.; von der Linde, D.; Oparin, A.; Meyer-ter-Vehn, J.; Anisimov, S. I.; *Phys. Rev. Lett.* 1998, 81, 224.
[37] Mingareev, I.; Horn, A.; *Appl. Phys. A* 2008, 92, 917-920.
[38] König, J.; Nolte, S.; Tünnermann, A.; *Opt. Exp.* 2005, 13, 10597-10607.
[39] Kabashin, A.V.; Meunire, M.; *J. Phys.: Conf. Ser.* 2007, 59, 354-359.
[40] Bonse, J.; Bachelier, G.; Siegel, J.; Solis, J.; *Phys. Rev. B* 2006, 74, 134106.
[41] Lindenberg, A.M et al.; *Phys. Rev. Lett.* 2008, 100, 135502.
[42] Kudryashov, S.I.; Bulgakov, A.V.; Bulgakova, N.M. (unpublished results).
[43] Ready, J.F.; *Effects of High-Power Laser Radiation*; Academic Press: NY, 1971.
[44] Anisimov, S. I.; Imas, Ya. A.; Romanov, G. S.; Khodyko, Yu. V. *Action of High-Power Laser Radiation on Metals;* Nauka: Moscow, 1970 (in Russian).
[45] Bäuerle, D. *Laser Processing and Chemistry*; Springer-Verlag: Berlin, 2000.
[46] Bulgakova, N.M.; Bulgakov, A.V.; *Appl. Phys. A* 2001, 73, 199-208.

[47] Russo, R. E.; Mao, X. L.; Liu, H. C.; Yoo, J. H.; Mao S.S., *Appl. Phys. A* 1999, 69, 887; Yoo, J. H.; Jeong, S. H.; Mao, X. L.; Greif, R.; Russo, R. E.; *Appl. Phys. Lett.* 2000, 76, 783.
[48] Paul, S.; Kudryashov, S.I.; Lyon, K.; Allen, S.D.; *J. Appl. Phys.* 2007, 101, 043106.
[49] Lyamshev, L.M.; Naugol'nikh, K.A.; *Sov. Phys. Acoust.* 1981, 27, 641-668; Bunkin, F.V.; Tribel'skiy, M.I.; *Sov. Phys. Usp.* 1980, 130, 193-239.
[50] Vitshas, A.F.; Dorozhkin, L.M.; Doroshenko, V.S.; Korneev, V.V.; Menakhin, L.P.; Terentiev, A.P.; *Sov. Phys. Acoust.* 1988, 34, 43.
[51] Bostanjoglo, O.; Niedrig, R.; Wedel, B.; *J. Appl. Phys.* 1994, 76, 3045-3048.
[52] Kudryashov, S.I.; *Proc. SPIE* 2006, 6106, 61061C.
[53] Vorobyev, A.Y.; Guo, C.; *Phys. Rev. B* 2005, 72, 195422.
[54] Kudryashov, S.I.; Joglekar, A.; Mourou, G.; Herbstman, J.F.; Hunt, A.J.; *Appl. Phys. Lett.* 2007, 91, 141111.
[55] Anisimov, S.I.; Luk'yanchuk, B.S.; *Usp. Fiz. Nauk* 2002, 172, 301-333.
[56] Kartashov, I.N.; Samokhin, A.A., Smurov, I.Yu.; *J. Phys. D* 2005, 38, 3703-3714.
[57] Kudryashov, S.I. (unpublished results).
[58] Jeschke, H.O.; Garcia, M.E.; Bennemann, K.H.; *Phys. Rev. Lett.* 2001, 87, 015003.
[59] Temnov, V.V.; Sokolowski-Tinten, K.; Stojanovic, N.; Kudryashov, S.; von der Linde, D.; Kogan, B.; Weyers, B.; Möller, R.; Seekamp, J.; Sotomayor-Torres, C.; *Proc. SPIE* 2002, 4760, 1032.
[60] Khodorkovskii, M.A.; Artamonova, T.O.; Murashov, S.V.; Shakhmin, A.L.; Belyaeva, A.A.; Rakcheeva, L.P.; Fonseca, I.M.; Ljubčik, S.V.; *Tech. Physics* 2005, 50, 1301-1304.
[61] Kozlov, B.N.; Kirillov, S.N.; Mamyrin, B.A.; *Proc. SPIE* 1997, 3093, 233-238.
[62] Amoruso, S.; Bruzzese, R.; Spinelli, N.; Velotta, R.; Vitiello, M.; Wang, X.; *Europhys. Lett.* 2004, 67, 404-410.
[63] Kudryashov, S.I.; Tikhov, A.A.; Bulgakov, A.V.; Bulgakova, N.M. (unpublished results).
[64] Kelly, R.; Miotello, A.; *Nucl. Instrum. Meth. Phys. Res. B* 1994, 91, 682-691.
[65] Körner, C.; Mayerhofer, R.; Hartmann, M.; Bergmann, H. W.; *Appl. Phys. A* 1996, 63, 123.
[66] Pakhomov, A.V.; Thompson, M.S.; Gregory, D.A.; *J. Phys. D* 2003, 36, 2067.
[67] Luft, A.; Franz, U.; Emsermann, A.; Kaspar, J.; *Appl. Phys. A* 1996, 63, 93.
[68] Price, D.F.; More, R.M.; Walling, R.S.; Guethlein, G.; Shepherd, R.L.; Stewart, R.E.; White, W.E.; *Phys. Rev. Lett.* 1995, 75, 252-255.
[69] Kudryashov, S.I.; Pakhomov, A.V.; Allen, S.D.; *Proc. SPIE* 2005, 5713, 508.
[70] Grigor'ev, I.S.; Meylikhov, E.Z. (ed.); *Fizicheskie velichini*; Energoatomizdat: Moscow, 1991.
[71] Batanov, V.A.; Bunkin, F.V.; Prokhorov, A.M.; Fedorov, V.B.; *Sov. Phys. JETP* 1973, 36, 311.
[72] Andreev, S.N.; Orlov, S.V.; Samokhin, A.A.; *Phys. Wave Phenom.* 2007, 15, 67-80.
[73] Yoo, J. H.; Jeong, S. H.; Greif, R.; Russo, R. E.; *J. Appl. Phys.* 2000, 88, 1638.
[74] Dabby, F.W.; Paek, U.C.; *IEEE J. Quant. Electron.* 1972, 8, 106.
[75] Anisimov, S.I.; Bonch-Bruevich, A.M.; El'yashevich, M.A.; Imas, Ya.A.; Pavlenko, N.A.; Romanov, G.S.; *Sov. Phys.- Tech. Phys.* 1967, 11, 935.

[76] Samokhin, A.A.; *Reports of General Physics Institute*; Nauka: Moscow, 1988, V.13 (in Russian).
[77] Kelly, R.; Miotello, A. *Appl. Surf. Sci.* 1996, 96-98, 205-215.
[78] Ubblohde, A.R.; *The Molten State of Matter*; Wiley, 1978.
[79] Landau, L.D.; Lifshitz, E.M. *Hydrodynamics*; Nauka: Moscow, 1986, V.6.
[80] Sokoloswski-Tinten, K.; Bialkowski, J.; Cavalleri, A.; Boing, M.; Schuler, H.; von der Linde, D.; *Proc. SPIE* 1998, 3343, 46-57.
[81] Landau, L.D.; Lifshitz, E.M. *Statistical Physics*; Nauka: Moscow, 1995, V.5, Part 1.
[82] Martynyuk, M.M.; *Russ. J. Phys. Chem.* 1996, 70, 1194-1197.
[83] Eberhart, J.G.; Schnyders, H.C.; *J. Phys. Chem.* 1973, 77, 2730-2735.
[84] Vidal, F.; Johnston, T.W.; Barthelemy, O.; Chaker, M.; Le Drogoff, B.; Margot, J.; Sabsabi, M.; *Phys. Rev. Lett.* 2001, 86, 2573-2576.
[85] Andreev, S.N.; Demin, M.M.; Mazhukin, V.I.; Samokhin, A.A.; *Bull. Lebedev Phys. Inst.* 2006, 3, 12-15.
[86] Lu, Q.; *Phys. Rev. E* 2003, 67, 016410.
[87] Zhigilei, L.V.; *Appl. Phys. A* 2003, 76, 339-350.
[88] Karabutov, A.A.; Kubyshkin, A.P.; Panchenko, V.Ya.; Podymova, N.B.; *Quant. Eelectron.* 1995, 22, 820-824.
[89] Sokolowski-Tinten, K.; Bialkowski, J.; Cavalleri, A.; von der Linde, D.; *Appl. Surf. Sci.* 1998, 127-129, 755-760.
[90] Ionin, A.A.; Kudryashov, S.I.; Sagitov, S.I.; Seleznev, L.V.; Sinitsyn, D.V.; Tikhov, A.A.; Zvorykin V.D. (unpublished results).
[91] Song, K.H..; Xu, X.; *Appl. Surf. Sci.* 1998, 127-129, 111-116.
[92] Meyyappan, M.; *Science and Technology of Carbon Nanotubes*: CRC Press: Boca Raton, FL, 2004.
[93] Palik, E.D. *Handbook of Optical Constants of Solids*; Academic Press: Orlando, 1985.
[94] Kudryashov, S.I.; Bulgakova, A.V.; Bulgakova, N.M.; Zvorykin V. D. (unpublished results).
[95] Reitze, D. H.; Ahn, H.; Downer, M. C.; Phys. Rev. B 1992, 45, 2677-2693.
[96] Ionin, A.A.; Kudryashov, S.I.; Seleznev, L.V.; Sinitsyn, D.V.; Zvorykin V.D. (unpublished results).
[97] Musella, M.; Ronchi, C.; Brykin, M.; Scheindlin, M.; *J. Appl. Phys.* 1998, 84, 2530-2534.
[98] Glushko, V.P. (ed.); *Thermodynamic Properties of Individual Substances*; Nauka: Moscow, 1979, V.2, book 1.
[99] Afanas'ev, Yu. V.; Krokhin, O. N.; *Zh. Eksp. Teor. Fiz.* 1967, 52, 966.
[100] Gusev, V.E.; Karabutov, A.A.; *Laser Optoacoustics*; AIP: NY, 1993.
[101] Zhigilei, L.V.; Garrison, B.J.; *J. Appl. Phys.* 2000, 88, 1281-1298.
[102] Kudryashov, S.I.; Lyon, K.; Allen, S.D.; *Phys. Rev. E* 2006, 73, 055301.
[103] Shori, R.K.; Walston, A.A.; Stafsudd, O.M.; Fried, D.; Walsh, Jr., J.T.; *IEEE J. Select. Topics Quant. Electron.* 2001, **7**, 959.
[104] Bunkin, F.V.; Kolomensky, A.A.; Mikhailevich, V.G.; Nikiforov, S.M.; Rodin, A.M.; Sov. Phys. Acoust. 1986, 32, 21.
[105] Emmony D.C., Geerken T., Klein-Baltnik H., *J. Acoust. Soc. Am.* 73, 220 (1983).
[106] MacDonald, K.F.; Fedotov, V.A.; Pochon, S.; Soares, B.F.; Zheludev, N.I.; Guignard, C.; Mihaescu, A.; Besnard, P.; *Phys. Rev. E* 2003, 68, 027301.

[107] Testud-Giovanneschi, P.; Alloncle, A.P.; Dufresne, D.; *J. Appl. Phys.* 1990, 67, 3560.
[108] Hatanaka, K.; Kawao, M.; Tsuboi, Y.; Fukumura, H.; Masuhara, H.; *J. Appl. Phys.* 1997, 82, 5799.
[109] Kudryashov, S.I.; Allen, S.D.; *J. Appl. Phys.* 2004, 95, 5820.
[110] Kotaidis, V.; Plech, A.; *Appl. Phys. Lett.* 2005, 87, 213102; Kotaidis, V.; Dahmen, C.; von Plessen, G.; Springer, F.; Plech, A.; *J. Chem. Phys.* 2006, 124, 184702.
[111] Lang, F.; Leiderer, P.; *New J. Phys.* 2006, 8,14-20.
[112] Tomita, S.; Andersen, J.U.; Gottrup, C.; Hvelplund, P.; Pedersen, U.V.; *Phys. Rev. Lett.* 2001, 87, 073401.
[113] Kaplan, A.F.H.; Mizutani, M.; Katayama, S.; Matsunawa, A.; *J. Phys. D* 2002, 35, 1218-1228.
[114] Landau, L.D.; Lifshitz, E.M. *Theory of Elasticity*; Nauka: Moscow, 1987,V.7.
[115] Mescheryakov, Y.P.; Bulgakova, N.M.; *Appl. Phys. A* 2006, 82, 363-368.
[116] Akhmanov, S.A.; Koroteev, N.I.; Shumay, I.L. *Nonlinear Optical Diagnostics of Laser-Excited Semiconductor Surfaces*, in International Handbook of Laser Science and Technology; Harwood Academic Publishers: Chur, 1989, V.2.
[117] Kudryashov, S.I.; Kandyla, M.; Roeser, C.; Mazur, E.; *Phys. Rev. B* 2007, 75, 085207.
[118] Martin, P.; Guizard, S.; Daguzan, Ph.; Petite, G.; D'Oliveira, P.; Meynadier, P.; Perdrix, M.; *Phys. Rev. B* 1997, 55, 5799.
[119] Yu, P.Y.; Cardona, M.; Fundamentals of Semiconductors: Physics and Materials Properties; Springer: Berlin, 2002, Ch. 3.
[120] Ionin, A.A.; Kudryashov, S.I.; Sagitov, S.I.; Seleznev, L.V.; Sinitsyn, D.V.; Tikhov, A.A.; Zvorykin V.D. (unpublished results).
[121] Shank, C.V.; Yen, R.; Hirliman, C.; *Phys. Rev. Lett.* 1983, 51, 900.
[122] Tom, H.W.K.; Aumiller, G.D.; Brito-Cruz, C.H.; *Phys. Rev. Lett.* 1988, 60, 1438.
[123] Callan, J.P.; Kim, A.M.-T.; Huang, L.; Mazur, E.; *Chem. Phys.* 2000, 251, 167.
[124] Hillyard, P.B.; Gaffney, K.J.; Lindenberg, A.M. et al.; *Phys. Rev. Lett.* 2007, 98, 125501.
[125] Fritz, D.M.; Reis, D.A.; Adams, B. et al.; *Science* 2007, 315, 633.
[126] Sokolowski-Tinten, K.; von der Linde, D.; *Phys. Rev. B* 2000, 61, 2643-2650.
[127] Ivanov, D.S.; Zhigilei, L.V.; Phys. Rev. Lett. 2003, 91, 105701.
[128] Ivanov, D.S.; Zhigilei L.V.; *Phys. Rev. B* 2003, 68, 064114.
[129] Kudryashov, S.I.; Emel'yanov, V.I.; *JETP Lett.* 2001, 73, 666-670.
[130] Bulgakova, N.M.; Stoian, R.; Rosenfeld, A.; Hertel, I.V.; Campbell, E.E.B.; *Phys. Rev. B* 2004, 69, 054102.
[131] Dachraoi, H.; Husinsky, W.; Betz, G.; *Appl. Phys. A* 2006, 83, 333-336.
[132] Lenner, M.; Kaplan, A.; Palmer, P.E.; *Appl. Phys. Lett.* 2007, 90, 153119.
[133] Kudryashov, S.I.; Emel'yanov, V.I.; *JETP* 2002, 94, 94.
[134] Bulgakova, N.M.; Zhukov, V.P.; Vorobyev, A.Y.; Guo, C.; *Appl. Phys. A* 2008, 92, 883-889.
[135] Morozov, A.A.; *Appl. Phys. A* 2004, 79, 997-999.
[136] Furusawa, K.; Takahashi, K.; Kumagai, H.; Midorikawa, K.; Obara, M.; *Appl. Phys. A* 1999, 69, S359-S366.
[137] Her, T.-H.; Finlay, R.J.; Wu, C.; Deliwala, S.; Mazur, E.; *Appl. Phys. Lett.* 1998, 73, 1673-1675.

[138] Sylvestre, J.-P.; Kabashin, A.V.; Sacher, E.; Meunier, M.; Luong, J.H.T.; *J. Am. Chem. Soc.* 2004, 126, 7176-7177.

[139] Ionin, A.A.; Kudryashov, S.I.; Seleznev, L.V.; Sinitsyn, D.V.; Tikhov, A.A.; Zvorykin V.D. (unpublished results).

[140] Herbstman, J.F.; Hunt, A.J.; Yalisove, S.M.; *Appl. Phys. Lett.* 2008, 93, 011112.

[141] Matsuda, O.; Wright, O.B.; Hurley, D.H.; Gusev, V.E.; Shimizu, K.; *Phys. Rev. Lett.* 2004, 93, 095501.

[142] Graham, R.A. Solids Under High-Pressure Shock Compression; Springer: Berlin, 1993.

[143] Ionin, A.A.; Kudryashov, S.I.; Seleznev, L.V.; Sinitsyn, D.V.; Tikhov, A.A.; Zvorykin V.D. (unpublished results).

[144] Tix, C.; Simon, G.; *Phys. Rev. E* 1994, 50, 453.

[145] Kudryashov, S.I.; Pakhomov, A.V.; Allen, S.D.; *Proc. SPIE* 2005, 5713, 508.

[146] Phipps, Jr., C.R.; Turner, T.P.; Harrison, R.F.; York, G.W.; Osborne, W.Z.; Anderson, G.K.; Corlis, X.F.; Haynes, L.C.; Steele, H.S.; Spicochi, K.C.; *J. Appl. Phys.* 1988, 64, 1083; see also Phipps, C.P.; Luke, J.R.; Lippert, T.; Hauer, M.; Wokaun, A.; *Appl. Phys. A* 2004, 79, 1385.

[147] Beinhorn, F.; Ihlemann, J.; Luther, K.; Troe, J.; *Appl. Phys. A* 2004, 79, 869.

[148] Glenzer, S.H.; Rozmus, W.; MacGowan, B.J.; Estabrook, K.G.; De Groot, J.D.; Zimmermann, G.B.; Baldis, H.A.; Harte, J.A.; Lee, R.W.; Williams, E.A.; Wilson, B.G.; *Phys. Rev. Lett.* 1999, 82, 97; for broader bibliography, see Ref. 48.

[149] Suarez, C.; Bron, W.E.; Juhasz, T.; *Phys. Rev. Lett.* 1995, 75, 4536; Hohlfeld, J.; Müller, J.G.; Wellershoff, S.-S.; Matthias, E.; *Appl. Phys. B* 1997, 64, 387.

[150] Hanada, Y.; Sugioka, K.; Gomi, Y.; Yamaoka, H.; Otsuki, O.; Miyamoto, I.; Midorikawa, K.; *Appl. Phys. A* 2004, 79, 1001.

[151] Gizzi, L.A.; Guilietti, A.; Willi, O.; Riley, D.; *Phys. Rev. E* 2000, 62, 2721; for broader bibliography, see Ref. 48.

[152] McCluney, R., *Introduction to Radiometry and Photometry*; Artech House: Boston, 1994.

[153] Lee, P.H.-Y.; Rosen, M.D.; *Phys. Rev. Lett.* 1979, 42, 236; Albritton, J.R.; Langdon, A.B.; *Phys. Rev. Lett.* 1980, 45, 1794.

[154] Kudryashov, S.I.; Lyon, K.; Allen, S.D.; *J. Appl. Phys.* 2006, 100, 124908.

[155] Russo, R.E.; Mao, X.L.; Liu, H.C.; Yoo, J.H.; Mao, S.S.; *Appl. Phys. A* 1999, 69, S887-S894.

[156] Kudryashov S.I., Tikhov, A.A. (unpublished results).

[157] Kudryashov, S.I.; Allen, S.D.; *J. Appl. Phys.* 2002, 92, 5627.

[158] N.G. Basov, O.N. Krokhin, and G.V. Sklizkov, *Reprint FIAN USSR* 1970, 52, 71 (in Russian).

In: Phase Transitions Induced by Short Laser Pulses
Editor: Georgy A. Shafeev
ISBN: 978-1-60741-590-9

Chapter 2

PHASE TRANSITIONS INDUCED BY NANOSECOND LASER PULSES IN CONFINED GEOMETRY

***A.A. Karabutov*[1]*, A.G. Kaptilniy*[2] *and A.Yu. Ivochkin*[1]**

[1] International Laser Center of M.V. Lomonosov Moscow State University, Leninskie Gory 1, Moscow, 119991, Russia

[2] Joint Institute for High Temperatures, Russian Academy of Science, Izhorskaya str. 13/19, Moscow, 125412, Russia

ABSTRACT

The results of the study of high-energy states and phase transitions for the case of lead and mercury as an example for the laser irradiation in large ranges of intensities are presented. In order to realize conditions of highly-effective pressure generation process at the conditions of local thermodynamic equilibrium the irradiated surface was mechanically confined by a plate of transparent dielectric. Dynamics of the thermodynamic state of the metal is analyzed by amplitude and shape of the pressure pulse (in the case of lead) which propagates away from the heated surface and by change of reflectivity of the irradiated metal surface. For the case of mercury the temperature measurements were additionally performed.

The changes in the pressure pulse shape made it possible to register phase transitions in lead: melting and boiling, in mercury – boiling at the irradiated surfaces at pressure up to $P_{\max} \sim 0.1$ GPa. It is shown that before the start of the lead melting and mercury boiling process the amplitude of pressure is proportional to the intensity of laser radiation. When the laser irradiation intensity steps over the melting (boiling) threshold the shape of the pressure pulse tends to be proportional to the integral of the adsorbed energy. The moment of the deformation of the form of the pressure pulse determines the start of the phase transition. During the realization of the high-energy states at the heating zone the density considerably decreases and that leads to considerable decrease of reflectivity as well as electroconductivity in the thin subsurface metal layer.

1. Introduction

The investigation of high-energy states and first-order phase transitions of materials induced by high power laser irradiation pulse is a problem of great importance both for the fundamental science and for its applications. High power (10^9 W/cm^2) laser irradiation makes it possible to convert the material into the area of high-energy states. It is a promising and an effective method of investigation of properties of matter in a wide area of temperatures and pressures. The process path in the space of thermodynamic parameters is determined both by the properties of the laser pulse (pulse shape, energy) and by the equation of state of the material itself. The matter of special interest is the near-critical area and the transition interval of the parameters – «metal - dielectric» ($T > 5000$ K). In this case it is an area of dense typically degenerated matter which is characterized by great coulomb interaction.

The determination of critical parameters of metals is an important fundamental task for the obtaining wide-area equations of state. The theoretical estimations give the values of tens of thousands degrees of K and tens of thousands degrees of atm. At this area of pressures and temperatures the usage of classical static methods is connected with great difficulties and in most cases is impossible. Due to this fact the critical parameters are experimentally determined only for part of the alkali metals and for mercury [1].

The interaction of short high power laser pulse with the material makes it possible to overcome limitations for the achieved values of near- and supercritical parameters – temperatures and pressures. For metals supercritical states can be obtained using laser irradiation with pulse energy of only a few Joules. Meanwhile it may well be that the duration of the laser pulse τ_l is less then the time of the decay of the metastable state τ_m and the matter can convert to a near-spinodal state. In this case the determination of rather high thermodynamic–parameters (temperature, pressure, density) which continuously change during laser impulse and at the same time are irregular in space, becomes the main task. Besides the development (design) of special methods of diagnostics both of the phase transitions in the matter and of finer effects, for example the effects that follow the deep immersion into the metastable area, crossing the spinodal into the labile state is essential. The solution of these problems will give information about the specific aspects of material behavior in the near-critical area and the necessary data for construction of the wide-range equations of the state of the matter [2].

One should note that for the interaction of laser radiation with metals the typical time of thermodynamic equilibrium establishment is, in the first place, the time of the electron-phonon relaxation ending, which is the 10^{-11}-10^{-12} s of the order of magnitude. This condition is not observed when the ultra-short laser pulses (of femto- and picosecond duration) are used $\tau_l \sim 10^{-14}$-10^{-12} s [3], but is valid for laser pulse of nanosecond duration $\tau_l \sim 10^{-8}$ s. In the last case the system behavior can be described using thermodynamic terms such as temperature and pressure. Besides, the possibility of generation of high values of dynamical pressure in the condensed matter is determined by prevail of pressure generation process dependent on heat diffusion process (τ_l) over the pressure relaxation process dependent on the speed of sound. In the case of metals this condition gives $\sqrt{\frac{\chi}{\tau_l}} \sim c_0$ (where χ - heat diffusion coefficient, c_0 – speed of sound in metal) and is valid for laser pulses of sub-nanosecond duration $\tau_l \sim 10^{-10}$ s. Due to this fact in micro- and especially in millisecond ranges of laser pulse duration $\tau_l \geq 10^{-6}$ s the efficiency of inducing high values of dynamical pressure decreases considerably. This is due to the effects of relaxation and interference of acoustical pulses which reflect many times from for sample borders within 10^{-6} -10^{-3} s. Thus the nanosecond range of laser pulse duration is optimal. On figure 1 the scheme of the temporal scales which determine the laser heating process is presented.

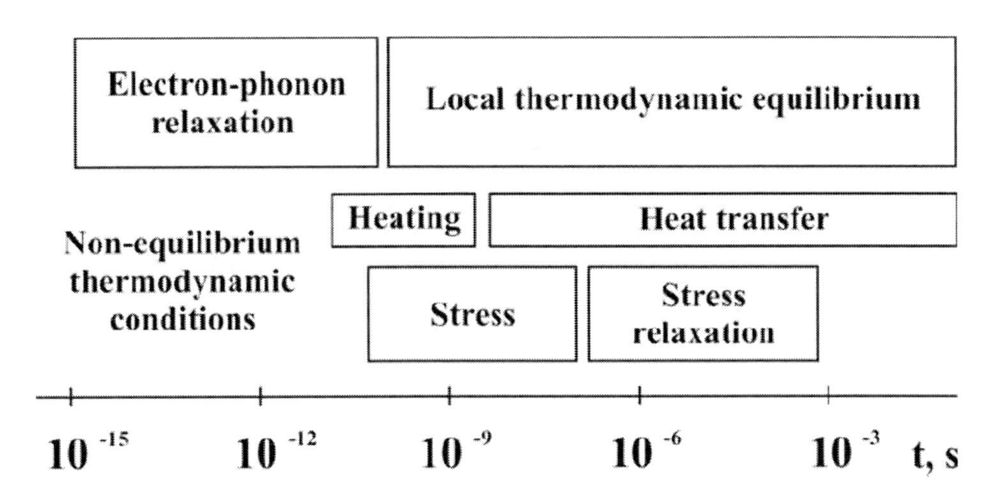

Figure 1. Scheme of the temporal scales which determine the laser heating process.

However, the study of phase transitions using thermodynamic methods has some peculiarities. The phase transition itself, in contrast to phase equilibrium is a non-equilibrium process, which is characterized by the time of relaxation, which determines the time of the establishment of equilibrium between the phases τ_a. However, if the duration of the laser pulse $\tau_l < \tau_a$ in this case the initiation of metastable states is possible, when the system is stable against small disturbances. Metastable states are characterized by great deviations of the phases from the equilibrium. In general cases the kinetic description of the above-mentioned processes is needed.

The creation of adequate models and the knowledge of physical processes which take place during laser interaction with matter are also needed for the progress in practical applications. The phase transitions play the main role in processes of laser ablation which is used for the surface treatment, thin films creation process and chemical composition determination [4]. The generation of shock waves in the process of laser-matter interaction is used for the surface hardening, besides this mechanism can be used as the shock wave source to study chemical reactions [6].

The dynamics of laser irradiation interaction with matter is mainly determined by two factors:

1. Laser pulse energy density;

2. The mechanical coupling conditions between the absorbing surface of metal and the ambient medium [7].

The influence of the laser pulse duration, wavelength on the interaction process is not considered in the current work.

1. Considering the influence of laser irradiation of nanosecond duration with metal surfaces one may distinguish three regimes of pressure generation process. At low laser energy densities the linear thermoelastic regime of pressure generation [7] is realized. It is characterized by small changes in surface temperature and the absence of phase transitions. In this regime the surface temperature and the amplitude of the induced pressure wave linearly depend on the irradiation intensity. With increase of absorbed energy value the material heating becomes considerable and the phase transitions can take place. In this case the contribution of phase transitions into the surface temperature and pressure becomes crucial – the non-linear regime is realized. At greater laser energy densities the ionization of material takes place and the plasma plume is formed, the process in which determines both the temperature and the pressure – the plasma regime of pressure generation is realized.
2. The mechanical coupling conditions between the absorbing metal surface and the ambient medium considerably influence the dynamics of each of the above-mentioned regimes. The surface can be acoustically free or confined. This condition determines the amplitude and the temporal shape of the generated pressure pulse [7]. In most works on laser-matter interaction [3] the free metal surface is studied. The irradiation of metal surface through the finely confined plate of transparent dielectric which prevents free expansion of the heated material makes it possible to realize high-efficient generation of pressure. In this case the ablation in a general case of this word is absent and the diagnostics of the metal state during the laser impact could be completed using non-contact methods, for example, determining the surface temperature using its thermal radiation.

The thremoelastic regime of pressure generation is realized at small energy densities of laser radiation of nanosecond duration – up to $I = 10^5$ W/cm^2. It was studied theoretically and experimentally in many papers. The analytical expression for temperature and pressure fields in material for an arbitrary laser pulse and for arbitrary mechanical coupling conditions at the metal surface was obtained [7]. It was showed that for the case of confined surface the pressure at the metal surface resembles the temporal form of the incident laser pulse and in the case of the free surface the efficiency of pressure generation process is much smaller. The theoretical calculations were proved by many experimental works (see [7] and literature cited there). In work [8] the direct comparison of the pressure generation efficiency between the cases of free and confined metal surface was conducted. It was shown that for the later case the efficiency of pressure generation is by a factor of a hundred larger. However the absolute values of pressure obtained in the linear regime are not great – tenths of atmospheres for a heating of tenths of degrees. For greater values of laser irradiation intensity the non-linear effects take place due to temperature dependence on thermophysic properties of the irradiated material and due to phase transitions.

The nonlinear regime of pressure generation is studied much less. Most papers are devoted to the laser irradiation interaction with the free surface. In this case the main process which determines the temperature and pressure in the irradiation absorption layer is the metal evaporation [3]. The evaporation intensity greatly depends on the surface temperature, so with the increase of laser irradiation energy density the contribution of the evaporation pressure becomes more and more substantial. For moderate values of laser irradiation energy densities (I = 10^5-10^9 W/cm^2) the contribution of the evaporation pressure into the recoil pressure is the main factor. In this case the intensive evaporation of metal leads to the movement of the border between the solid and gaseous phases into the metal and to the cavity formation. This regime is of much interest for technological applications. However in the process of ablation the temperature field is the main factor. A large number of works were devoted to the study of it [9-12]. The pressure distribution and its generation effectiveness play the secondary role. The analytical solution was received which couples the velocity of the movement of the border between phases and the value of absorbed laser irradiation flux in the case of neglecting thermodynamics of flying vapor [3]. The evaporation pressure was studied in the works [13-14]. In this regime of interaction the surface temperature is a few thousand Kelvins and pressure is a few hundreds of atmospheres.

In a narrow range of intensities of laser radiation the comparative contribution into the temperature and pressure fields besides evaporation makes process of melting of metal. Such a regime was studied numerically in works [15-16]. The distributions of temperature and pressure in samples and also the profiles of acoustic waves that propagate into the depth of the sample were obtained. The experimental study of influence of phase transitions on the shape of the optoacoustic signals for semiconductors was conducted in paper [17]. It was shown that the pressure signals are sensitive to the change of the phase state of material.

The usage of a confined surface (impedance surface border) in the regime with phase transitions also leads to a considerable increase of efficiency of pressure generation process. This has been shown in works [18-20]. However, neither analytical nor numerical studies on pressure generation in this regime were conducted.

At laser energy densities over I = 10^9 W/cm^2 (for the case of the acoustically free surface) the character of the running process changes. The internal energy of vapor begins to considerably exceed the heat of evaporation. The increase of the vapor temperature leads to its ionization and a rapid increase of irradiation absorption coefficient. As a result, the screening of the surface of the material takes place. In this case the internal energy of the forming plasma increases more. Due to this fact the hydrodynamic movement of plasma begins to play an important role. In the case of intensive nanosecond pulses the profile of the density of laser plasma in the case of the free surfaces forms at the times of about $5 \cdot 10^{-10}$ s [4]. During this time the irradiation of the laser pulse is adsorbed at the edge of the flying plasma (plume). At the conditions of the stationary regime when the transient processes at the front of the laser pulse have come to the end and the profile of plasma density has been formed, the laser radiation starts to be absorbed near the region with critical density where the speed of the plasma movement achieves the speed of sound. The heated layer that expands leads to a propagation of a shock wave which gives a considerable energy into the depth of the metal. This leads to the evaporation of metal in the wave of unloading and then due to thc attenuation of the shock wave to the melting and mechanical destruction.

So in the case of the free surface at the intensity which is necessary to obtain pressures of several kilobars and temperature of about 1 eV the formation of the plasma plume will inevitably take place. It will oppose temperature measurements using thermal radiation. One may prevent it using the confined metal surface.

In the case of the confined surface plasma formation, its expansion and generation of shock waves will also take place, but in this case the pressure amplitude will be much higher. It was suggested to use the mechanical confinement of the surface in order to increase the efficiency of pressure generation for the first time in the work [21]. In this work, pressure with amplitude of 30 kbar was obtained. In this case laser irradiation of considerable intensity ($2 \cdot 10^9$ W/cm^2) was used. The generation of pressure took place during the expansion of the heated plasma, which was confined by a transparent material. The pressure profiles were registered by a quartz piezo-transducer. In the same article simple estimations for the maximum reachable value of pressure which satisfactorily describe the experimental data were obtained. The estimations were based on the balance of energy neglecting the ionization losses. Thereupon the series of works devoted to the study of the pressure generation process in the confined geometry followed. The application was the shock-wave hardening of materials. In the work [6] the pressures of several tenths of kilobars were obtained using samples of different metals with different types of confining transparent media. The numerical modeling of the processes that took place was conducted. According to it the maximum value of pressure is limited by value of 100 kbars. The further growth of pressure at laser irradiation fluxes over $4 \cdot 10^9$ W/cm^2 is limited by a considerable reflection from the layer of dense plasma and (or) by an optical discharge of the transparent media. The results of the numerical modeling were in a good agreement with experimental data. In order to increase the values of maximum achievable values of pressure it was suggested to use constructive interferention of pressure signals reflected from the material with higher acoustical impedance. In this work the case of putting a titanium film confined by a quartz glass on a molybdenum surface was considered. In this case the calculated value of achievable pressures were over 100 kbars. The other suggested method of increase of pressure was the irradiation of the confined metallic target from both sides. In this case the pressure in the center of the target was increased by the factor of 2 due to overlay of two waves.

In the work [22] the interaction of laser radiation of picosecond duration (~ 10 ps) with copper foams (40-250 μm) confined between two plates of quartz glass was studied for the first time. The pressure was registered using an optical method – the rear side of the target was used as a mirror in one of the Michelson interferometer's arm. The pressure of 15 kbar was experimentally registered. However in the case of ultrashort excitation the shock wave greatly attenuates with distance. Considering this fact the pressures of about 300 kbar were obtained.

So in plasma regions the generation of pressures up to hundreds of kilobars is possible. However it is possible to obtain kilobar level of pressures which corresponds to critical parameters of metals in non-plasma regime using much lesser levels of laser energy densities – in the regime with phase transitions under the acoustical confined conditions.

Summarizing it should be noted that the suggested method of laser impact on acoustically confined surface of the material under study could be effectively used in several important directions:

- experimental study of near-critical area of materials, phase diagram, determination of critical parameters of metals;
- the study of vapor branch of the phase diagram of metals;
- the study of metal-semiconductor-dielectric phase transition in the supercritical area of the phase diagram;
- generation and study of deep metastable and labile state (in the area below the critical point). When the τ_l ~ 10^{-8} s the duration of the laser pulse is lesser than the metastable state decay time and the high intensity of the laser radiation gives the way of simple realization of such states [10];
- generation of high-intensive shock waves for the study of shock-plastic deformation of metals. In this case the pressure overfall at the shock wave front induced by laser pulse must exceed the yield strength of the material under study [7,11];
- the study of optical properties of materials on a wide range of temperatures ~ 1 eV and densities ~ 1 g/cm^3. In this area the absorption properties of materials are almost unknown due to considerable difficulties both in conduction of theoretical calculation and in the stating of experimental research.

The Aim of the Work

The aim of the current work is the experimental study of the possibility of obtaining states of metals with temperatures and pressures which are close to critical and the study of the dynamics of phase transitions during the impact of laser radiation of nanosecond duration at the metal surface confined by a layer of transparent dielectric. This aim requires the development of a method for the simultaneous measurement of temperature, pressure and the reflectivity of acoustically confined surface of the sample during its radiation by laser pulses of nanosecond duration.

2. The Coupled Problem of Thermoelasticy of an Absorbing and Transparent Media During Impact of Nanosecond Laser Radiation

In the simplest, so called thermoelastical regime of pressure generation the laser irradiation fluxes are quite small, the heating of the absorbing medium is insignificant, the thermodynamical properties change little and phase transitions are absent (in the case of the acoustically free surface the evaporation process can be neglected). In this case it is possible to obtain distributions of pressure and temperature in the system by solving the coupled problem of thermoelasticity in a media with external heat sources.

For simple interpretation of the obtained results the problem during the experiment should be one-dimensional if it is possible. This condition is valid if the transverse size of the heated area which is equal to the diameter of the spot of the incident laser irradiation a is large in comparison with:

- heat diffusion depth in the transverse direction (during the time of the laser impact τ_l);
- the irradiation absorption depth α^{-1} (α - laser radiation absorption coefficient in metal);
- wavelength of the excited acoustical pulse.

During the absorption and thermalization of laser irradiation in the skin-layer of metal the heat diffusion length is determined by high level of thermoconduction $\delta_T = \sqrt{\chi \tau_l}$ (χ - thermoconductivity coefficient of metal) [23]. For example, for lead for $\tau_l \sim 10^{-8}$ s, $\delta_T \sim$ 500 nm, this value is ten times greater than the absorption depth of the incident laser pulse (skin-layer depth is $\delta_L \sim$50 nm) so the heating with great accuracy can be treated as a surface heating. Besides, the heat diffusion depth is inessential in comparison with the size of the heated area ~1 mm and greater (spot of laser irradiation at the target) so the problem can be solved in one-dimensional approximation. Let's consider the closed system of equations which describes the processes of heat conduction and thermoelastical excitation of acoustical waves in a system consisting of a transparent dielectric and an absorbing metal during an impact of laser irradiation. The Z axis is directed perpendicularly to the border surface of two medium, the metal occupies area $z > 0$, transparent dielectric - $z < 0$:

$$\begin{cases} \dfrac{\partial}{\partial t}\left(\rho_{0i} c_{Pi} T_i\right) = \dfrac{\partial}{\partial z}\left(\lambda_i \dfrac{\partial T_i}{\partial z}\right) + \dfrac{\alpha(1-R(t))e^{-\alpha z} q(t)\theta(z)}{\rho c_p} & i = \begin{cases} gl, & z \in (-\infty, 0] \\ m, & z \in [0, +\infty) \end{cases} \\ \dfrac{\partial^2 P_i'}{\partial z^2} - \dfrac{\partial^2}{\partial t^2}\left(\dfrac{1}{c_{Li}^2} P_i'\right) = -\dfrac{\partial^2}{\partial t^2}\left(\beta_i^* \rho_{0i} T_i\right) & \theta(z) = \begin{cases} 0, & z \in (-\infty, 0] \\ 1, & z \in [0, +\infty) \end{cases} \end{cases} ,(1)$$

where index i denotes transparent or absorbing medium, T_i - temperature, P_i' - increase of pressure, ρ_{0i} - density, $c_{Pi}(T)$ - heat capacity at constant pressure, $\lambda_i(T)$ - thermo conduction coefficient, $c_{li}(T)$ - propagation speed of longitudinal acoustic waves, $\beta_i^* = \beta_i \cdot (1 - 4\dfrac{c_{ti}^2}{3c_{li}^2})$, $\beta_i(T) = -\dfrac{1}{\rho}\left(\dfrac{\partial \rho}{\partial T}\right)_P$ - thermal expansion coefficient, $R(T)$ - reflectivity of metal surface, $q(t)$ - intensity of incident laser irradiation pulse. The equation for the increase of pressure $P_i' = P_i - P_{0i}$ is formulated in the acoustical approach. For materials in condensed state the acoustical approach (i.e. the description of shock wave as acoustical [23]) is valid when the acoustical Max number $M = v / c_0 << 1$, where $v = P_a' / (\rho \cdot c_0)$ - amplitude of oscillating velocity in acoustical wave, P_a' - amplitude of pressure, c_0 - sound velocity in the material. In the current work $M_{\max} \cong 3 \cdot 10^{-3}$ and the usage of the acoustical assumption is valid for all of the range of studied parameters. This makes it possible to measure pressure by registering the amplitude and the shape of the acoustical wave with a wide-band piezo-transducer. The system of equations (1) should be supplemented with border conditions. Let's consider $\sqrt{\chi\tau} >> \alpha^{-1}$; in this case the heating can be treated as surface heating. In this case there is also the condition of equal temperature heat fluxes, pressures and oscillating velocities at the border between two mediums:

$$\begin{cases} T_{gl} = T_m \\ \lambda_{gl}\dfrac{\partial T_{gl}}{\partial z} = -\lambda_m \dfrac{\partial T_m}{\partial z} \end{cases} \qquad \begin{cases} P'_{gl} = P'_m \\ \dfrac{1}{\rho_{gl}}\dfrac{\partial P'_{gl}}{\partial z} = \dfrac{1}{\rho_m}\dfrac{\partial P'_m}{\partial z} \end{cases} . \tag{2}$$

The thermophysical parameters used in the system (1)-(2) in a general case depend on temperature, close to the phase transitions area. In the thermodynamical parameters space this dependence becomes considerable. However for low intensity fluxes of laser irradiation the heating of the material is insignificant and this dependence can be neglected. Later this regime will be called linear. In this case it is possible to obtain the distributions of temperature and pressure in the system for a laser pulse of an arbitrary shape.

2.2. The Temperature Field in Metal and in Transparent Dielectric in the Approximation of Constant Thermophysical Properties of Material

As it can be seen from Eq. (1) pressure doesn't appear in the equation of the temperature field, so there is the possibility of independent solution of the heat problem and subsequent usage of the obtained solution as an external source in the equations for pressure. For constant thermo conductivity coefficient λ, heat capacity c_p and reflectivity of metal R the heat problem is linear. Due to this fact it is possible to obtain the solution for the arbitrary temporal shape of laser pulse using Fourier transformation.

For Fourier coefficients of temperature in transparent and absorbing medium it is possible to obtain the following expressions:

$$\tilde{T}_i(\omega,z) = \frac{\tilde{q}_0(\omega)(1-R)(\omega\chi_m)^{-1/2}}{\rho_{0m}c_{pm}(1+R_T)}\exp\left((-1)^k\sqrt{\frac{i\omega}{\chi_i}}z\right) \qquad \begin{cases} k=0, \ \text{при} \ i=gl \\ k=1, \ \text{при} \ i=m \end{cases} ,\tag{3}$$

$$\tilde{T}_i(\omega,z) = \frac{1}{2\pi}\int_{-\infty}^{\infty} T(t,z)e^{-i\omega t}dt ,$$

$$\tilde{q}_0(\omega) = \frac{1}{2\pi}\int_{-\infty}^{\infty} q_0(t,z)e^{-i\omega t}dt ,$$

$$R_T = \frac{\rho_{0gl}c_{pgl}\sqrt{\chi_{gl}}}{\rho_{0m}c_{pm}\sqrt{\chi_m}}$$

is the parameter which determines the ratio of the heat fluxes into the absorbing and transparent mediums. During the deduction of Eq. (3) it was assumed that the condition $\sqrt{\chi_m \tau_l} >> \alpha^{-1}$ is valid (the heat diffusion depth during the time of laser impact is much greater than the skin-layer depth).

Using Eq. (3) it is possible to obtain the distribution of temperature in the transparent and absorbing media as a result of convolution of laser pulse with Green function:

$$\begin{cases} T_m(z,t) = \int\limits_0^{\infty} \dfrac{(1-R_T) q_0(t-\tau)\exp\left(-z^2/4\chi_m\tau\right)}{(1+R_T)\rho_{0m} c_{pm}\sqrt{\pi\chi_m\tau}} d\tau \\ T_{gl}(z,t) = \int\limits_0^{\infty} \dfrac{(1-R_T) q_0(t-\tau)\exp\left(-z^2/4\chi_{gl}\tau\right)}{(1+R_T)\rho_{0m} c_{pm}\sqrt{\pi\chi_m\tau}} d\tau \end{cases} \quad (4)$$

Eq. (4) for $z = 0$ gives the dynamics of temperature change at the metal surface during the laser impact.

2.3. The Distribution of Pressure in Metal and in Transparent Dielectric without Account of Thermal Dependence of Thermophysical Parameters

Using the solution of the heat conductivity equation it is possible to obtain the distribution of pressure in both media. For this it is also possible to use Fourier method, in this case for the spectral component of pressure in the absorbing medium there is the following expression:

$$\tilde{P}'_m(\omega,z) = \frac{(1-R_T)\tilde{q}(\omega)\beta^*_m c_{lm}}{c_{pm}(N+1)(1+R_T)}\left[\frac{b}{1-i\omega/\omega_{gl}}\left(1-\sqrt{i\omega/\omega_{gl}}\right)+\frac{1+N\sqrt{i\omega/\omega_m}}{1-i\omega/\omega_m}\right]\exp\left(-\frac{i\omega}{c_{lm}}z\right) - \frac{\beta_m\tilde{q}(\omega)\sqrt{i\omega\chi_m}}{c_{pm}(1+R_T)\left(1-\dfrac{i\omega}{\omega_m}\right)}\exp\left(-\sqrt{\frac{i\omega}{\chi_m}}z\right) \quad (5)$$

$\omega_m = \dfrac{c_{lm}^2}{\chi_m}$, $\omega_{gl} = \dfrac{c_{lgl}^2}{\chi_{gl}}$, $N = \dfrac{\rho_{0m} c_{lm}}{\rho_{0gl} c_{lgl}}$ - ratio of the acoustical impedances of the transparent and absorbing media,

$$b = \frac{\beta^*_{gl}\sqrt{\chi_{gl}}}{\beta^*_m\sqrt{\chi_m}}.$$

The Eq. (5) can be simplified. For laser pulses of nanosecond duration for metals and even more so for dielectric the following condition is valid: $\frac{\omega}{\omega_{m,gl}} << 1$. It corresponds to the fact that the area of disturbance during the time of about duration of laser pulse $c_l \tau_l$ is much greater than the heat diffusion depth during the same time. So saving the quantities of the same infinitesimal order one obtains the following:

$$\tilde{P}'_m(\omega, z) = \frac{(-R)\tilde{q}(\omega)\beta^*_m c_{lm}(+1+N\sqrt{i\omega/\omega_m})}{c_{pm}(N+1)(+R_T)}\exp\left(-\frac{i\omega}{c_{lm}}z\right) -$$

$$-\frac{\beta^*_m \tilde{q}(\omega)\varsigma_{lm}\sqrt{i\omega/\omega_m}}{c_{pm}(+R_T)}\exp\left(-\sqrt{\frac{i\omega}{\chi_m}}z\right) \qquad (6)$$

The expression (6) consists of two items. The first item corresponds to the plane acoustic wave which propagates into the depth of absorbing media with velocity c_{lm}. The second item accounts for extra addition to pressure in the area of heat generation. It is seen from the Eq. (6) that the major role in the pressure generation process at the sample surface plays the value of ratio of acoustical impedances N. If the metal surface is acoustically free ($N \to \infty$) the first and second items in (6) can be of the same infinitesimal order. If $(N \to 0)$ the irradiated surface is mechanically loaded by a layer of transparent dielectric and in this case the generation of pressure is much more effective (in $\sqrt{\frac{\omega}{\omega_{m,gl}}}$ times) in comparison with free surface, that was experimentally approved in [8].

If the metal surface is confined $(N \sim 1)$, than the second item in (6) can be neglected. So the acoustical wave which propagates into the depth of material resembles the temporal form of pressure pulse at the metal surface with accuracy of values of order $\sqrt{\frac{\omega}{\omega_{m,gl}}}$. It gives the possibility to measure pressure at the sample surface by registering the acoustic wave in the medium depth. With the account of the above-mentioned assumptions using (6) the spectrum of the excited acoustical wave running out of the heat generation zone can be presented in the following form:

$$\tilde{P}'_m(\omega, z) = \frac{(-R)\tilde{q}(\omega)\beta^*_m c_{lm}(+1)}{c_{pm}(N+1)(+R_T)}\exp\left(-\frac{i\omega}{c_{lm}}z\right). \qquad (7)$$

The same expression can be obtained also for the wave which propagates into the transparent medium:

$$\tilde{P}'_{gl}(\omega,z)=\frac{(1-R)\tilde{q}(\omega)\beta^*_m c_{lm} N (G+1)}{c_{pm}(N+1)(1+R_T)}\exp\left(-\frac{i\omega}{c_{gl}}z\right). \tag{8}$$

It follows from (7) that the shape of the acoustical pulse resembles the temporal form of the laser irradiation:

$$P_m(t)=\frac{(1-R)q(t-z/c_{lm})\beta^*_m c_{lm}(G+1)}{c_{pm}(N+1)(1+R_T)}. \tag{9}$$

Expressions (4), (9) for the time dependencies of temperature and pressure at the surface of the sample $(z=0)$ give, in a parametric form, the curve of the process of laser heating of metal surface in $P-T$ coordinates. At figure 2 the temporal dependencies of temperature and pressure at the metal surface for Gaussian laser pulse with FWHM duration of τ_l = 8.5 ns obtained using Eq. (4),(9) are presented. At figure 3 the heating process diagram of metal in $P-T$ coordinates is presented.

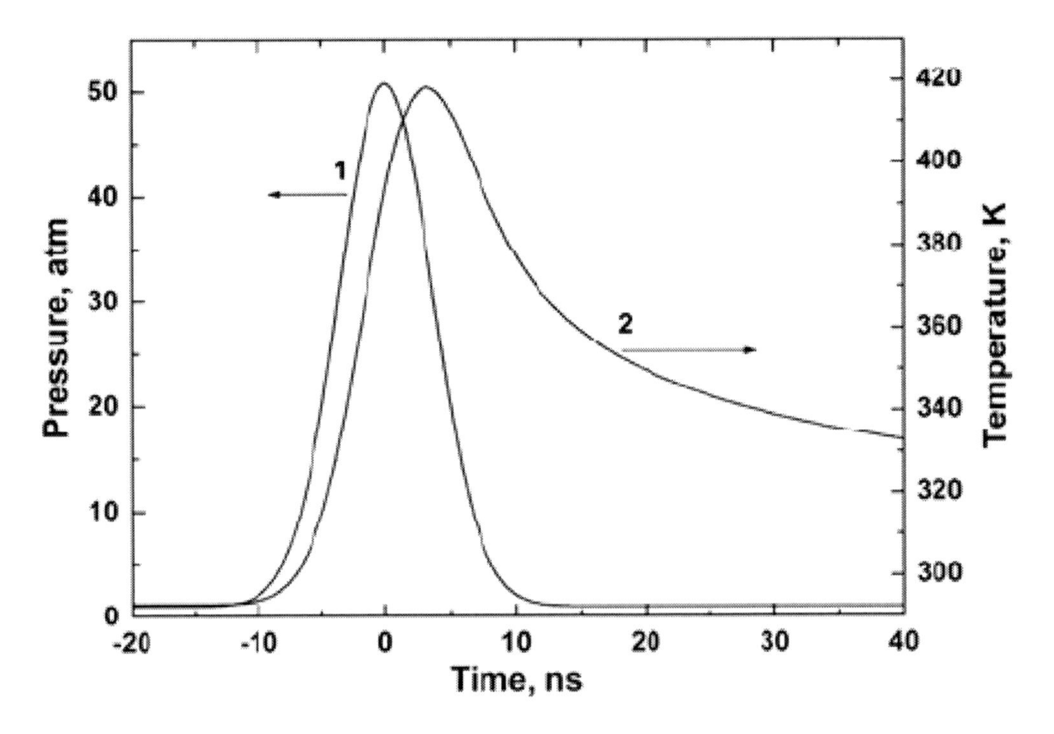

Figure 2. Pressure (1) and temperature (2) at the metal surface (lead) under a layer of transparent dielectric subjected to laser irradiation with E = 10 mJ/cm^2.

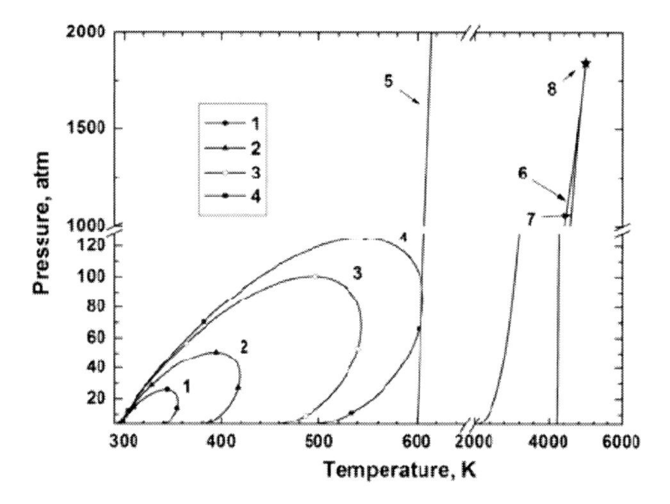

Figure 3. The curves of the heating process (lead) in *P-T* coordinates for different values of absorbed laser irradiation energy density: 1 – E = 5 mJ/cm^2, 2 – E = 10 mJ/cm^2, 3 – E = 20 mJ/cm^2, 4 – E = 25 mJ/cm^2, 5 – melting curve, 6 – binodal, 7 – spinodal, 8 – critical point.

2.4. The Influence of Change of Thermo Physical Properties of Metal on Distribution of Temperature and Pressure in Transparent and Absorbing Media

In formulation of linear thermoelastic regime the temperature dependence of thermophysical properties of transparent and absorbing medium were neglected. If the surface heating is considerable, when the thermodynamic parameters approach the spinodal and especially when phase transitions take place, the temperature dependence of thermophysical properties can not be neglected. However in this case the equations become nonlinear, and there is no analytical solution ; only numerical solution is possible.

In the first approximation the thermo-physical properties can be treated as linear functions of temperature. In Table 1 the values of temperature corrections for the case of lead are presented. The temperature dependence of properties of transparent medium in first approximation can be neglected.

Table 1. Temperature dependence of thermophysical properties of lead

Parameter	Temperature coefficient, K^{-1}
Density ρ	$8{,}6\cdot10^{-5}$
Absorptive power $1-R$	$2{,}14\cdot10^{-3}$
Heat conduction coefficient λ	$-3\cdot10^{-4}$
Heat capacity c_p	$3{,}8\cdot10^{-4}$
Thermal-expansion coefficient β	$9{,}9\cdot10^{-4}$
Sound velocity c	$-5{,}4\cdot10^{-4}$

The temperature dependence of reflectivity is calculated using the data on temperature dependence of conductivity of lead using the following formula:

$$\frac{1}{1-R}\frac{\partial(1-R)}{\partial T}=-\frac{1}{2\sigma_0}\frac{\partial \sigma}{\partial T}.$$

From the data presented at Table 1 it is seen that the temperature dependence of reflectivity makes the major contribution.

The qualitative notion about nonlinear transformation of acoustical pulses due to the temperature dependence of reflectivity of the sample can be obtained considering the model problem in which the temporal form of the laser pulse is a step-like function of time.

Let's consider the impact of the laser irradiation with the power density $I(t)=I_0\cdot f(t)$ on the metal surface, where

$$f(t)=\begin{cases}0, & t<0;\\ 1, & t>0.\end{cases} \tag{10}$$

In the case of constant reflectivity the surface temperature of the sample can be calculated using Eq. 4:

$$\begin{cases}T=T_0, \quad t<0;\\ T=T_0+\dfrac{2I_0(1-R_0)\sqrt{t}}{\sqrt{\pi\rho c_p\lambda}}, \quad t>0.\end{cases} \tag{11}$$

Eq. (4) becomes invalid when the temperature approaches the value $T_m=\left(\dfrac{1}{R_0}\dfrac{dR}{dT}\right)^{-1}\approx 700$ K (for the lead). In this case one should solve the heat conduction equations directly taking into account the temperature dependence of the reflectivity coefficient.

$$\begin{cases}\dfrac{\partial T}{\partial t}=\chi\dfrac{\partial^2 T}{\partial x^2}, \quad x>0;\\ \lambda\dfrac{\partial T(0,t)}{\partial x}=-I_0\cdot\left[1-R_0\left(1+\dfrac{1}{R_0}\dfrac{dR}{dT}T(0,t)\right)\right], \quad T<T_m=-\left(\dfrac{1}{R_0}\dfrac{dR}{dT}\right)^{-1}.\end{cases} \tag{12}$$

The system of equations (12) is the heat conduction equation with the border condition of the third order. Its solution can be written in the following expression [24]:

$$T(x,t) = \frac{(1 - R_0)}{R_0 k}\left[erfc\frac{x}{2\sqrt{\chi t}} - e^{\frac{I_0}{\lambda}R_0 kx + \left(\frac{I_0}{\lambda}R_0 k\right)^2 \chi t} erfc\left(\frac{x}{2\sqrt{\chi t}} + \frac{I_0}{\lambda}R_0 k\sqrt{\chi t}\right)\right], \quad (13)$$

where $k = \frac{1}{R_0}\frac{dR}{dT}$, which is valid for times $t < t_c = \frac{0.42\lambda\rho\, c_p}{I_0^2 R_0^2 k^2}$ (for lead) when the reflectivity becomes zero.

At figure 4 the temporal dependences of temperature at the lead surface are presented for $I_0 = 4.5 \cdot 10^6$ W/cm^2 for the constant reflectivity (linear regime) and when R depends on T (nonlinear case). At figure 5 the temporal dependence of absorbed laser irradiation power density which resembles the temporal form of the generated acoustical wave is presented.

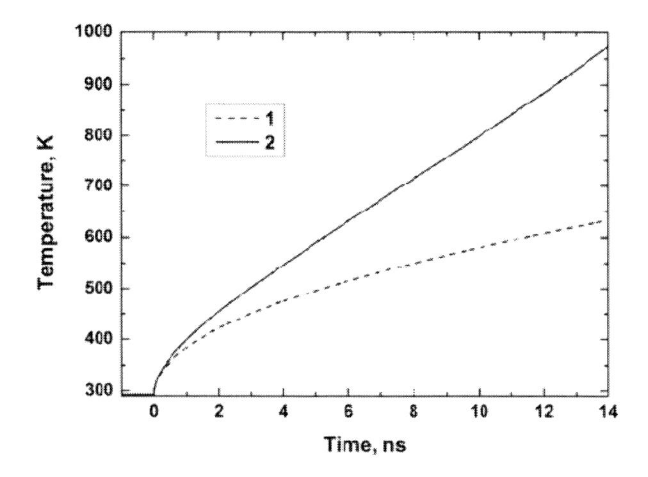

Figure 4. Temporal dependence of surface temperature at linear (1) and non-linear (2) regimes.

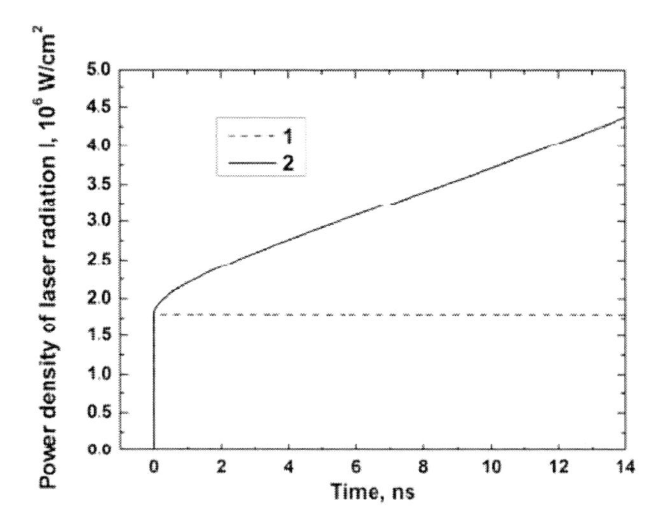

Figure 5. Temporal dependence of power of absorbed laser radiation at linear (1) and non-linear (2) regimes.

From figures 4-5 it is clear that in nonlinear regime the temperature rises considerably greater in comparison with the linear regime. The pressure pulse resembles the temporal form of absorbed laser irradiation power density, which for the case of Gaussian pulse changes in the way that it becomes non-symmetrical and it maximum shifts to greater times.

2.5. The Results of Numerical Modeling of Temperature, Pressure and Density Fields During the Impact of Laser Radiation on the Confined Metal Surface Taking into Account the Temperature Dependence of Thermophysical Parameters of Materials

Below the results of the numerical modeling using finite-element program FEMLAB of the coupled problem of thermo-elasticity taking into account the temperature dependence of thermo-physical parameters (from Table 1) are presented. Lead has been taken as an absorbing material, quartz glass K8 – as a transparent one. At figure 6 $P-T$ diagrams of processes of laser heating obtained numerically are presented. The calculations took into account the above-mentioned thermal nonlinearity. Different curves correspond to different laser irradiation fluencies. The maximum value of laser fluence of 16 mJ/cm^2 was taken in order to obtain the peak temperature near the lead melting point (600 K). Dotted lines denote the diagram which doesn't take into account the thermal nonlinearity for laser fluence of 16 mJ/cm^2.

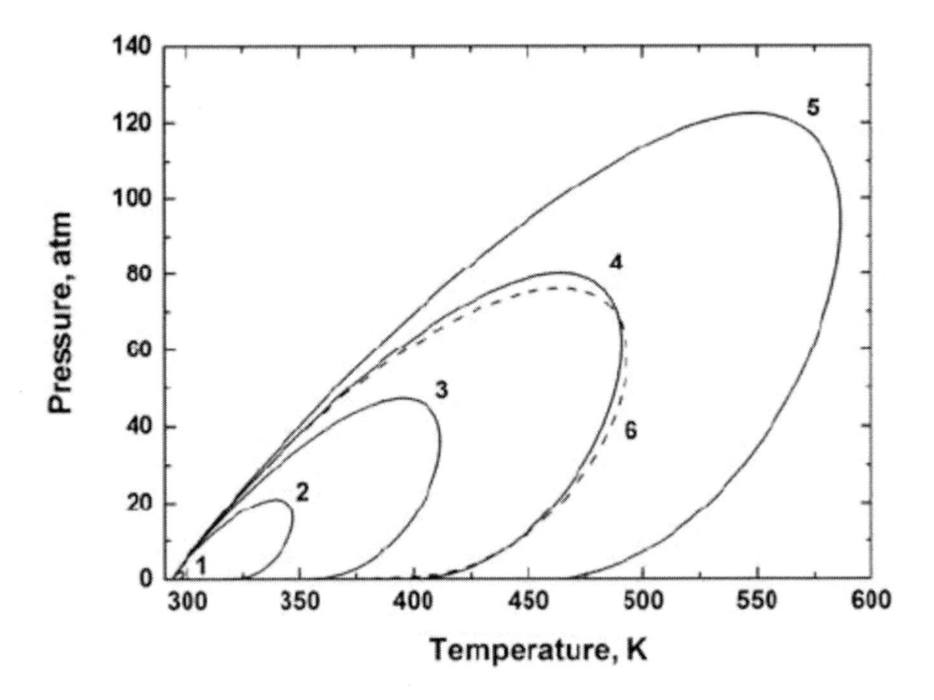

Figure 6. *P-T* diagrams of heating process with an account of thermal nonlinearity for different values of laser radiation energy density: 1 – 0.4 mJ/cm^2, 2 – 4 mJ/cm^2, 3 – 8 mJ/cm^2, 4 – 12 mJ/cm^2, 5 – 16 mJ/cm^2, 6 (dotted line) – 16 mJ/cm^2 (without thermal nonlinearity).

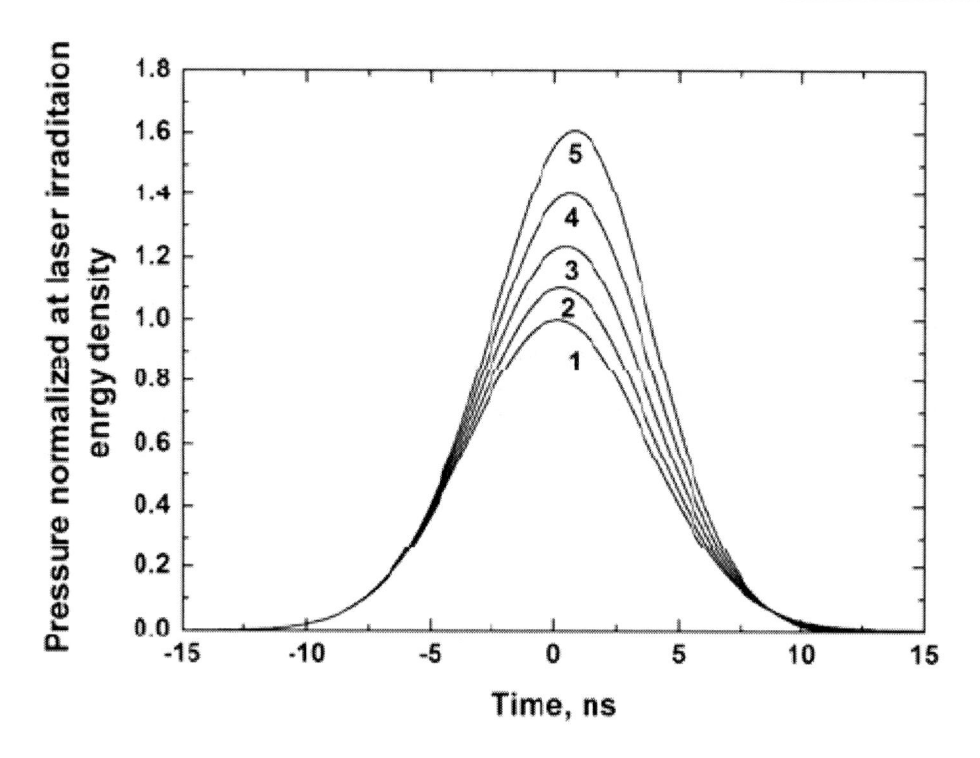

Figure 7. Temporal dependencies of pressure normalized at energy density of incident laser radiation: 1 – 0.4 mJ/cm^2, 2 – 4 mJ/cm^2, 3 – 8 mJ/cm^2, 4 – 12 mJ/cm^2, 5 – 16 mJ/cm^2.

At figure 7 the pressure pulses normalized on the laser fluence are presented. For fluences below $E_0 = 0.4$ mJ/cm^2 with or without thermal nonlinearity the calculations give the same result – thermal nonlinearity is inessential. For fluxes over $E_0 = 0.4$ mJ/cm^2 the contribution of thermal nonlinearity is considerable. At figure 8 the normalized temporal profiles of pressure for $E_0 = 16$ mJ/cm^2 with and without nonlinearity are presented. It is seen that due to thermal nonlinearity the pressure maximum shifts slightly to the right and the dependence becomes nonsymmetrical. This can be explained by the fact that the maximum of the intensity of the absorbed laser radiation also shifts in time due to increase of reflectivity with temperature. At figure 9 the temporal dependency of the pressure generation efficiency η during the laser impact calculated using the following formulae:

$$\eta = \frac{P(E_0)/E_0}{P(E_0^{lin})/E_0^{lin}} \tag{14}$$

are presented. Here $E_0^{lin} = 0.4$ mJ/cm^2. At figure 10 the dependence of pressure generation efficiency at maximum upon laser energy density E_0 is presented.

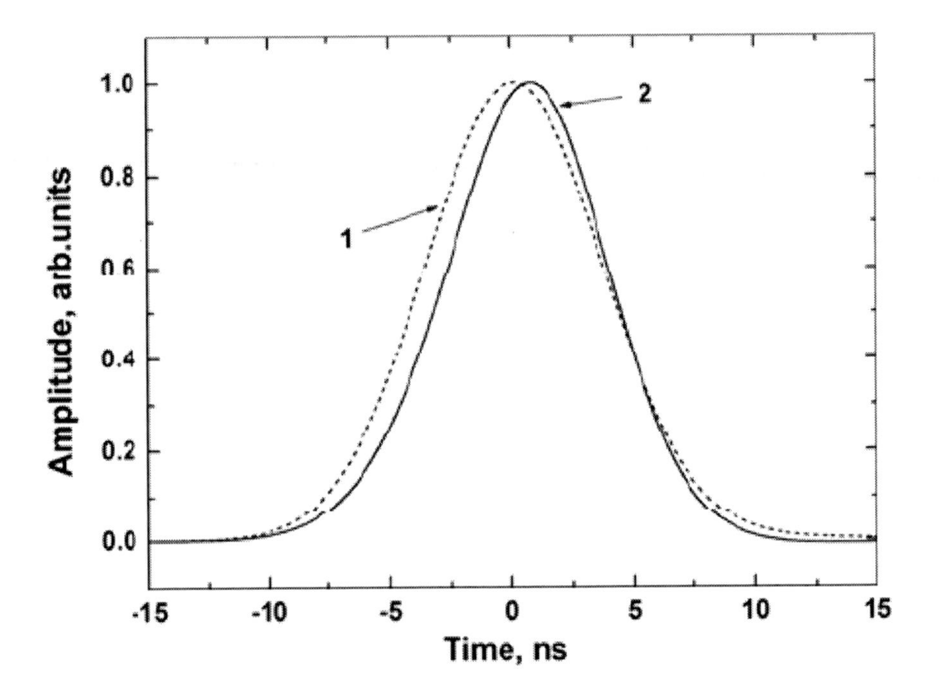

Figure 8. Temporal dependencies of pressure at the surface with (2) and without (1) account of thermal nonlinearity normalized at maximum for E = 16 mJ/cm^2.

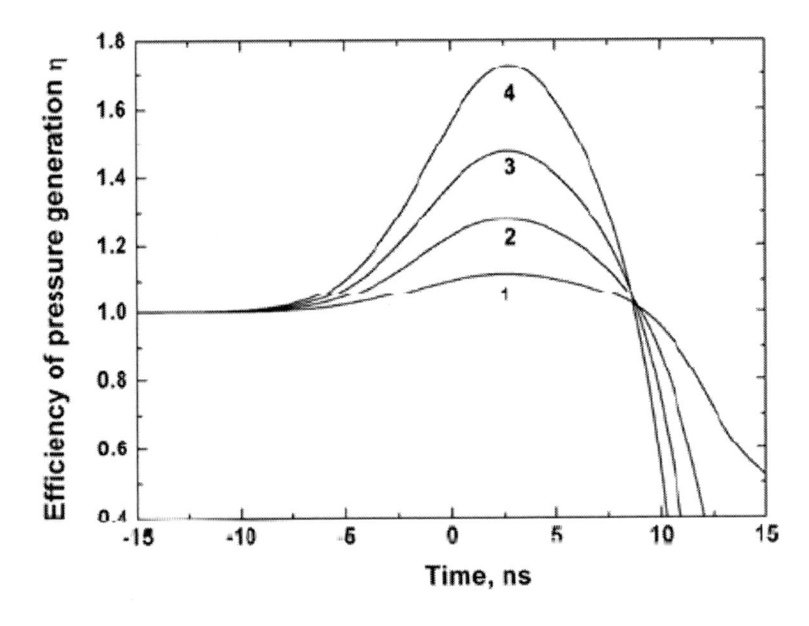

Figure 9. The efficiency of pressure generation process for different values of laser radiation energy density: 1 – 4 mJ/cm^2, 2 – 8 mJ/cm^2, 3 – 12 mJ/cm^2, 4 – 16 mJ/cm^2.

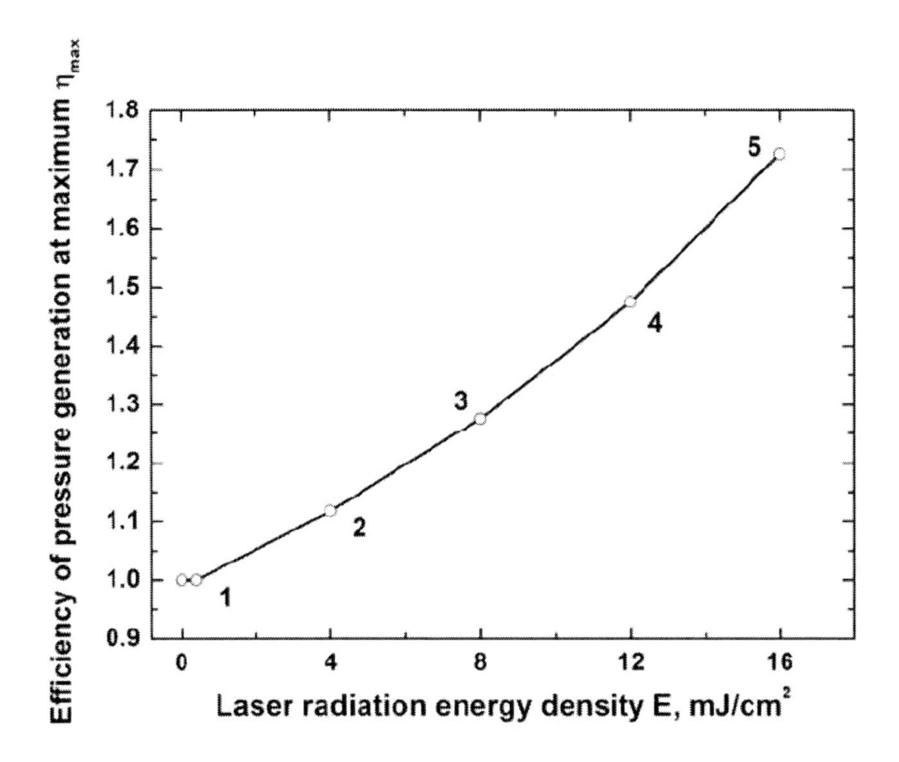

Figure 10. Dependence of pressure generation efficiency at maximum upon absorbed laser radiation energy density: 1 – E = 0.4 mJ/cm^2, 2 – E = 4 mJ/cm^2, 3 – E = 8 mJ/cm^2, 4 – E = 12 mJ/cm^2, 5 – E = 16 mJ/cm^2.

The temporal form of the pressure pulse changes little due to thermal nonlinearity; that much is clear from figure 8. There are considerable changes only in the signal amplitude which nonlinearly increases with the absorbed laser energy density (Figure 10).

3. The Influence of Phase Transitions on the Distributions of Temperature, Pressure and Density During the Laser Irradiation Impact on the Metal Surface Confined by Transparent Dielectric

3.1. The Problem Formulation for the Case of Phase Transition – Melting

If the phase transition takes place the problem complicates considerably. Several areas with different phase states of material appear. The borders between them are moving. Such problems are initially nonlinear and there is no analytical solution for them.

At first let's consider the simplest case of phase transition – material melting. We consider the melting process to be heterogeneous – taking place at the seeds of the liquid phase, i.e. at the melting border. In this case the position of the border between the phases and its velocity can be determined. Let's consider that the melting temperature doesn't depend on pressure; besides, we will neglect the possibility of the appearance of metastable states.

In this case in order to describe the process of melting under the impact of laser pulse, taking into account hydrodynamic processes, it is necessary to solve the full hydrodynamic system of equations in two media – solid and liquid. The boundary between these two media is moving. Lets denote X_{ls} - coordinate of the solid-liquid boundary, V_{ls} - its velocity. Then the full system of hydrodynamic equations for three domains: quartz glass, liquid and solid metal phase can be written in the following form:

$$\left[\frac{\partial \rho}{\partial t}+\frac{\partial(\rho u)}{\partial z}=0\right]_I \qquad \left[\frac{\partial(\rho u)}{\partial t}+\frac{\partial(\rho u^2)}{\partial z}=-\frac{\partial P}{\partial z}\right]_I, \tag{15-16}$$

$$\left[\frac{\partial}{\partial t}\left[\rho\left(\mathrm{E}+\frac{u^2}{2}\right)\right]+\frac{\partial}{\partial z}\left[\rho u\left(\mathrm{E}+\frac{u^2}{2}\right)\right]=-\frac{\partial(Pu)}{\partial z}+\lambda\frac{\partial T}{\partial z}+\alpha I_0(t)e^{-\alpha x}\right]_I. \tag{17}$$

where index $I = gl, ml, ms$ denotes quartz glass, liquid and solid phase correspondingly. Eq. (15-17) should be supplemented with border and initial conditions and also with equation of state.

At the glass-metal boundary the conditions of equal oscillating velocities, pressures and heat fluxes are the following:

$$[u_{gl}=u_{ml}]_{x=0}, \qquad [p_{gl}=p_{ml}]_{x=0}, \left[\lambda_{gl}\frac{\partial T_{gl}}{\partial x}=\lambda_{ml}\frac{\partial T_{ml}}{\partial x}\right]_{x=0}. \tag{18}$$

At the solid-liquid boundary X_{sl} the laws of mass, impulse and energy conservation in the presence of phase transition must be fulfilled:

$$[\rho_s V_{ls}=\rho_l(u_s-u_l+V_{ls})]_{x=X_{ls}(t)}, \tag{19}$$

$$[\rho_s V_{ls}^2+p_s=\rho_l(u_s-u_l+V_{ls})^2+p_l]_{x=X_{ls}(t)}, \tag{20}$$

$$[\lambda_s\frac{\partial T_s}{\partial x}-\lambda_l\frac{\partial T_l}{\partial x}=L_m\rho_s V_{ls}]_{x=X_{ls(t)}}. \tag{21}$$

3.2. Numerical Solution of the Heat Conduction Problem (Stefan Problem) for a Melting Phase Transition

Because the melting temperature doesn't depend on pressure the heat problem can be solved independently from the acoustical problem. The problem of heat conduction in two-domain areas, taking into account the position of the phase transition boundary, is the classical Stefan problem. There is the analytical solution of the Stefan problem only for the case of constant temperature. Even for the constant heat flux there is only an approximate solution in the form of series. The numerical solution of the Stefan problem is complicated by continuous movement of boundary between phases. The position of this boundary must be tracked during the problem solving. In order to overcome this difficulty the special approximate methods were developed. They allow reduction of the problem to the simply connected domain. This is the so called enthalpy methods [25-26].

The thermal conductivity problem can be reformulated introducing the second unknown variable – enthalpy H. In this case the problem reduces to the simply connected domain:

$$\frac{\partial H}{\partial t} = \frac{\partial}{\partial z}\left(\lambda \frac{\partial T}{\partial z}\right), \quad z \in [0, \infty), \tag{22}$$

where enthalpy H is a step-like function of temperature:

$$H = \begin{cases} \rho_s c_p^s (T - T_m), T < T_m, \\ \rho_l c_p^l (T - T_m) + \rho_s L_m, T > T_m. \end{cases} \tag{23}$$

The reduced problem (21-22) can be solved by the finite-difference method. Below, the numerical solution of the Stefan problem for the case of heating of lead surface by a Gaussian laser pulse is presented. At figure 11 the temporal dependency of surface temperature for $E_0 = 32$ mJ/cm^2 is presented. At the beginning of phase transition a fracture is formed at the thermogram. At figure 12 the special distribution of temperature for different times is shown. Using this data it is possible to determine the temporal dependency of position of boundary between phases $x(t)$ and also its velocity.

So the temporal dependence of temperature at the metal surface during melting has a peculiarity – a fracture which corresponds to the melting temperature. This peculiarity can be suitable for determination of the moment of the start of phase transition using an experimentally obtained thermogram.

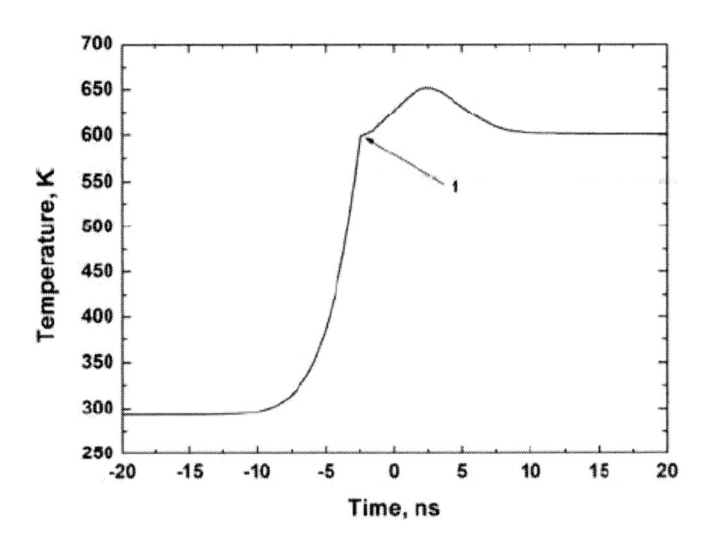

Figure 11. Temporal dependency of temperature of sample surface at the presence of phase transition: 1 – the start of melting.

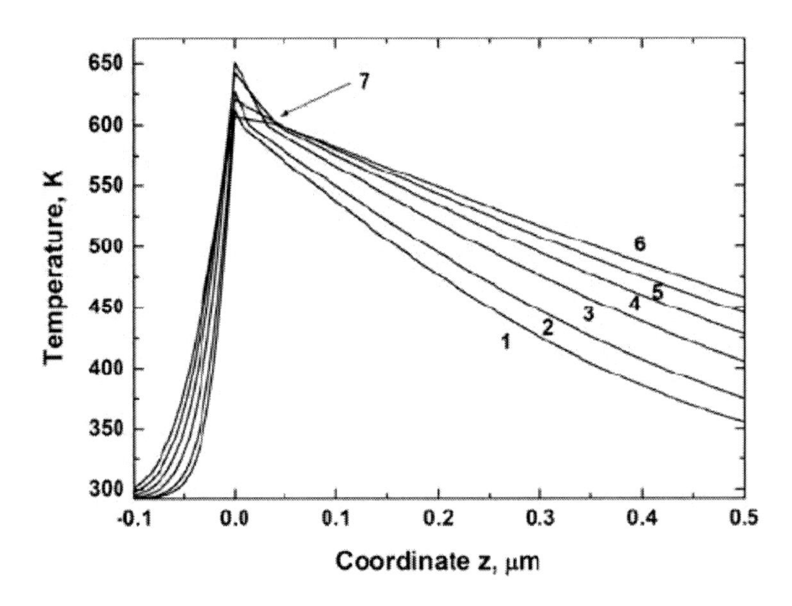

Figure 12. Profiles of temperature distribution for different moments of time: 1 – -2 ns, 2 – 0 ns, 3 – 2 ns, 4 – 5 ns, 5 – 6 ns, 6 – 8 ns, 7 – the movement the phase transition boundary.

3.3. The Influence of Melting on the Pressure Generation Efficiency for the Acoustically Confined Irradiated Surface

In the current work the melting of “normal” metals – for which the specific volume of melted material is greater than the specific volume of the solid phase is considered. The “volume defect” $\Delta\rho = \rho_s - \rho_l$ (the difference between densities of liquid and solid phases) which appear during melting leads to a rapid increase of pressure at the conditions when the possibility of metal expansion is limited by a transparent dielectric.

Let's consider the domain $z>0$, the boundary $z=0$ let's consider to be rigidly fixed. The absorption of laser radiation takes place at this boundary and leads to a heating and melting of metal at the moment $t=0$. Let's consider that the front of phase transition moves with constant velocity V_{ls}. Let's denote ρ_l, ρ_s - densities of liquid and solid phases correspondingly, $\Delta\rho = \rho_s - \rho_l$ - "defect of volume", c_l, c_s -sound velocities in liquid and solid phases.

Let's consider the state of the system at the moment t. By this time the system consists of free parts: liquid phase of mass $V_{ls}t\rho_s$, deformed solid area of length $c_s t$ and undisturbed area $z>(V_{ls}+c_s)t$ (the acoustical wave doesn't reach it by the moment t). Let's denote the contraction of deformed solid part of the system by Δl_s, then the absolute elongation of the liquid part will be $\Delta l_l = -\Delta l_s$. The relative elongation (contraction) of the solid domain will be $\varepsilon_s = \dfrac{\Delta l_s}{c_s t}$. The deformation of the liquid domain can be calculated in the following way. The liquid of mass $V_{ls}t\rho_s$ will occupy the area with length $V_{ls}t\dfrac{\rho_s}{\rho_l}$. So the full absolute elongation will be $V_{ls}t + \Delta l_l - V_{ls}t\dfrac{\rho_s}{\rho_l}$ and the relative deformation $\dfrac{V_{ls}t(\rho_l-\rho_s)+\Delta l_l\rho_l}{V_{ls}t\rho_s}$.

Next, let's formulate the condition of equal pressure in both phases. Pressure is the result of multiplication of relative deformation and quantity ρc^2: $P=\varepsilon\rho c^2$. For pressure P we obtain the following expression:

$$P = \frac{\Delta\rho\rho_s c_s V_{ls}}{\rho_l\left(1+\dfrac{\rho_s^2 c_s V_{ls}}{\rho_l^2 c_l^2}\right)}. \tag{24}$$

The melting front velocity is much smaller than the speed of sound $V_{ls} << c_l$ so the expression can be simplified.

$$P = \Delta\rho\cdot c_s V_{ls}\frac{\rho_s}{\rho_l} \tag{25}$$

Considering the melting phase transition in Eq. (25) we can assign $\dfrac{\rho_s}{\rho_l}\approx 1$. Then

$$P = \Delta\rho\cdot c_s V_{ls}. \tag{26}$$

Let's estimate the melting front velocity for the case of laser pulse with step-like temporal form:

$$I(t) = \begin{cases} 0, & t < 0; \\ I_0, & t > 0. \end{cases} \tag{27}$$

The energy absorbed by the moment t is spent on heating of the solid phase of length $X_{ls}(t)$ to melting temperature, and also on latent heat capacity of phase transition L_m. In this case we neglect the temperature gradient in the liquid phase. Using the energy conservation law one can obtain:

$$\int_0^t I(t')dt' = (c_p \Delta T_{sl} + L_m) \cdot \rho_0 \cdot X_{ls}(t). \tag{28}$$

Differentiating (28) by time one can obtain the expression for the melting front velocity:

$$V_{sl}(t) = \frac{I(t)}{\rho_0 (c_p \Delta T_{sl} + L_m)}. \tag{29}$$

Putting (29) in (26) one can obtain expression for pressure induced by melting:

$$P_{sl} = \frac{c_s \Delta\rho_{sl}}{\rho_0 (c_p \Delta T_{sl} + L_m)} I(t). \tag{30}$$

Let's make quantitative estimations. For the lead $\frac{\Delta\rho}{\rho_0} = 3.5 \cdot 10^{-2}$, $L_m = 2.5 \cdot 10^4$ J/kg, $c_p \cdot \Delta T_{sl} = 4 \cdot 10^4$ J/kg, $c_s = 2.16 \cdot 10^3$ m/s. For laser energy density let's take the threshold value of starting of melting $I \sim 3 \cdot 10^{10}$ J/m². For pressure P_{sl} in this case one can obtain the estimation $P_{sl} \sim 350$ atm.

Let's compare the different contributions into pressure generation process: thermoelastic contribution and contribution due to phase transitions. For thermoelastic contribution the following formulae which can be obtained using Eq. (8) is valid:

$$P_{lin} = \frac{\beta c_s}{c_p} I(t). \tag{31}$$

For the ratio of contributions one can obtain the following expression:

$$\frac{P_{sl}}{P_{lin}} = \frac{\Delta\rho_{sl} c_p}{\beta\rho_0 \left(c_p \Delta T_{sl} + L_m\right)}. \quad (32)$$

For lead $\frac{P_{sl}}{P_{lin}} = 0.7$. So during the melting the contribution of "defect of volume" in pressure generation process is comparable with the thermoelastic contribution, and in absolute values the additional pressure is about several hundreds of atmospheres.

3.4. The Metal Boiling Induced by Pulse of Laser Radiation at Acoustically Confined Surface

The description of boiling in the subsurface area of metal under the layer of transparent dielectric is very complicated. The dependence of vapor temperature at the front of the liquid-vapor phase transition on pressure according to Clapeyron-Clausius law is exponential:

$$T_b(p) = \frac{T^{(0)}}{\ln(p^{(0)}/p)}, \quad (33)$$

where $T^{(0)}$, $p^{(0)}$ - some characteristic properties of materials which can be obtained from the approximation of table data. For melting the dependence of melting temperature on pressure could be neglected in a wide range of pressures. For boiling the dependence (33) leads to the impossibility of separation of heat and acoustical problems. They must be solved jointly.

During the laser impact on confined metal surface simultaneously with temperature increase there is the increase of pressure at the surface. This higher value of pressure corresponds to a higher boiling temperature. The melted metal doesn't boil, but the dynamical shift of boiling point takes place [19]. The boiling occurs at the rear front of the pressure pulse inside the unloading wave, when the acoustic wave which propagates into the depth of metal takes away the excess of pressure. During the rapid decrease of pressure the boiling process (due to homogeneous nucleation) has no time to develop in this case the deep immersion into metastable state – overheated liquid, is possible. Its subsequent behavior will be determined by kinetics of nucleation process.

If the speed of pressure decrease is greater, the state of material can approach the border of the absolute thermodynamic instability area – spinodal. Crossing the spinodal into the labile area leads to an explosive boiling of liquid – spinodal decomposition [5]. For boiling the clear liquid-vapor phase boundary in contrast to heterogeneous melting can be absent. The material in the subsurface area will represent the two-phase liquid-vapor mixture. So the description of the process of boiling by expressions analogous to (15)-(21) is impossible. The possible way of solution of the problem of description and modeling of non-equilibrium phase transitions during the laser impact on metals and semiconductors is the molecular dynamics method [27].

4. The Experimental Method of Simultaneous Measurements of Pressure, Temperature and Reflectivity of Metals Confined by a Layer of Transparent Dielectric During Impact of Laser Irradiation of Nanosecond Duration

4.1. Calculation of Optimal Targets Parameters

As it was showed in the previous paragraphs the mechanical loading of the irradiated metal surface on the one hand leads to the increase of pressure generation efficiency and on the other hand provides the possibility of temperature measurements using its thermal radiation. During the production of samples-targets, the series of factors must be taken into account. These factors, besides providing efficiency of the pressure generation process, give the possibility to determine the value of pressure registering the acoustical wave which propagates into the depth of the metal. It was suggested to use the construction in which the layer of metal is confined between two plates of optical glass as a target (Figure 13).

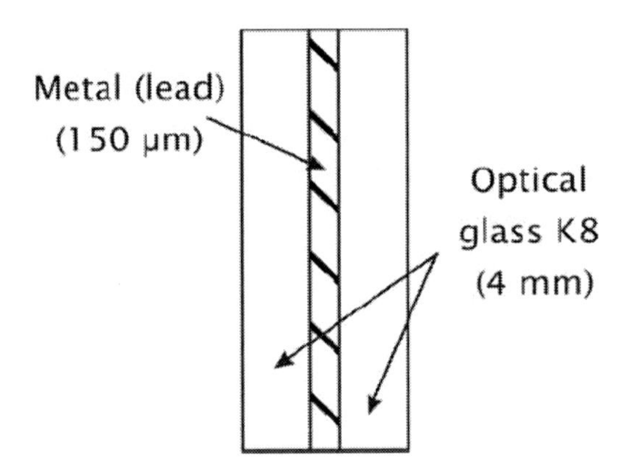

Figure 13. Construction of target.

While choosing the optimal geometrical dimensions the following conclusions were taken into account. At first it is necessary to avoid the distortion of the shape of the acoustical signal due to its interference with pulses reflected from the boundaries of quartz-metal. To prevent it the thickness of the metal layer l_m should be sufficiently large so that the duration of the double run of the acoustical wave through this layer $\frac{2l_m}{c_{lm}}$ exceeds the duration of the registered acoustical pulses. For metal thickness of 150 μm the time window for lead is ~ 150 ns and that is sufficient, because in linear regime the duration of the acoustical pulses doesn't exceed 10 ns (FHWM). The optical glasses also should be thick enough so that the reflection of sound in them does not distort the registered acoustical signal.

On the other hand, the acoustical pulses of the finite amplitude, propagating in a medium, are subjected to the influence of diffraction. The diffraction leads to a decrease of the amplitude of low-frequency components of the signal spectrum. It is possible to estimate the influence of diffraction using the following equation for the characteristic diffraction frequency:

$$\nu = \frac{Lc}{\pi a^2}. \tag{34}$$

The frequencies of the acoustical signal spectrum greater than ν at the distance of L from the source of radius a are non-subjected to the influence of diffraction.

In the target construction the optical glasses of 4 mm thickness were used. For the diameter of the acoustical pulse of 2 mm the diffraction frequency is 1.8 MHz. So the diffraction distorts the acoustical pulse during its propagation in glass, however, even more distortion as a rule takes place during its registration in a piezo-transducer (diffraction in the volume of piezo-transducer itself).

4.2. The Scheme of Pressure Measurement in Acoustic Wave Propagating in Metal and Its Connection with Pressure at the Irradiated Surface of the Sample

The pressure in acoustic wave which propagates into the metal depth is registered using lithium niobate piezo-transducer. The acoustical contact between the transducer surface and the rear side of the target was ensured. As a piezo-transducer the crystal of lithium niobate of z-cutting was used. The concentric electrodes were applied to its rear surface. The lithium niobate thickness was 6 mm. The estimation of the threshold value of diffraction for the propagation of the signal in the material of piezo-transducer using (34) is $\nu = 3.8$ MHz. So, the signal is considerably distorted during its propagation at the piezo-transducer due to diffraction effects. Besides that there is the contribution of non constant frequency sensitivity of the piezo-transducer to the signal distortion.

Spectrum of the registered signal $S(\omega)$ can be presented as a result of multiplication of the spectrum of the initial signal $S_0(\omega)$, the function of spectral sensitivity of piezotransducer $K_{trans}(\omega)$ and also the transmission function of the diffraction distortion $K_{diff}(\omega)$:

$$S(\omega) = S_0(\omega) K_{tras}(\omega) K_{diff}(\omega). \tag{35}$$

If $K_{trans}(\omega)$ and $K_{diff}(\omega)$ are known the shape of the initial signal can be restored. In order to determine $K_{trans}(\omega)$ and $K_{diff}(\omega)$ the piezo-transducer calibration is needed. Such calibration can be conducted in the following way. For low values of laser irradiation intensities, as it was mentioned before, the shape of the acoustical impulse resembles the temporal profile of the laser pulse. The absolute value of excited pressure can be calculated with the knowledge of the absorbed laser energy density using Eq. (8). So in a linear regime of impact with the knowledge about the form of the laser pulse the spectrum of the initial pressure signal $S_{0lin}(\omega)$ is known. Registering the distorted signal it is possible to find its spectrum $S_{lin}(\omega)$. The ratio of the spectrums gives the ability to obtain the transmission function of the whole acoustic section $K_{trans}(\omega)K_{diff}(\omega)$:

$$K_{trans}(\omega)K_{diff}(\omega) = S_{0lin}(\omega)/S(\omega). \tag{36}$$

After the calibration with the knowledge of $K_{trans}(\omega)K_{diff}(\omega)$ and the spectrum of the distorted pressure signal it is possible to restore the initial signal $P_0(t)$ using equation:

$$P_0(t) = \int_{-\infty}^{\infty} \frac{S(\omega)}{K_{trans}(\omega)K_{diff}(\omega)} e^{i\omega t} d\omega . \tag{37}$$

As it was shown earlier (p. 2.3) the profile of the acoustic wave which propagates into the sample depth resembles the temporal form of the pressure at the sample surface. So using (37) it is possible to determine the dynamics of the pressure change at the surface of the sample.

4.3. The Scheme of the Measurements of Temperature of the Sample Surface and Calibration of Optical Pyrometer

During the impact of intense laser radiation at the acoustically confined metal surface the ablation and the attendant processes of plasma generation which screen the heated metal surface are concentrated in a very thin subsurface layer of a few tenths of nanometers. This layer is transparent for radiation and that makes it possible to conduct pyrometrical measurements of the surface temperature with nanosecond temporal resolution.

The spectral radiation density (spectrum density of power of thermal radiation, radiated from the elementary surface area into an elementary solid angle) of the black body heated to the temperature T is determined by the Plank's law:

$$b_0(\lambda,T) = \frac{4\pi^2\hbar c^2}{\lambda^5} \frac{1}{\exp\left(\frac{2\pi\hbar c}{\lambda kT}\right) - 1}. \tag{36}$$

Thermal radiation of real bodies differs from the black body radiation. For them the spectral radiation density is equal to the result of multiplication of spectral radiation density of black body $b_0(\lambda,T)$ and the spectral emittance $\varepsilon(\lambda,T)$ (later on if it is not stated specially we will consider spectral quantities i.e. that depend on wavelength omitting the term "spectral"):

$$b(\lambda, T) = \varepsilon(\lambda, T) \cdot b_0(\lambda, T). \tag{37}$$

For nontransparent bodies the connection between emittance $\varepsilon(\lambda, T)$ and the reflectivity $R(\lambda, T)$ is determined by the Kirchhgoff's law:

$$\varepsilon(\lambda, T) = 1 - R(\lambda, T). \tag{38}$$

For the most part of materials there is no data on dependence of reflectivity upon wavelength although it is known that the thermal radiation of metals is well described by a radiation of "grey body." In this case the emittance is constant in the whole optical spectral range. Besides, the emittance greatly depends on a body's temperature and on its aggregative state. So in order to determine the surface temperature of the metal simultaneous measurements of thermal radiation and reflectivity changes are needed.

In the current work, in order to measure the surface temperature of metal using its thermal radiation an optical pyrometer with nanosecond temporal resolution was developed. Thermal radiation from the sample was collected using short-focused lenses (focus distance 65 mm) and was transported into a quartz optical fiber (of 1 mm thickness). After passing the fiber coming through the series of filters the radiation was focused on the photodetector surface. High-speed silicon PIN-diodes Thorlabs DET-210 with temporal resolution ~ 1 ns, and the maximum sensibility of 0.45 A/W an 750 nm wavelength were used.

The photodetector signal is determined by the power of the incident thermal radiation flux:

$$V = R \iiint \varepsilon(\lambda, T) b_0(\lambda, T) K(\lambda) W(\lambda) dS d\Omega d\lambda. \tag{39}$$

Here V [V] – signal of the photodetector, W [A/W] – its sensitivity, R [Ohm] – the resistance at which the photodetector is loaded, and $K(\lambda)$ - transmittance coefficient of the whole optical section of the pyrometer including filters. The integration is performed at the part of the area of laser spot which is projected by optical system onto the sensitive surface of the photodetector and at the part of the solid angle in limits of which the optical system collects the radiation.

In order to determine the temperature using the thermal radiation of the surface the calibration of the optical section of pyrometer is needed. In the current work such calibration was performed using a calibrated source of radiation – black body model.

4.4. The Scheme of the Measurements of Reflectivity at the Wavelength of the Incident Laser Radiation

On the one hand the measurements of the reflectivity of the sample is needed in order to determine the temperature. However, it is interesting in itself. In order to determine the reflectivity of the sample at the wavelength of laser radiation its surface was radiated with an angle of incidence of 45 degrees (Figure 14). The measurement of the initial value of reflectivity of the sample was performed at low level laser pulse energy in order to exclude the dependence of reflectivity upon temperature. The lead targets with mirror-like surface were used. The reflectivity of the targets before the laser impact was ~ 0.6. In order to obtain the value of reflectivity the energies of incident and reflected laser pulses were simultaneously measured using two transducers for pulsed laser energy measurement and the ratio of this integral values were considered as a initial value of reflectivity. In order to determine the change of the reflectivity of the sample relative to its initial value during the pulsed laser impact two photodetectors were used. They registered the incident and the mirror-like reflected radiation. The ratio of these signals, registered by photodetectors of the incident and reflected radiation, is proportional to the surface reflectivity at each moment of time. The signals from both photodetectors are shifted in time relatively to each other due to time delays between the channels of the oscilloscope. Since later the ration of the signals is calculated, the time delays and the different sensitivities of photodetectors should be compensated. Due to this reason before each shot with considerable energy the calibration of transducers was performed – the determination of difference in sensibility and time delay.

5. Experimental Study of Behavior of Metals under the Layer of Transparent Dielectric During the Impact of Laser Irradiation of Nanosecond Duration

5.1. Selection of the Target Material

In order to check the developed method the experiments with lead and mercury were carried out. The data on equation of state for these metals – theoretical calculations and experimental data (particularly for lead) are widely enough presented in literature [28-30] and so it is convenient to choose these metals as references. Due to low value of melting temperature of lead, small energies of laser radiation are needed for the study of phase transitions, besides the latent heat of phase transition is of the same order of magnitude as the energy for the heating of metal to the temperature of phase transition, so the phase transition plays the determinative role in the dynamics of temperature and pressure at the target surface. From the practical point of view the low value of melting temperature makes the target producing process easy although it hampers the temperature measurement.

The melting temperature of lead is $T_{melt} = 600.6$ K, boiling temperature at normal conditions $T_{boil} = 2018$ K. For lead there is a semi-empirical equation of state based on the model of soft spheres which satisfactorily describes experimental data for temperatures up to 4000 K [31]. Experimental data on critical parameters of lead are absent but according to theoretical calculations $T_{cr} = 4980$ K, $P_{cr} = 1.84$ kbar, $\rho_{cr} = 3.25$ g/cm^3. According to another data [32]: $T_{cr} = 5395$ K, $P_{cr} = 1.07$ kbar, $\rho_{cr} = 2.37$ g/cm^3. Such considerable deviation is due to the complexity of calculations and also due to the lack of experimental data. The same situation takes place for other metals with the exception of alkali metals for which the critical parameters were measured in static conditions and the deviation of calculations and the experiment doesn't exceed 10% [32].

5.2. The Scheme of the Experimental Setup

The scheme of the experimental setup is presented at figure 14. As a source of laser radiation solid-state Nd:YAG laser (1) working in a Q-switched regime with the following parameters of radiation: wavelength 1.06 um, pulse duration ~ 10 ns, maximum pulse energy ~ 1 J was used. The laser radiation was attenuated by a series of neutral filters (4); after that it was focused by a collecting lenses (f=30 cm) (5) onto the target surface (10) into a spot of 1-3 mm in diameter. The change of the attenuation coefficient and of the distance between the lenses and the target surface allowed us to vary the incident laser radiation energy density in a wide range of values. The uniform distribution of intensity in the laser spot was achieved by a homogenizer (7) (polished glass plates) positioned at half the distance between the lenses and the target. The temporal form of the incident laser pulse was registered with a silicon PIN-diode (6) (rise-time ~ 1 ns). A portion of radiation was directed onto it using a glass plate (8). The registration of the temporal form of the reflected radiation was made with the analog photodiode (11).

The system of registration of thermal radiation consisted of collecting lenses, optical fiber (12), filter from a BGG – glass (14) and photodetector (16). The thermal radiation was focused using lenses (9) onto the end of the fiber glass (12). The radiation outgoing from the fiber was collimated by a lenses (13) and in a parallel beam a BGG-glass (14) a filter was put. Using it, the laser radiation of the wavelength of 1.06 um was cut and after that the radiation was focused by lenses (15) on a photo-transducer of thermal radiation (16). So the photodetector (16) registered the intensive spectrum of thermal radiation.

The measurement of the acoustical signal from the sample was conducted with a wide-band acoustical piezo-transducer from lithium niobate (10). The electrical signal from the transducer was registered by a digital saving oscilloscope Tektronix TDS 3034. The signals from the photodiodes were registered by another four-channel oscilloscope TDS 2024. The synchronization of both of oscilloscopes was conducted using trigger photodiode (2).

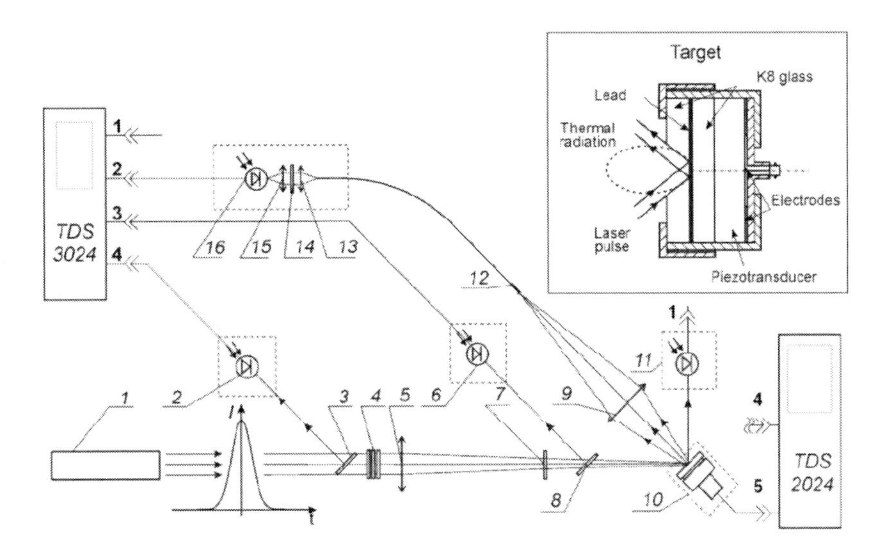

Figure 14. Scheme of the experimental setup.

5.3. Profiles of the Acoustical Pulses Induced by Laser Impact in Metals

Two reference metals: mercury (Hg) and lead (Pb) were chosen for the experiments. The measurements were conducted in a wide rage of incident laser energy density (from 1 mJ/cm^2 to 500 mJ/cm^2, about 40 regimes).

At figures 15a and 16a the temporal profiles of pressure pulses of lead and mercury for different values of laser irradiation energy densities are presented. The profiles are normalized on the maximum value of each curve.

At figures 15b and 16 b the absolute values of obtained pressures in a semi-logarithmical scale are presented. The maximum value of obtained pressure for lead was about 100 MPa (~1000 atm) and for mercury 750 MPa (~ 7500 atm).

For low values of laser pulse energy density (up to ~ 4 mJ/cm^2 for lead and ~ 12 mJ/cm^2 for mercury) the temporal form of pressure corresponds to a linear thermoelastic regime of pressure generation described at paragraph 2 (p. 2.3). According to eq. (8) the pressure pulse propagating into the depth of the metal resembles the temporal form of the laser pulse that is approved by the obtained experimental results.

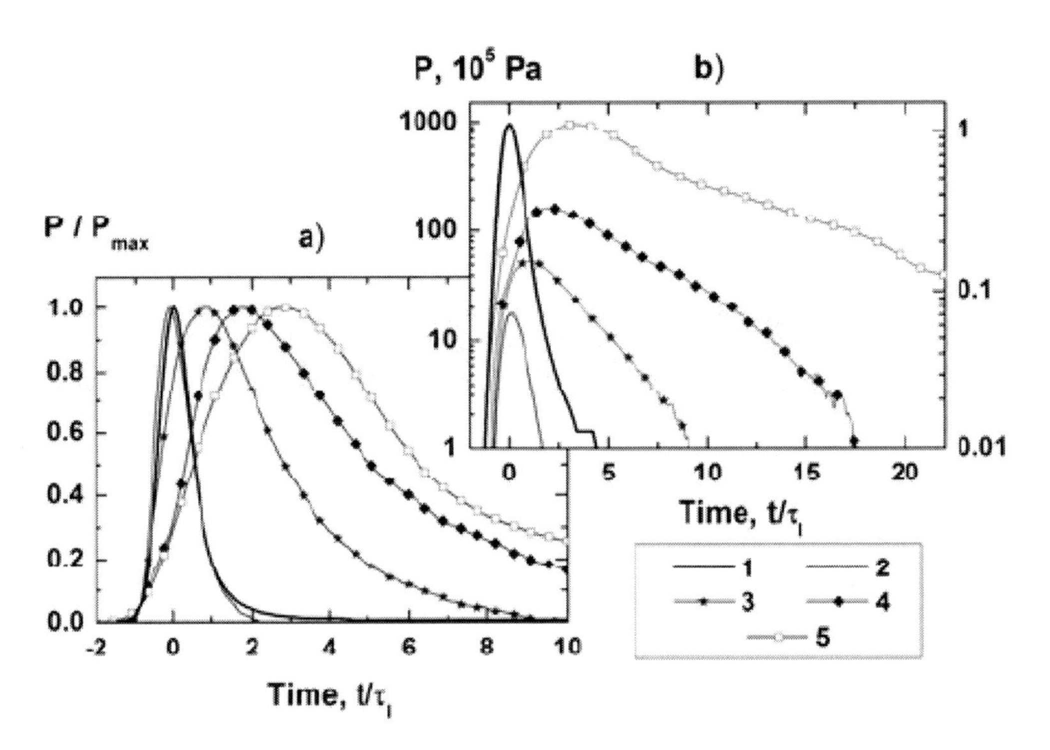

Figure 15. Temporal dependencies of pressure for lead a) – pressure is normalized at its maximum value, b) – absolute values of pressure. 1 – laser pulse, 2 – E = 2.4 mJ/cm^2, 3 – E = 10 mJ/cm^2, 4 – E = 24 mJ/cm^2, 5 – E = 230 mJ/cm^2.

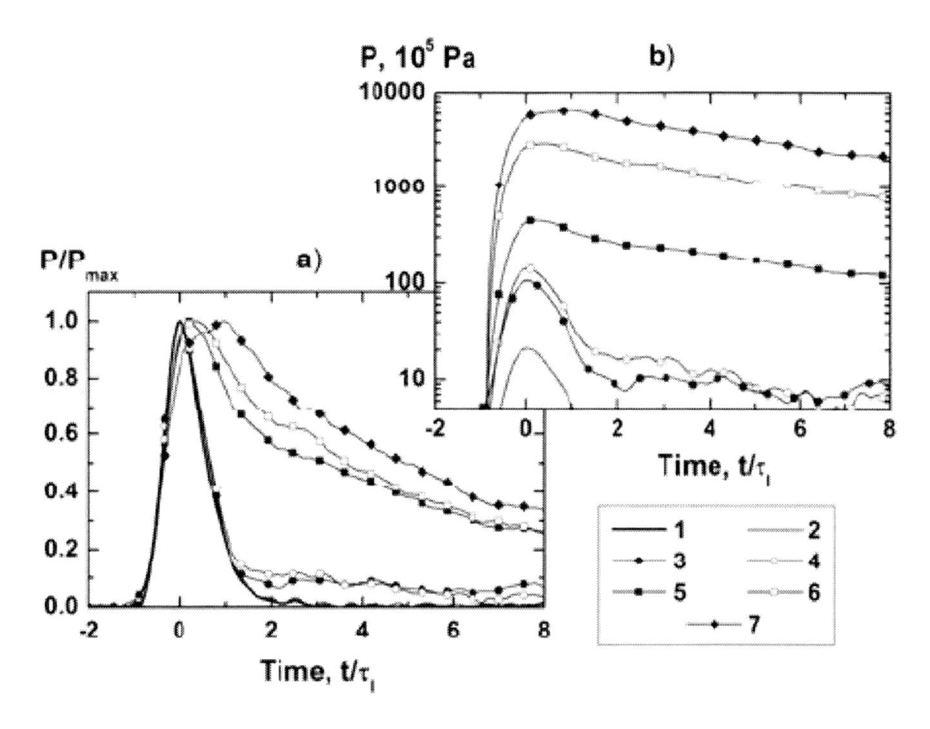

Figure 16. Temporal dependencies of pressure for mercury a) – pressure is normalized at its maximum value, b) – absolute values of pressure. 1 – laser pulse, 2 – E = 6 mJ/cm^2, 3 – E = 32 mJ/cm^2, 4 – E = 38 mJ/cm^2, 5 – E = 83 mJ/cm^2, 6 – E = 200 mJ/cm^2, 7 – E = 270 mJ/cm^2.

5.4. The Influence of Phase Transitions of Melting and Boiling on the Temporal Shape of the Acoustic Pulses

As it is shown on figures 15 a and 16 a at laser energy densities over 4 mJ/cm^2 for lead and 12 mJ/cm^2 for mercury the temporal form of the pressure pulses begins to distort. In comparison to the linear regime, the widening of the pressure pulses takes place. In this range of laser energy densities the material heating is considerable that leads to the appearance of thermal non-linearity due to temperature dependence of thermophysical parameters of absorbing and transparent medium. However as it was shown before by numerical calculations (p. 2.5) the thermal non-linearity leads to a non-linear dependence of signal amplitude upon the absorbed laser energy density, but the change of the signal shape is small. The considerable widening of the pressure pulses can be explained by the start of the melting process at the metal surface. The irradiation energy density at which a considerable widening of the pressure pulses is observed (8-16 mJ/cm^2 for lead) corresponds to the result of calculation of threshold value of laser energy density of heating of lead up to melting temperature 16.5 mJ/cm^2 (taking into account thermal non-linearity). At this range of parameters of state of the heated metal the mechanism of pressure generation changes. The mechanism of "defect of volume" (mentioned above) due to phase transition in the heated zone and appearance of a new phase with smaller density starts to exert considerable contribution in the pressure generation process. From the definite moment which is explicitly connected with the threshold value of integral of absorbed energy of the laser pulse, the separation of the leading edges of the laser intensity pulse and the pressure pulse takes place (figure 15 a (Pb), curve 3, figure 16 a. (Hg) curve 5). This moment determines the beginning of the phase transition and from this moment the current value of pressure amplitude becomes proportional to the integral of the absorbed energy. The analog results were obtained earlier [18,20].

The conducted estimations of the contribution of the embedded energy and the latent heat of the phase transition allow suggesting that boiling of mercury and melting and boiling of lead takes place. The assumption about boiling of mercury is approved by a rather high value of pressure (figure 16 a, curves 5-7) after the end of the laser impact that can be explained by slow (~10 τ_l = 90 ns) vapor condensation. It also should be noted that boiling of mercury also takes place in regimes 3, 4 during the unloading of metal – dropping of pressure and the rear edge of the laser pulse (figure 16 a, curves 3-4 at $t/\tau_l > 2$).

In the case of relatively high levels of absorbed energy in lead a very rapid transition from heating of the solid phase through melting to boiling (the energy deposition which is necessary for melting is much lower) takes place. At figure 15 a curves 4,5 the separation of front edges (of laser and pressure pulses) hypothetically due to boiling begins almost from the start of irradiation. For the exact definition of this phase transition as boiling the change of density must be known. Also the general regularity is observed: with increase of level of absorbed energy besides the above-mentioned widening, the maximums of pressure pulses shift onto the area of larger times (figure 16 a, Pb).

5.5. The Results of Temperature Measurements

Simultaneous measurements of temperature and pressure were carried out in two regimes for mercury. At figure 17a the absolute values of temperatures are shown. The dynamics of the relative change of reflectivity at the laser irradiation wavelength during the time of the laser impact (with angle of incidence of 45 degrees) for lead in relative units (R/R_0) are presented at figure 17b.

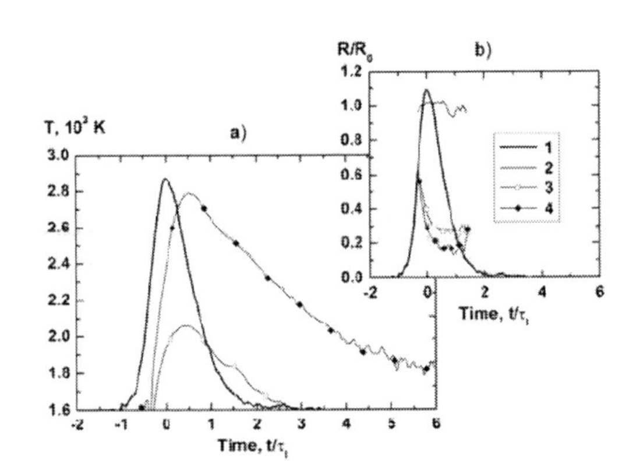

Figure 17. Temporal dependencies of temperature (a) and reflectivity (b) for mercury: 1 – laser pulse, 2 – E = 6 mJ/cm^2, 3 – E = 200 mJ/cm^2, 4 – E = 270 mJ/cm^2.

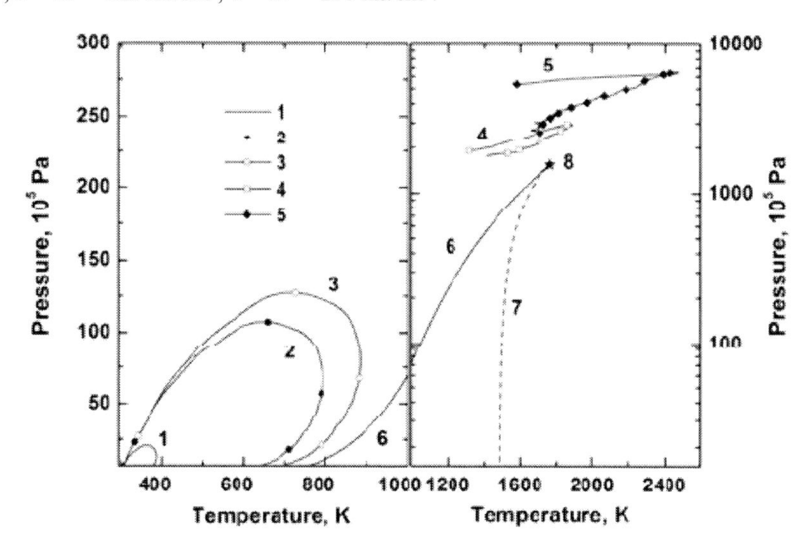

Figure 18. *P-T* diagram of laser heating of mercury: 1 – E = 6 mJ/cm^2, 2 – E = 32 mJ/cm^2, 3 – E = 38 mJ/cm^2, 4 – E = 200 mJ/cm^2, 5 – E = 270 mJ/cm^2, 6 – binodal, 7 – liquid spinodal, 8 – critical point.

The obtained results of pressure and temperature measurements allow plotting a $P-T$ diagram of the mercury heating process during irradiation of its surface under a layer of transparent dielectric. *At* left part of figure 18 the results of the calculation of the P-T diagram for the linear regime using eq. (4,9) (the same way as at figure 3) are presented. At the right side of the picture the experimental data is presented (curves 4, 5). In these regimes the supercritical states of mercury which are close enough to the critical point were obtained. The down branches of both of regimes, which corresponds to the rear edge of the laser pulse – the unloading process, are parallel to the critical isochore, that gives the prospect of obtaining near-critical states accordantly choosing the parameters of laser irradiation.

5.6. The Dynamics of the Reflectivity Change of the Surface During the Impact of Laser Radiation

The dynamics of the relative change of reflectivity at the laser irradiation wavelength at the time of the laser impact with angle of incidence of 45 degrees for mercury and lead in relative units (R/R_0) are presented at figure 17 b and figure 19 correspondingly. The initial value of reflectivity coefficient for mercury was R_0 = 0.50 and for lead R_0 = 0.60. With an increase of incident laser irradiation power density the reflectivity (at times the scale being comparable with the duration of the laser pulse) decreases more than five times. This considerable decrease of reflectivity and correspondingly the increase of absorbed energy during the duration of the laser pulse τ_L demands the binding account of this effect in all calculations during the processing of experimental data. At figure 20 the estimation of temporal dependence of conductivity obtained using data on reflectivity change and the Drude model [33]:

$$\sigma = \frac{4c}{\lambda_0 (1-R)^2}, \qquad (41)$$

is presented. Here σ - conductivity, c – speed of light, λ_0 - laser irradiation wavelength. In the limits of this model the order of the conductivity decrease is the same as the decrease of reflectivity. The considerable decrease both of the conductivity and reflectivity can be explained by the substantial density decrease for high levels of absorbed energy that takes place with an increase of temperature in a thin subsurface layer of metal.

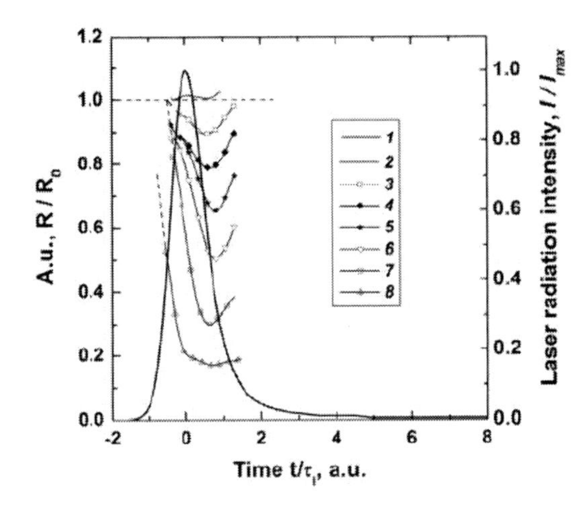

Figure 19. Dynamics of the reflectivity change during the impact of laser pulse (lead): 1 – laser pulse, 2 – E = 1.62 mJ/cm^2, 3 – E = 3.9 mJ/cm^2, 4 – E = 8.1 mJ/cm^2, 5 – E = 11.3 mJ/cm^2, 6 – E = 16.2 mJ/cm^2, 7 – E = 39 mJ/cm^2, 8 – E = 370 mJ/cm^2, τ_l = 8.5 ns.

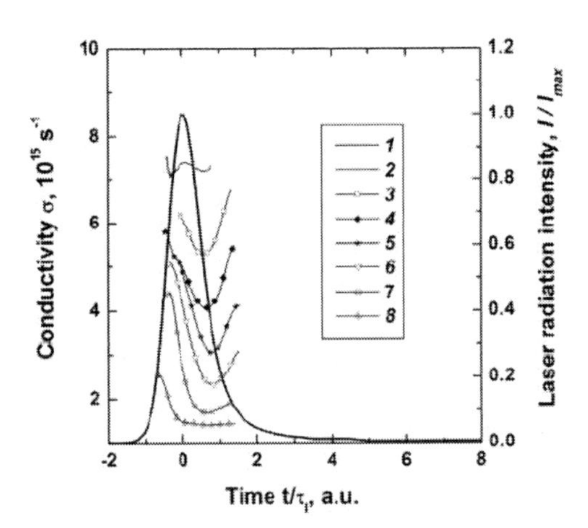

Figure 20. Dynamics of the conductivity change during the impact of laser pulse for different values of laser radiation energy density (lead): 1 – laser pulse, 2 – E = 1.62 mJ/cm^2, 3 – E = 3.9 mJ/cm^2, 4 – E = 8.1 mJ/cm^2, 5 – E = 11.3 mJ/cm^2, 6 – E = 16.2 mJ/cm^2, 7 – E = 39 mJ/cm^2, 8 – E = 370 mJ/cm^2, τ_l = 8.5 ns.

CONCLUSION

The possibility of obtaining high-energy states of metals with temperatures ~ 1 eV and pressures up to 10 kbar and higher by the impact laser pulses of nanosecond duration of moderate intensity up to 10^8 W/cm^2 at the metal surface under a layer of transparent dielectric is founded.

The method of inducing high energy states of metals, when the irradiated surface of metal is covered by a transparent dielectric layer, has been worked out.

The method of experimental research with laser irradiation impact was suggested, when it was possible to simultaneously measure pressure, temperature and reflecting ability of metal surface within nanosecond resolution.

Experimental research of pressure with laser irradiation impact on the surface of lead and mercury target was carried out. At low density of laser energy the pressure pulse repeats the form of laser pulse. At higher laser energy density pressure pulse becomes wider. It can be explained by the beginning of phase transition on the metal target surface.

Before melting (boiling) the pressure is proportional to the current laser intensity, but after the melting (boiling) threshold the pressure becomes proportional to the absorbed energy of laser pulse. The moment when the "deformation" of the leading edge of the pressure pulse starts determines the beginning of the phase transition.

Experimental research showed decrease of reflectivity of metal surface at high energy laser pulse density about five times, as compared to normal conditions.

This effect depends on considerable decrease of density in a thin absorbing surface layer of metals.

Simultaneous measurements of pressure and temperature allowed us to plot a P-T diagram of the metal heating process and also to prove the obtaining of the supercritical states of mercury.

The presented results show the efficiency of using confined metal surface for inducing high-energy states of metals and studying phase transitions.

It should be noted that the suggested method can be successfully applied in three important areas of study:

- for inducing and studying of metastable and unstable states in the sub-critical area of the phase diagram;
- for intense shock waves generation to study the shock-plastic deformation of metals;
- for the study of optical properties in a wide range of temperatures (~ 1 eV) and densities (~ 1 g/cm^3).

REFERENCES

[1] Fortov V.E. (editor.) Encyclopedia of low-temperature plasma. V. 3-1. Thermodynamical properties of low-temperature plasma. Moscow.: *Fizmatlit,* 2004, 541 pp. [in Russian]

[2] Fortov V.E., Iakubov I.T. The Physics of Non-ideal Plasma. *World Scientific,* 1999, 403 pp.

[3] Anisimov S.I. et al. The action of high-power laser beams on metals. Moscow.: *Nauka,* 1970, 272 pp. [In Russian]

[4] Krokhin O.N. Laser plasma. Physics and applications. Moscow: MIFI, 2003, 400 pp. [In Russian]

[5] Skripov V.P. Metastable liquids. *John Wiley & Sons,* 1974, 272 pp.

[6] Fairand B.P., Clauer A.H. Laser generation of high-amplitude stress waves in materials // *J. Appl. Phys. V.* 50(3), P. 1497, (1979)

[7] Gusev V.E., Karabutov A.A. Laser optoacoustics. *American Institute of Physics,* 1993, 336 pp.

[8] von Gutfeld R.J., Melcher R.L. 20-MHz acoustic waves from pulsed thermoelastic expansion of constrained surfaces // *Appl. Phys. Lett. V.*30(6), P. 257, (1977)

[9] Gross M.S., Black I., Muller W.H Computer simulation of the processing of engineering materials with lasers – theory and first applications // *J. Phys. D. V.* 36, P. 929, (2003)

[10] Xie J., Kar A. Mathematical modeling of melting during laser materials processing // *J. Appl. Phys. V.* 81(7), P. 3015, (1997)

[11] Tokarev V.N., Kaplan A.F. An analytical modeling of time dependent pulsed laser melting // *J. Appl. Phys. V.* 86(5), P. 2836, (1999)

[12] Kovalev A.A., Zhvavyi S.P., Zykov G.L. Dynamics of laser-induced phase transitions in cadmium telluride // *Semiconductors. V.* 39(11), P. 1299, (2005)

[13] Samokhin A.A. First-order phase transitions during the impact of laser radiation at absorbing media // *Proceeding of General Physics Institute, V.*13, P. 3, (1989) [In Russian]

[14] Samokhin A.A., Uspenskiy A.B. Evaporation of material under the impact of laser radiation // *Preprint of Physical Institute of Academy of Sciences,* V. 143, (1979) [In Russian]

[15] Breslavkiy V.P., Mazhukin V.I. Algorithm of numerical solution of hydro dynamical variation of Stephan problem using dynamically adapting meshes // *Mathematical modeling. V.*3(10), P. 104 (1991) [In Russian]

[16] Mazhukin V.I., Nikiforova N.M., Samokhin A.A. Photoacoustic effect upon material melting and evaporation by laser pulses // *Proceeding of General Physics Institute. V.* 60, P. 108 (2004) [In Russian]

[17] Veselovki I.A., Zhiryakov B.M., Popov N.I., Samokhin A.A. Photoacoustic effect and phase transitions in semiconductors and metals during pulsed laser irradiation impact // *Proceeding of General Physics Institute, V.* 13, P. 108 (1989) [In Russian]

[18] Karabutov A.A., Kubyshkin A.P., Panchenko V.Ya., Podymova N.B., Savateeva E.V. Optoacoustical investigation of the melting of indium by a laser pulse under a confined surface // *Quantum electronics, V.* 28(8), P. 670 (1998)

[19] Karabutov A.A., Kubyshkin A.P., Panchenko V.Ya., Podymova N.B. Dynamic shift of the boiling point of a metal under the influence of laser radiation // *Quantum electronics.* V. 25(8), P. 789 (1995)

[20] Karabutov A.A., Kaptil'nyi A.G., Kubishkin A.P. Investigation of indium and graphite melting induced by the dynamic laser action // *Izvestiya Akademii Nauk. Ser. Fizicheskaya. V.* 63(10), P. 1934 (1999) [In Russian]

[21] Anderholm N.C. Laser-generated stress waves // *Appl. Phys. Lett. V.* 16(3), P. 113 (1970)

[22] Schoen P.E., Campillo A.J. Characteristics of compressional shocks resulting from picosecond heating of confined foil // *Appl. Phys. Lett. V.* 45(10), P. 1049 (1984)

[23] Zel'dovich Ya.B., Raizer Yu.P. Physics of Shock Waves and High-Temperature Hydrodynamic Phenomena. New-York: Dover Publications, 2002, 944 pp.

[24] Lykov A.V. The theory of thermal conductivity. Moscow: Vyssh. Shkola, 1967, 599 pp.

[25] Samarski A.A., Moiseenko B.D. Express scheme for multidimensional Stephan task// *Journ. of Comp. Math. and Math. Mod. V.* 5(5), C.816 (1965) [In Russian]

[26] Budak B.M., Solov'eva E.N., Uspenskiy A.B. Difference method with coefficients smoothing for Stephan problem solution // *Journ. of Comp. Math. and Math. Mod. V.* 5(5), C.828-840 (1965) [In Russian]

[27] Lorazo P., Lewis L.J., Meunier M. Thermodynamic pathways to melting, ablation, and solidification in absorbing solids under pulsed laser irradiation // *Phys. Rev. B. V.* 73, P. 134108 (2006)

[28] Hixson R.S., Winkler M.A., Shaner J.W. Sound speed in liquid lead at high temperatures // *Int. Journ. of Thermophysics. V.* 7(1) (1986)

[29] Mehdipour N., Boushehri A., Eslami H. Prediction of the density of molten metals // *J. of non-crystalline solids. V.* 351, P. 1333 (2005)

[30] Partouche-Sebban D., Hixson R.S., Borror S.D. et al. Measurement of the shock-heated melt curve of lead using pyrometry and reflectometry // *J. Appl. Phys., V.* 97, P. 043521 (2005)

[31] Young D.A. A soft-sphere model for liquid metals // *Lawrence Livermore Lab. Report* UCRL-52352, Livermore, 1977

[32] Basin A.S. The main parameters of critical point of metals with close-packed crystal structure // Chemistry and computer modeling. *Butler's announcements.* V. 10, P. 83 (2002)

[33] Born M., Wolf E. Principles of optics. Cambridge, 2002. 720 pp.

In: Phase Transitions Induced by Short Laser Pulses
Editor: Georgy A. Shafeev
ISBN: 978-1-60741-590-9

Chapter 3

PHASE TRANSFORMATIONS IN THE UV LASER IRRADIATION OF MOLECULAR CRYOGENIC SOLIDS: IMPLICATIONS FOR LASER-BASED TECHNIQUES

Olga Kokkinaki and Savas Georgiou
Institute of Electronic Structure and Laser,
Foundation for Research and Technology – Hellas, 71110,
Heraklion, Crete, Greece

ABSTRACT

Laser-induced phase transformations have been extensively studied for metals and semiconductors. In contrast, very little has been reported on the laser-induced phase transformation of molecular systems. Here, we review pertinent experimental work on simple condensed molecular solids. Distinctly different structural processes are induced in different fluence ranges. Specifically, with increasing laser fluence, devitrification, glass annealing and melting and homogeneous bubble formation are shown to occur. At even higher fluences, bubble growth results in the unselective material ejection, namely ablation. Thus, in the nanosecond case, ablation is due to "explosive boiling" due to overheating of the matrix. In particular, the dynamics of laser-induced vitrification/devitrification as well as the nucleation and growth of the bubble cannot be accounted quantitatively by classical nucleation models. These differences may be ascribed to the very high heating and cooling rates involved in laser irradiation. We discuss the implications of these results for laser-based techniques such as matrix-assisted pulsed laser deposition and matrix-assisted laser detection techniques. The capability of pulsed laser irradiation for inducing phase transformations in molecular systems provides the potential for high time-resolved studies of the dynamics of molecular glasses and for obtaining fundamental information on the dynamics of bubble formation.

I. INTRODUCTION

In the case of semiconductors and metals, pulsed laser irradiation has been demonstrated to result in well defined phase transformations and structural changes [[1]-[7]]. These changes have found important industrial applications, particularly in microelectronics (e.g. for Si annealing, optical recording, etc). In parallel, their investigation has provided fundamental information on the kinetics of phase transformations in these systems. In fact, a major motivation for these studies has been the examination of phase transformations, such as melting and solidification/crystallization processes, under far from thermal equilibrium conditions. In contrast, the examination or use of pulsed laser-irradiation for inducing phase transformations in molecular/organic systems has been quite limited. In polymers, various morphological changes are observed upon laser irradiation [[1]], but these relate to complex processes, including debris deposition, decomposition, other chemical modifications etc. Thus, despite their interest for applications, these effects in polymers do not yield direct information on the dynamics of phase transformations. Indeed, one of the major reasons that have hindered the study of laser-induced phase transformations in molecular/organic systems is the complication due to the fact that, in parallel, chemical processes may occur. In fact, for a long time, the interaction of laser pulses with organics has been exclusively described within photochemical mechanisms. However, even within the proposed thermal models, discussion has mainly focused on the analytical description of the thermal decomposition, whereas attention on the phase transformations has been rather more limited. In contrast, the discussion on phase transformations has been limited to the possibility of explosive boiling in simple molecular systems (to be discussed in detail below) and the morphological changes indicative of melting and resolidification in the UV irradiation of some polymers.

Second, the examination of laser-induced phase transformations in molecular systems is hindered by a number of difficulties. In the case of semiconductors, the studies take advantage of the large optical and conductivity changes in these materials upon melting or order-disorder transitions, thus enabling temporally monitoring of the laser-induced processes. On the other hand, simple organics are generally amorphous or powdery and thus highly optically scattering. Furthermore, their optical properties (i.e., refractive index, absorption, etc.) do not change much upon phase change (e.g. melting). Even for glassy polymers (which exhibit low optical scattering), plausible melting upon laser irradiation has been almost exclusively assessed by post-irradiation phenomenological optical examination of the surface or indirectly through monitoring the diffusion of dopants as demonstrated by studies by Bounos et al. [[8]].

Combined, these two factors limit the potential of optical techniques for monitoring phase transformations of molecular substrates. Yet, as in the case of metals and semiconductors, the study of laser-induced phase transformations may provide significant insight into the dynamics of molecular systems. In particular, understanding the dynamics of molecular glasses is a topic of intense current interest in the area of physical chemistry [[9]-[11]]. As discussed below, the use of frozen or condensed samples permits the selection of simple enough compounds in which complications due to photodissociation and chemical reactivity are altogether obviated. In addition, at least for molecular systems that easily vitrify, some of the experimental difficulties noted above can be overcome and thus laser-induced transformations can be probed optically or even spectroscopically.

An important reason for the examination of the laser-induced phase transformations relates to the elucidation of the mechanism(s) of laser ablation of molecular solids and for the systematic optimization of the laser material processing schemes/applications. Laser ablation constitutes the basis of powerful techniques [[1]] in a wide spectrum of applications, ranging from the micron/ sub-micron patterning and structuring of organic/ polymeric substrates in microelectronics [[12], [14]], to the fabrication of new materials by processing within liquids [[15], [16]], to polymer/ biopolymer characterization in analytical chemistry (Matrix-Assisted-Laser-Desorption-Ionization of biopolymers, MALDI) [[17]], to tissue excision in medicine [[18]] and conservation of painted artworks [[19]]. Ablation also constitutes the basis of Pulsed Laser Deposition (PLD) [[20]]. In particular, a variation of PLD, "Matrix Assisted Pulsed Laser Evaporation" (MAPLE) has successfully met the challenge of depositing organic/ polymeric materials with minimal thermal or chemical decomposition. To this end, in MAPLE, the polymers/ biomolecules are dissolved in an absorbing frozen solvent or solution, instead of being directly irradiated in the bulk, so that the "violent nature" of laser interaction may be largely reduced (at least, the photochemical deleterious modifications to the polymer). The deposition of a wide range of organic macromolecules (e.g. carbohydrates, nanotubes), polymers/ biopolymers or even of larger biological structures (e.g. viruses, proteins, cells, tissue components) in intact and functional form has been demonstrated [[21]].

Well before the advent of MAPLE, we initiated the study of cryogenic solids for the detailed examination/ elucidation of the processes in laser-organics interactions [[22]]. Elucidation of the processes in laser interaction with molecular materials calls for the use of simple compounds. However, under ambient conditions, simple organic compounds are largely gaseous or liquid. Thus, for simulating the solid state, we turned to the study of the van der Waals solids formed by the condensation of vapours of these compounds on low temperature substrates. Given the physicochemical simplicity of these systems, photodesorption/ ejection processes can be probed in detail. Furthermore, the structure of the condensed solids can be varied systematically (e.g. glassy or crystalline), thus enabling assessment of its influence on the ejection processes. In addition, there is extensive information on the photophysics/ photochemistry of these compounds [[23]], thus, enabling accurate and quantitative evaluation of the processes and effects that are involved upon laser irradiation.

In the following, we review studies on the laser-induced phase transformations of molecular cryogenic solids. In particular, upon irradiation with increasing laser fluence (F_{LASER}), devitrification, glass annealing, and melting occur (Section III). Clearly, these are compatible with a thermal process and this is further confirmed by the examination of the desorbate intensities, which, in fact, show that a simple thermal process is valid up to F_{LASER} about twice the melting point (Section IV). At even higher F_{LASER}, explosive boiling is initiated resulting in the massive unselective ejection of a layer of material that can be identified with ablation (Sections V and VI). The implications of these results for laser-based techniques such as MALDI and MAPLE are discussed (Section VII). The plausible influence of chemical processes on the evolution of phase transformations in the UV irradiation of photolabile systems is addressed in Section VIII. In addition, implications of these results for nanoscience/ technology are briefly discussed (Section IX).

II. Employed Methodologies

A common phenomenological description of laser desorption/ ejection relies on the examination of the etching depth (or ejected amount) as a function of F_{LASER}. Such a graph for the 248 nm irradiation of neat frozen $C_6H_5CH_3$ solid is shown in figure 1. A sharp increase of the desorbate intensity is observed above ~100 mJ/cm^2. The onset of the sharp increase of the etching depth/ ejected amount is considered to correspond to the threshold fluence for ablation. However, in most cases, the dependence is quite smooth, so that it is both difficult to specify accurately and also questionable if it represents the onset of new processes. In fact, the terms "desorption" and "evaporation" in the MAPLE [[21]] and MALDI [[16]] acronyms suggest a simple thermal desorption process. To illustrate the problems of relying on such graphs, a closer examination of the plot for $C_6H_5CH_3$ reveals another increase of the desorption intensity at ~ 45 mJ/cm^2. In fact, as shown below, in semi-logarithmic (Arhhenius-type) plot of the desorption intensity vs. the inverse of the estimated temperatures, the slopes of the linear fittings differ below and above this F_{LASER}, whereas no such specific change in slopes is observed above and below the indicated ablation threshold.

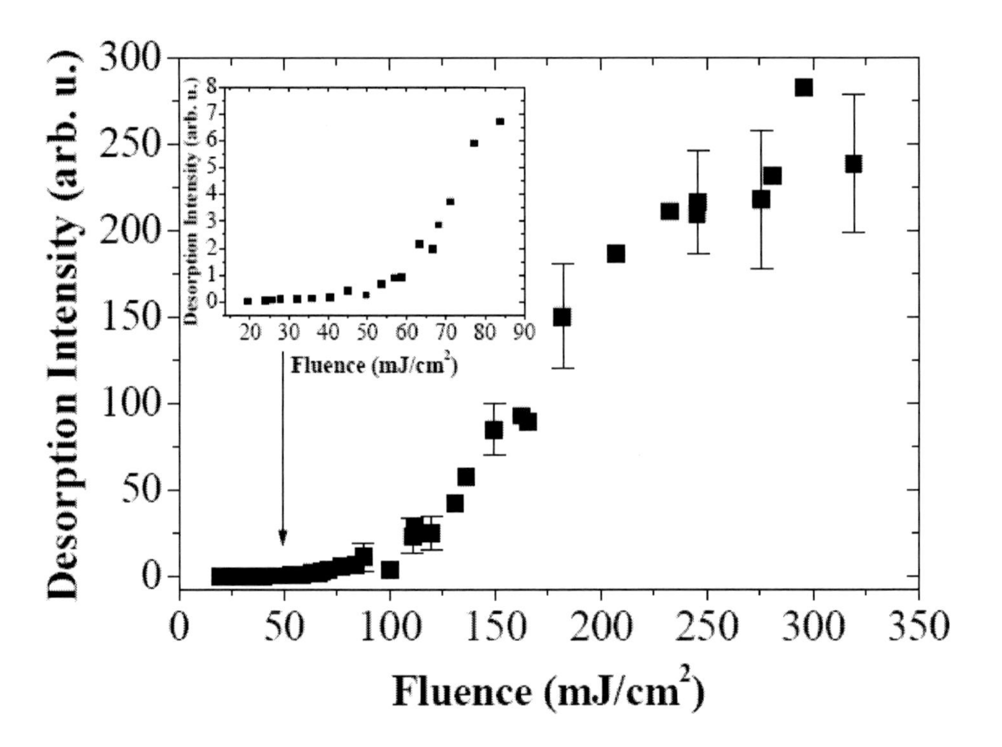

Figure 1. Intensity of the ejected toluene $C_6H_5CH_3$ recorded from freshly deposited $C_6H_5CH_3$ films as a function of the incident F_{LASER} (λ=248 nm). The error bars represent 2σ, as determined from at least 6-7 different measurements of each datum point.

Several studies rely on the examination of the desorbate translational distributions for obtaining information on the photo-ejection processes. In particular, the translational distributions are often fitted to ("shifted") Maxwell-Boltzman ones, as a way of assessing the "thermal" or "non-thermal" nature of the laser-induced desorption/ejection process [[23]]. However, the desorbate translational distributions may be severely modified by post-desorption collisions (even for desorption of as low as 0.1 monolayers). Therefore, the characteristics of the translational distributions may not directly reflect the processes occurring in the substrate that lead to material ejection. A number of analytical models have been developed to account for the perturbation of the postdesorption collisions, but their validity/applicability especially in the ablation regime has been questioned by molecular dynamics simulations. A detailed discussion of these issues has been presented in [[23]].

Ideally, for examining whether or not a thermal or other mechanism is applicable, we would like to examine systems in which all the parameters are kept the same (e.g., absorption coefficient, chemical constitution, etc), with only the binding energy between molecules varied in a systematic way. Of course, in practice, such systems are not available, but the objective can be attained by comparing the dependence of the ejection efficiency/ signal of dopants dispersed within a matrix [[24]-[26]]. To this end, the matrix is always the same (toluene-$C_6H_5CH_3$) and the employed dopants differ in their binding energies to the matrix. The binding energy can be "varied" by employing as dopants compounds of increasing size within a homologue series. With increasing size, the number of pairwise additive interactions of the dopant with the matrix increases and so does its overall binding to the matrix. To ensure that the absorption coefficient is always the same, the chosen dopants do not absorb at the irradiation wavelength (248 nm), so that absorption is exclusively by the matrix and in all cases the dopant-to-matrix molar ratio is the same (1:5, determined to be the optimal in terms of S/N). Since the excitation/ deactivation processes are the same in the comparison, the relative ejection signals of the dopants provide direct information on the nature of the energy dissipation in the substrate and on the mechanisms of material ejection. For a thermal process, the desorption intensities of the dopants should correlate with their binding energy to the matrix. On the other hand, no such correlation should be observed for mechanisms such as photomechanical or photochemical.

In most of our work, as matrix we employ $C_6H_5CH_3$ (at 248 nm, absorption coefficient ~ 3700 cm^{-1} from in-situ measurements) for three main reasons. First, it is one of the simplest organic molecules and it may be a useful solvent in MAPLE. Second, it presents minimal photo-fragmentation (at 248 nm), thereby avoiding complications due to any photoreactivity [[27]]. Third, it has been extensively studied, thus the well-defined values [[28]-[30]] permit quantitative analysis of the results. As dopants, we include alkanes (e.g. c-C_3H_6, c-C_6H_{12}, $C_{10}H_{22}$) and ethers/alcohols (e.g. $(CH_3)_2(CH_2)_nO$, D_2O, $CH_3(CH_2)_nOH$) which are nearly transparent at 308 nm and 248 nm. Among all the examined dopants, we refer in particular to the results concerning dimethylether ($(CH_3)_2O$) and decane ($C_{10}H_{22}$). Their binding energies to the $C_6H_5CH_3$ matrix are respectively >0.4 eV/molecule and >0.8 eV/molecule, as determined by Thermal Desorption Spectroscopy (for reasons of brevity, we refer to $(CH_3)_2O$ as "volatile", while to $C_{10}H_{22}$ as "non-volatile" dopant) [[26], [31]]. The intensities of the ejecta are probed as a function of F_{LASER} via quadrupole mass spectrometry (i.e., neutral desorbates are detected by electron-impact ionization). The film thickness is $\geq$ 50 μm, i.e., much larger than the optical penetration depth. Further information on the experimental procedures can be found in the cited references.

In parallel with the mass spectroscopic studies, optical/imaging techniques have been employed in order to monitor the structural changes that the irradiated matrices undergo. As already mentioned, molecular substrates are generally amorphous/ powdery (thus, highly optically scattering), thereby limiting the potential of optical techniques in these substrates. In our studies, the above problems are alleviated by studying molecular/organic glasses. In particular, toluene is one of the simplest organic liquids known to supercool easily to the point of vitrification (T_g~117 K) [[11]]. Therefore, upon vapor condensation at lower temperatures, $C_6H_5CH_3$ may form a glass of high optical quality, so that the processes that are induced upon UV (248 nm) irradiation can be monitored via temporally resolved imaging/optical monitoring (HeNe or diode laser used for probing). From experimental standpoint, it is important to note that the quality of the glass produced depends on the rate of the deposition/condensation [[9]-[11]]. In fact, recent studies have shown that for very slow deposition rates $\leq$ 10 nm/min, ultrastable glasses can be formed [[32]]. In the present experiments, deposition rates ~10-100 times faster have been employed, first because otherwise the formation of sufficiently thick matrices (depth larger than the penetration depth) would require exceedingly long times and second, because the promotion of laser-induced phase transformations, in particular of devitrification and bubble formation, rely sensitively on the pre-existence of defects (as shown, due to the ultrafast heating/cooling rates involved).

III. LASER-INDUCED DEVITRIFICATION AND GLASS RE-ANNEALING

Figure 2 shows images recorded upon irradiation of *neat* $C_6H_5CH_3$ glass at two indicative UV F_{LASER}. For the quantitative evaluation, the intensity of the specularly reflected probe beam has been monitored (Figure 3). Both imaging and optical examinations demonstrate that distinct morphological changes are induced to the glass in different UV laser fluence ranges.

- at F_{LASER} <15 mJ/cm^2, no changes are detectable in the images, and the transmitted and reflected probe beam intensities remain constant, even after irradiation with more than ~2000 UV laser pulses. Generally, at such low F_{LASER}, desorption of sub-monolayers is observed; desorption is system-specific (usually, mediated by electronic excitation) [[33]] and differs from that observed at higher fluences. There is no direct relevance to the topic of phase transformations addressed in this review.
- In contrast, upon irradiation in the 15-30 mJ/cm^2 range, the intensities of both the specular reflection and of the transmitted probe decrease sharply upon irradiation of the $C_6H_5CH_3$ glass with successive UV laser pulses, evidently due to pronounced scattering of the probe beam. Under broadband illumination, the irradiated area exhibits the "whitish/milky" appearance typical of polycrystalline samples. Examination of the angular distribution of the probe beam scattering indicates that the crystals formed are of size $r \approx 70-100nm$ (Mie-type scattering). In the subsequent pulses, scattering increases evidently due to an increase in the size of the crystals and due to multiple scattering becoming important.

Importantly, for well annealed glasses, a much higher number of laser pulses is required in order to initiate observable devitrification. Vapor as-deposited glasses are generally characterized by a large free volume (for $C_6H_5CH_3$, estimated to be ~10% [[11]] of the extrapolated at these temperatures liquid volume) and defect formation cannot be avoided. In contrast, annealed glasses are of higher density and are largely free of defects. Thus, laser-induced devitrification is very sensitive to the existence of defects /microcrystallites that act as "nuclei".

- Surprisingly, upon irradiation at higher UV fluences (30-50 mJ/cm^2), annealing of the as-deposited glass takes place, with its reflectivity increasing with successive laser pulses by as much as ~30-50 % (Figure 3). The irradiated area remains of high optical quality, even after irradiation with more than 200 laser pulses.

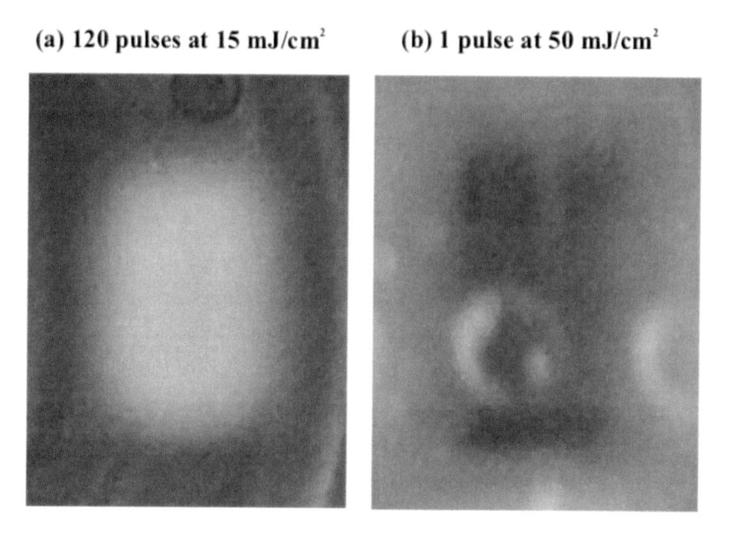

Figure 2. Photographs of the toluene glass following irradiation with the indicated number of pulses at (a) 15 mJ/cm^2, (b) 50 mJ/cm^2.

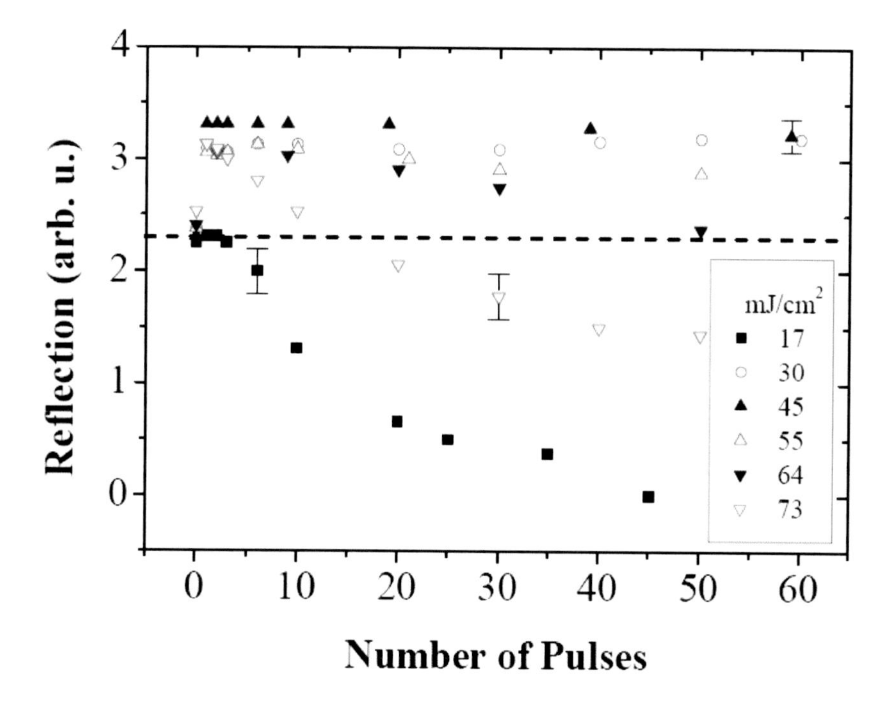

Figure 3. Pulse dependence of the intensity of the specularly reflected probe beam (as measured by a PMT) at various UV laser fluences. The values at 0 number of pulses represent the (initial) reflected intensity from the as-deposited films.

Glasses do not undergo a sharp first-order transition to melt, but, as commonly considered [[9]], a gradual "softening" towards the liquid. In competition with this process, a glass may undergo de-vitrification to the stable crystalline form. Indeed, in conventional (slow) heating experiments $C_6H_5CH_3$ glass is known to undergo crystallization (devitrification) at ≥130 K [[11]]. In the case of laser irradiation, the matrix surface temperature at the end of the laser pulse is accurately estimated by

$$T = T_0 + \frac{\alpha F_{LASER}}{\rho C_p} e^{-\alpha z} \tag{1}$$

where T_0=110 K, z is the depth from the surface (since heat diffusion for t<1 μs can be neglected, as the thermal diffusion time is $\tau_{thermal} \sim 1/\alpha^2 D_{th} \approx 65$ μs - $D_{th} = k_{th} / \rho C_p$ ρ is the molar density, C_p the molar heat capacity, D_{th} the thermal diffusivity [[30]] and $k_{th} \sim 0.16$ $Wm^{-1}K^{-1}$ - and losses due to desorption at F_{LASER}<50 mJ/cm^2 are minimal). For $C_6H_5CH_3$ glass and supercooled liquid C_p~ 135 J mol^{-1} K^{-1}, ρ_{glass}~0.011 mol/cm^3. Furthermore, glasses are of higher enthalpy than the crystal, which for vapor-deposited $C_6H_5CH_3$ glass (at 110 K) is estimated to be ~1.5-2.0 kJ/mole [[11]]. Based on these, the onset for laser-induced devitrification is estimated (Eq. (1)) at a surface temperature of the matrix of ~150 K (assuming that the excess enthalpy of the glass is released during devitrification). This is somewhat higher than the temperature (130 K) at which devitrification is observed in conventional thermogravimetric experiments. The difference seems small to warrant discussion, but it does turn out to have important implications in relationship with the elucidation of the laser induced processes.

Considering next the changes for F_{LASER} in the 30-50 mJ/cm^2, the surface temperature of the matrix (at the end of the laser pulse) is estimated to range between 180 and 240 K, i.e. melting is effected (melting point: 178.15 ^{0}K). Melting is further demonstrated by the examination of the dependence of the intensity of "volatile" dopant (e.g. $(CH_3)_2O$) on the number of successive laser pulses (Figure 4). For F_{LASER} <30-40 mJ/cm^2, the dopant signal is found to decrease with successive laser pulses. The total signal over ~ 50-100 pulses indicates that only dopant from the upper (surface) layer (~ 10 nm) of the film desorbs. In contrast, in the 50-100 mJ/cm^2 fluence range, the dopant signal per pulse is much higher and, in fact, it increases with successive laser pulses. The total signal corresponds to desorption of the dopant from at least ~ 100 nm depth. Diffusion of the dopant from such depths shows that the film viscosity is in the ~ 10^{-3} Pa s range, which corresponds to the viscosity of liquid toluene (for T=210-255 K) [[30]]. In contrast, the viscosity of (frozen) $C_6H_5CH_3$ at ~100 K is in the ~10^{12} Pa s range.

Subsequent cooling of the melt results in annealed glass (figure 2(b) and 3). The extent of crystallization vs. glass formation is known to depend on the cooling rate, with faster cooling promoting glass formation [[9]]. Upon pulsed laser irradiation, the initial cooling rate increases with F_{LASER} as $\frac{dT}{dt} \cong -\frac{a^3 D_{th} F_{LASER}}{\rho c_p}$ [[1]] (for heat-diffusion-dominated process). However, with time, the heat redistribution results in changes of the cooling rate. The temperature evolution in the substrate following irradiation is given by [[32]-[35]]

$$T(z,t) = T_0 + \frac{\alpha F_{LASER}}{2C_P} \times \exp(a^2 D_{th} t)$$

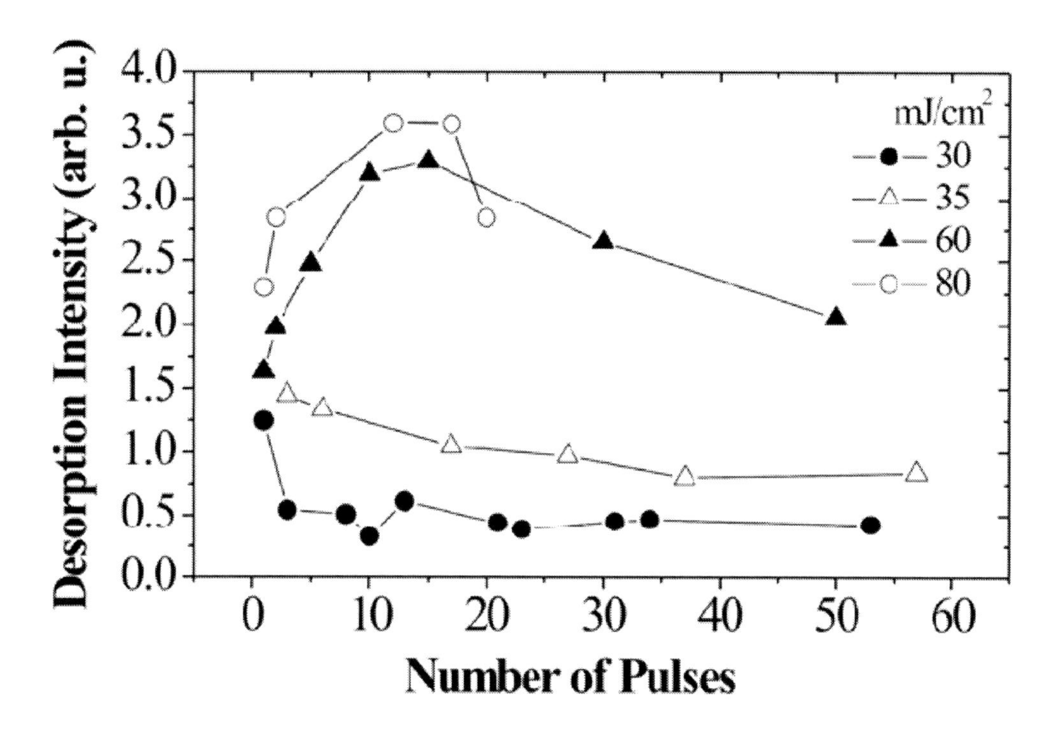

Figure 4. Desorption intensities of $(CH_3)_2O$ dopant and of $C_6H_5CH_3$ upon irradiation of the $(CH_3)_2O/C_6H_5CH_3$ mixture (1:5 molar concentration) with successive laser pulses at λ=248 nm. The intensities are corrected for the different relative ionization efficiencies of the two compounds in the mass spectrometer.

$$\times [\exp(-\alpha z) erfc(a \sqrt{D_{th} t} - \frac{z}{2\sqrt{D_{th} t}}) + \exp(az) erfc(a \sqrt{D_{th} t} + \frac{z}{2\sqrt{D_{th} t}})]$$

where z is the depth from the film surface (erfc: complementary error function). The equation represents the solution of the heat diffusion equation under the assumption that the initial temperature distribution in the substrate follows Beer's absorption profile. The calculations show that the cooling rate at the surface of the melt when reaching 180 K and below is slower than the cooling rate upon irradiation at $F_{LASER}<30$ mJ/cm^2 (Figure 5). Thus, at the very high cooling rates (10^6-10^7 K/s) attained upon laser irradiation, devitrification/vitrification are not specified by differences in the cooling rates. Instead, devitrification is determined by the existence or not of defects large enough to promote crystal growth on the relevant time scales, which explains why the process is highly dependent on the degree of annealing of the as-deposited glass.

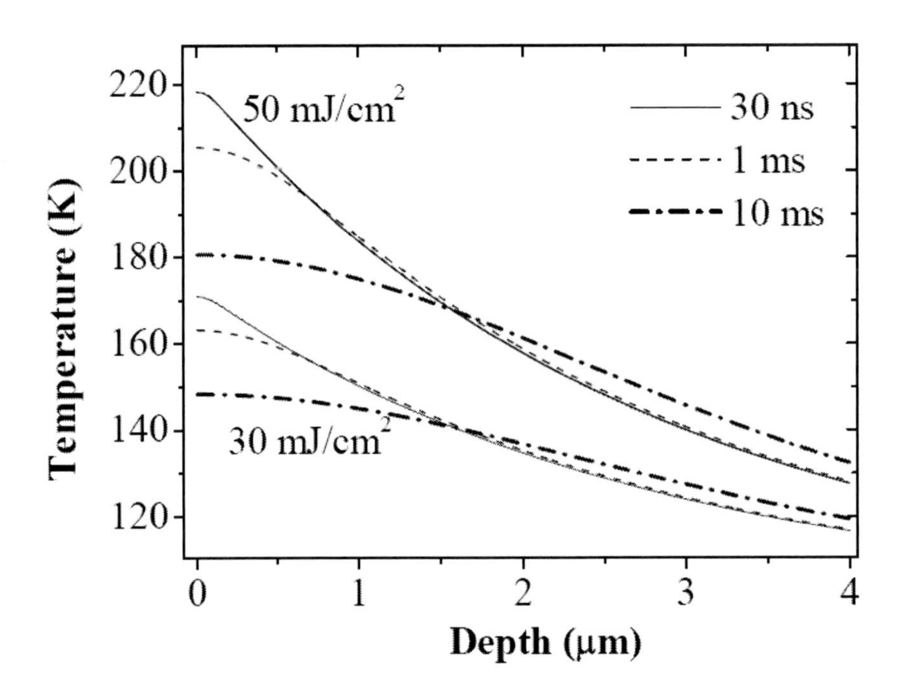

Figure 5. Temporal evolution of the temperature (as estimated by Eq. (2)) at various depths within the glass at F_{LASER} below 50 mJ/cm^2 at which energy losses due to desorption can be neglected.

Crystallization can be considered to involve two steps, namely nucleation i.e., formation of crystal nuclei of large enough size and the subsequent growth of formed or pre-existing nuclei [[36]-[40]]. Experimentally, the observation of very fast nucleation requires that, at the very least, one nucleus to be formed <1 μs time within the ~3 μm optical penetration depth, which correspond to nucleation rates J=3.3x10^{13}-3.3 x10^{11} m^{-3} s^{-1}. On the other hand, theoretically, the nucleation rate for crystal formation is estimated by the classical nucleation model, based on the assumption that nuclei have to grow up to a critical nucleus with radius r^*_{cr} and free energy ΔG^*_{cr} (in this section, the subscript "cr" represents crystallization). If this size is reached, the nuclei are going to grow into a (micro)crystal, whereas if r^*_{cr} is not reached, then they dissolve. The classical nucleation theory predicts that

$$r^*_{cr} = \frac{2\sigma_{cr,l}}{\Delta G_{cr,l}}, \quad \Delta G^*_{cr} = \frac{4\pi}{3}\sigma_{cr,l} r^{*\,2}_{cr}$$

and the nucleation rate (nuclei/per unit time/per unit volume) is then given by :

$$J_{cr} \approx \frac{N_V D_t}{d^2}\exp\left[-\frac{\Delta G^*_{cr}}{k_B T}\right] \text{ or } J_{cr} \approx \frac{N_V}{2\pi\tau}\exp\left[-\frac{\Delta G^*_{cr}}{k_B T}\right]$$

where $\Delta G_{cr,l}$ is the difference in the Gibbs energy between crystal and liquid (per unit volume), D_t is the molecular diffusivity at T, $\sigma_{cr,l}$ ~12 mJm^{-2} is the interfacial crystal-liquid tension per unit area [38], d represents a characteristic distance (approximately taken to be the molecular diameter), τ represents the structural relaxation time of mode α which essentially determines viscous diffusion $\tau = 6.3x10^{-13}\, s\exp(434K/(T-104K))$ [[32]]. $\Delta G_{cr,l}$ can be calculated from the reported Cp values or there are other approximate formulae for its estimation. At $T \approx T_{glass}$, $\Delta G_{cr,l}$ is quite high but J_{cr} is limited by the very small value of D_t, whereas at high T, D_t is large but the thermodynamic driving force is small. The maximum nucleation rate is estimated to be 2-4 orders of magnitude smaller than the experimentally indicated ones. Note that the above formulae represent the steady-state conditions (at a constant temperature), whereas for the very fast heating/cooling rates upon laser irradiation, the lag in nuclei induction can be significant; ($J = J_{cr}\, exp(-\tau/t); \tau \approx \pi^2 r_{cr}^2 / 4D_t$) [[37]] resulting in even lower nucleation rates. Thus, it is clear why devitrification on the relevant time and spatial scales can be only attained if there are sufficient and large enough pre-existing nuclei (i.e., larger than the critical size). Such defects are relatively abundant in the as-deposited glass.

For pre-existing defects or nuclei of size larger than the critical size, growth will be specified by the self-diffusion D_t and the thermodynamic “driving force”. Approximately, the velocity of the microcrystal growth is $u \approx \frac{D_t(T)}{d}\exp\left[-\exp(-\Delta G_{cr,l}/RT)\right]$ [[37]]. For the growth of crystallites of r~100 nm on 30 ns -1 μs, u must be ~10-0.1 ms^{-1}. In the 120-140 K temperature range (i.e., where devitrification is observed in thermogravimetric experiments), despite the high supercooling (i.e. very high $\Delta G_{cr,l}$), D_t is too low (10^{-13} m^2/s) to enable growth even of pre-existing nuclei on the relevant cooling time. In contrast, in the 150-170 K range, D_t ranges from 10^{-12} m^2/s to 10^{-10} m^2/s. However, even in this temperature range, u is estimated to be ~ 10^{-3} – 0.3 m/sec, i.e. not large enough to account for the growth of the pre-existing nuclei to the size of 100 nm. Thus, at present, laser-induced devitrification is not fully accounted by the known parameters. It is likely that at the very high heating/cooling rates, “domains” known to exist in glasses do not have time to re-adjust, so that the kinetics of devitrification cannot be described in a satisfactory way by the simple formulae of classical nucleation model. At any rate, it is clear that laser-induced devitrification provides new tools for assessing the dynamics of the process in molecular glasses.

On the other hand, for F_{LASER}>30 mJ/cm^2, melting results in the “annihilation” of defects and dissolution of any microcrystalline “nuclei” (i.e., since they become thermodynamically unstable). Based on the previous discussion about the need for pre-existing nuclei for promoting devitrification, it is evident why the subsequent cooling of the melt results in annealed glass even though its cooling rate is slower than that attained upon irradiation at 15-30 mJ/cm^2.

Besides its importance for the examination of vitrification/devitrification processes in molecular glasses, the above results also have direct importance for the elucidation of laser ablation. A fundamental implication of the "explosive boiling" mechanism is that upon irradiation even at fluences below the ablation threshold, the solid melts. Indeed, MD simulations [[42]] have observed a high degree of film disordering, strongly suggestive of this phase change. However, no direct experimental evidence of a solid-to-liquid phase transformation has been obtained so far, except for indirect observations via studies of the translational distributions of the desorbates and post-irradiation optical examination of the morphology of irradiated areas [[43]-[48]]. In the IR-ablation of neat C_6H_6 films, Braun and Hess [[44]] have shown that the desorbate most probable translational energy E_{TRANS}^{mp} does not increase monotonically with laser fluence, but most interestingly it shows a "phase-transition" like dependence. The characteristic "plateau" of the diagram appears at fluences suggestive of film temperatures close to the melting point of the compound. In view of the similarity of the (E_{TRANS} vs. F_{LASER}) diagram to a (P, T) thermodynamic diagram but not necessarily a valid correspondence, ablation, at least in IR, was claimed to be photothermal in nature. Evidently, the implicit assumption is that the most probable desorbate velocities reflect mainly the temperature of the film. However, the fitting of the time-of-flight spectra requires non-zero and rather large υ_{drift} values, which indicates that the translational distributions and thus the observation of a plateau in the E_{TRANS}^{mp} may be due to the effect of the gas-phase (postdesorption) collisions.

IV. Thermal Desorption and Onset of Explosive Boiling

As shown above, based mainly on optical examination, devitrification and melting of the matrix are demonstrated to occur upon irradiation in the 30-50 mJ/cm^2. At these very low F_{LASER}, the desoption signal is too low to enable detailed information. However, at F_{LASER} >50 mJ/cm^2, the desorption signal is high and thus we turn to the results of the mass spectroscopic examination.

The ejection intensities of the dopants (corrected for their ionization/ detection efficiency in the mass spectrometer vs. $C_6H_5CH_3$) and of toluene are plotted as a function of F_{LASER} (Figure 6) in the irradiation of the mixtures of toluene with $(CH_3)_2O$, C_6H_{12} and $C_{10}H_{22}$. It is evident that two fluence ranges can be delineated with distinctly different F_{LASER}-dependence of the ejection intensities.

At F_{LASER}<100-120 mJ/cm^2, the signal of the "volatile" dopants (e.g. $(CH_3)_2O$) is considerably higher than the $C_6H_5CH_3$ signal, even though its molar concentration in the film is 1/5 that of $C_6H_5CH_3$ (Figure 7). On the other hand, at these F_{LASER}, no signal is detected for the non-volatile dopants (e.g. for $C_{10}H_{22}$). Apparently, at low fluences, only dopants that are weakly bound to the matrix desorb, indicative of a thermal desorption process.

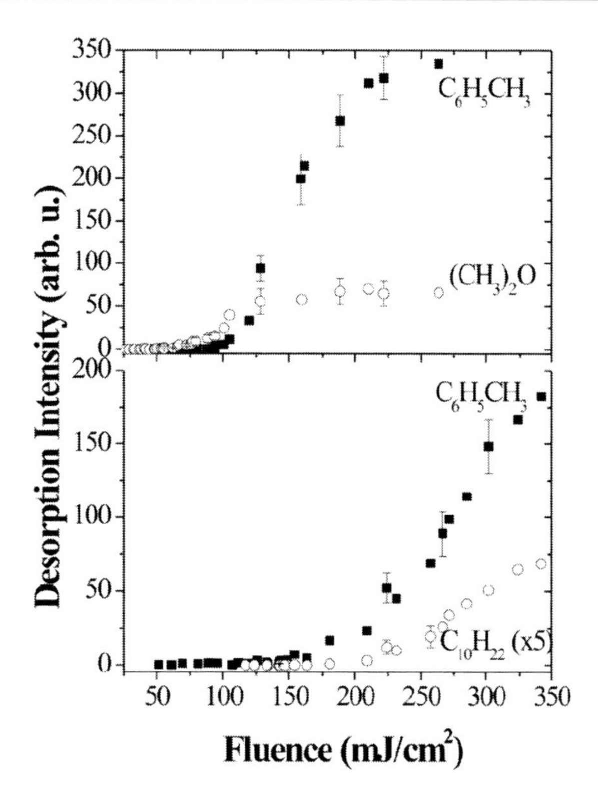

Figure 6. The intensities of the dopant and of $C_6H_5CH_3$ desorbing as a function of F_{LASER} upon irradiation of the mixtures $(CH_3)_2O$/ $C_6H_5CH_3$, C_6H_{12}/ $C_6H_5CH_3$ and $C_{10}H_{22}$/ $C_6H_5CH_3$ (1:5 molar concentration) at λ=248 nm. The intensities are corrected for the different relative ionization efficiencies of the two compounds in the mass spectrometer.

To establish accurately this point, an Arrhenius-type analysis of the data is presented (i.e., ln(signal) vs. 1/T, where the T is the peak (surface) temperature attained upon irradiation $T = T_0 + \frac{\alpha F_{LASER}}{\rho c_p} e^{-\alpha z}$ where T_0=120 ^{0}K) (Figure 8). The activation energy for desorption (ΔE_{des}) of the compounds can be determined from the slope of the linear fitting. The plots demonstrate that even below 100 mJ/cm^2, two sub-ranges can be delineated, in which the slopes differ distinctly. The change occurs at F_{LASER} close to the one specified for the melting of the glass, although in retrospect it seems that the change should better be correlated with the onset of bubble formation discussed below.

However, a much more accurate estimation of the evolution of the temperature is attained from:

$$\frac{\partial T}{\partial t} - \left(\frac{x_1 P_1(T)}{\sqrt{2\pi M_1 RT}} + \frac{x_2 P_2(T)}{\sqrt{2\pi M_2 RT}} \right) \frac{\partial T}{\partial z} = \frac{k_{th}}{\rho C_p} \frac{\partial^2 T}{\partial z^2} + \frac{\alpha I_{\mathrm{LASER}}}{\rho C_p} \quad (3)$$

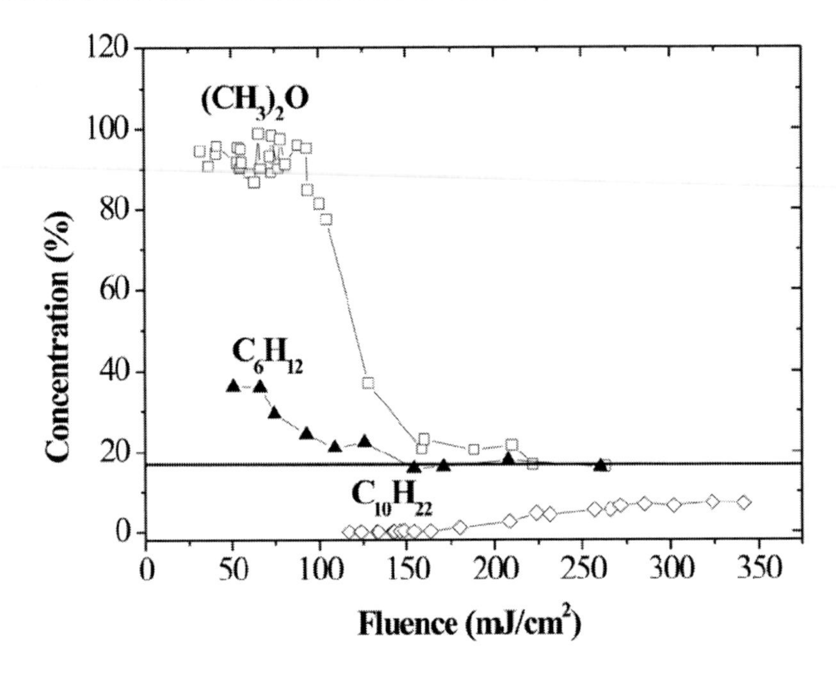

Figure 7. Replot of the data of the previous figure in the form of dopant concentration [i.e. $I_{dopant}/(I_{dopant} + I_{toluene})$] of $(CH_3)_2O$, C_6H_{12} and $C_{10}H_{22}$ dopants in the plume as a function of the F_{LASER} in the irradiation of their mixtures with $C_6H_5CH_3$. The horizontal line indicates the initial concentration of dopants in the sample.

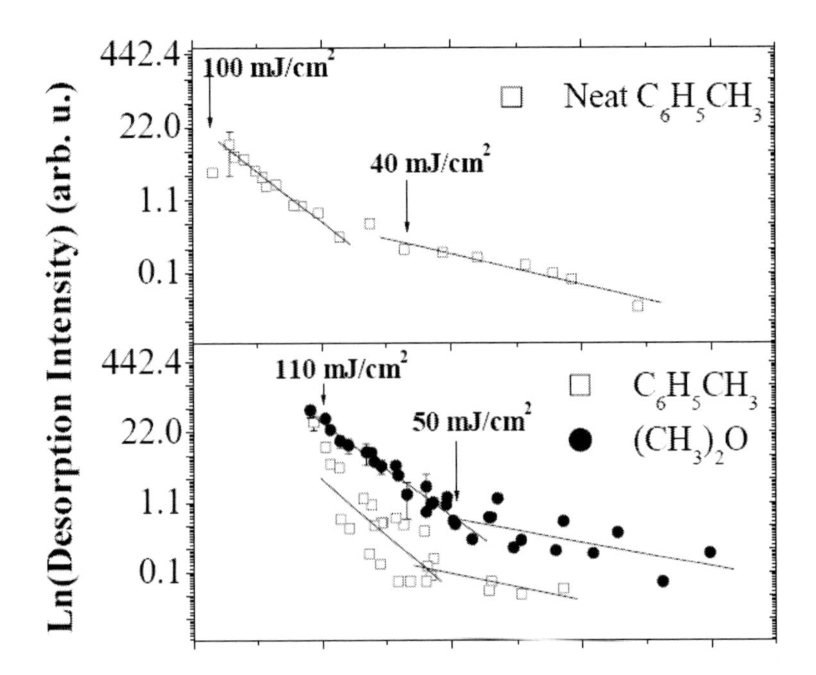

Figure 8. Semilogarithmic (Arrhenius type) plot of the $C_6H_5CH_3$ and of the dopant intensities in the irradiation of the mixtures vs. $1/T_{peak}$ (data points at fluences above the ablation thresholds are not included in the graphs).

with the boundary condition $-k_{th}\left(\frac{\partial T}{\partial z}\right)_{z=0} = -\frac{x_1 P_1 \Delta H_{LV,1}}{\sqrt{2\pi M_1 RT}} - \frac{x_2 P_2 \Delta H_{LV,2}}{\sqrt{2\pi M_2 RT}}$ where k_{th}: thermal conductivity, ρ, C_p have been defined above, the subscripts 1 and 2 denote toluene and dopant, respectively, $\Delta H_{LV} \equiv E_{TDS}$ the thermal desorption enthalpy as determined by TDS (Table 1), x_1=0.84 and x_2=0.16 represent, respectively, the molar concentrations of toluene and dopant (we assume that they do not change much from their initial values) and P pressure ($P \propto \exp\left\{-\frac{E_{TDS}}{RT}\right\}$). For a single component (i.e., neat toluene solid), Eq. (3) reduces to the well established one that has been extensively used in ablation studies [[49]]. We note that if evaporative cooling is neglected and instead the peak temperature is simply estimated by Eq. (1), the estimated activation values are quite low. It is likely that in MALDI studies, some discrepancies between the activation values derived for the laser-induced ejection and thermodynamic values may be partly ascribed to the use of the simple formula for temperature estimations.

For all systems, there is good correspondence between the E_{LASER} determined as above and the TDS activation values, thus establishing unambiguously that the laser-induced desorption process in this fluence range corresponds to thermal desorption.

Table 1. The results of the Arrhenius-type analysis of the studied systems

System	Molar average[1] C_p (J mol^{-1} K^{-1})	Detected Compound	E_{LASER}[2] (Intermediate Fluences)[2] (kJ/mol)	E_{TDS}[3] (kJ/mol)
Neat Toluene	135 (glass) 150 (liq.)	$C_6H_5CH_3$	35	40 ± 5
Dimethylether/ Toluene	130 (glass) 142 (liq.)	$(CH_3)_2O$	30	20 ± 5
		$C_6H_5CH_3$	39	35 ± 5
Cyclopropane/ Toluene	125 (glass) 140 (liq.)	c-C_3H_6	15	~12-15
		$C_6H_5CH_3$	35	30 ± 5
Cyclohexane/ Toluene	140 (glass) 153 (liq.)	c-C_6H_{12}	42	33± 5
		$C_6H_5CH_3$	39	40 ± 5
Decane/ Toluene	163 (glass) 175 (liq.)	$C_{10}H_{22}$	-	(77)[4]
		$C_6H_5CH_3$	48	45 ± 10

[1] Molar C_p values assumed for the estimation of the temperatures in Eq. (3). For the mixtures, the Cp values represent the molar (1:5) average values for the Cp of the dopant and that of toluene. The error in the values depends on the accuracy of the procedure in estimating the peak temperatures via Eq. (3), as well as in the accuracy in the parameters employed (in particular, the use of molar average values presumes weak interaction between dopant and matrix). Neglecting uncertainties in the employed quantities, the error in the E_{LASER} estimations in the mixtures is ~10 kJ/mol.

[2] The activation energies determined from the Arrhenius plot for the intermediate laser fluence range.

[3] Activation energies for desorption of the compounds as determined from fittings of the rising edge of the TDS curves.

[4] $C_{10}H_{12}$ starts desorbing only at high temperatures after $C_6H_5CH_3$ has largely desorbed. The indicated value derives from literature data based on phase equilibria of $C_{10}H_{12}/C_6H_5CH_3$ mixtures.

V. ABLATION

In contrast, at higher fluences, both the weakly and the strongly-bound-to-the-matrix dopants are found to eject in the gas phase. In fact, for all systems, the dopant-to-matrix signal ratios reach values close to the film stoichiometry, although in the case of the strongly bound dopants, deviations are observed (Figure 7) [[25]]. Clearly, at these F_{LASER}, the ejection intensity of the dopants does not correlate with their binding energy to the matrix. For the "nonvolatile" dopants (e.g. $C_{10}H_{22}$), the ejection intensities relatively to that of the matrix remain almost constant with successive laser pulses (Figure 9) [[26]]. Furthermore, the dependence of the "nonvolatile" dopant ejection efficiency on F_{LASER} for irradiation above the ablation threshold is almost identical to that of the matrix, i.e. for both $C_6H_5CH_3$ and $C_{10}H_{22}$,

$$S_{ejected} \propto \ln\left(\frac{F_{LASER}}{F_{thr}}\right)$$

where $S_{ejected}$ represents the intensity of the corresponding compound and F_{thr} represents a threshold value. Taken together, the results show that these fluences entail the unselective expulsion of a volume of material, i.e., independently of the binding energy of the dopants. Besides the above, there are a number of differences in the ejection dynamics of "volatile" dopants (i.e., dopants desorbing at $F_{LASER} \geq 20$ mJ/cm^2) above vs. below the indicated threshold.

These characteristics are not due to a change in the absorption process, since the absorptivity is measured to remain constant for F_{LASER} well above the ablation thresholds. Thus, the strikingly different ejection features at high F_{LASER} from those at low fluences unambiguously demonstrate the operation of different ejection mechanisms in the corresponding ranges. Accordingly, the ablation threshold represents a physically significant parameter. The onset of the ejection of the strongly-bound to the matrix dopants constitutes a direct experimental criterion for establishing the laser ablation threshold of molecular solids.

Excluding the simple thermal desorption scheme, three, at least, different mechanisms have been considered in the literature for laser-induced ablation, namely photochemical, photomechanical and phase explosion [[1], [18]]. All three can account for massive, unselective material ejection, but they differ considerably in their nature.

According to the photochemical mechanism, material ejection is due to the expulsion exerted by gaseous products and fragments produced by the photolysis of the parent molecule. This mechanism has been advanced in a number of cases e.g. for the UV ablation (at 248 nm) of aromatic compounds *in liquid state* (under ambient conditions) [[50], [51]]. In the case of $C_6H_5CH_3$, the photochemical mechanism can be rejected on the basis of a number of experimental observations, most importantly on the basis of the observation that no products are detected by mass spectrometry upon UV laser irradiation of $C_6H_5CH_3$ films.

A spallation/ photomechanical mechanism, in which material ejection results from the high thermoelastic pressures that develop in the substrate due to ultrafast heating upon irradiation with ultrashort laser pulses has sometimes been considered [[1], [18]]. This mechanism requires that the developed (tensile) pressure exceeds the "fracture" energy of the substrate/matrix. Alternatively, it has been considered that even in cases that the tensile pressure is not high enough to result in direct "cold" fracture, it may facilitate explosive boiling by promoting bubble formation at lower temperatures than required for a "pure" heating/thermal process [[18], [52]]. However, these possibilities can be rejected because of the weak stress confinement:

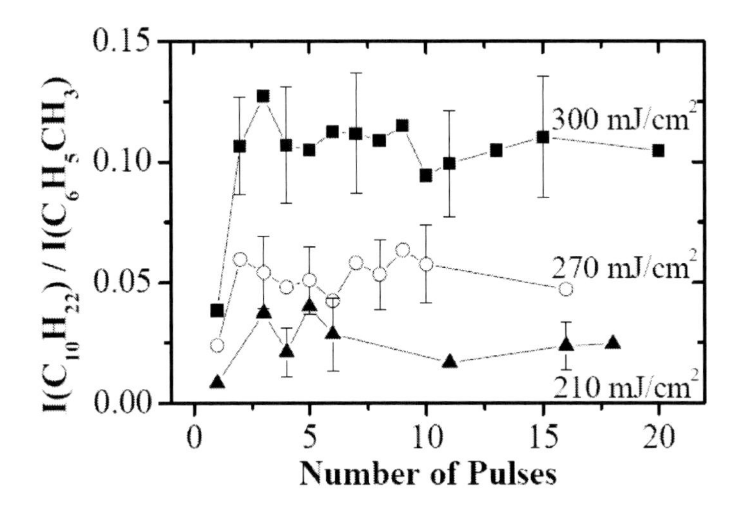

Figure 9. Desorption intensities of $C_{10}H_{22}$ and of toluene as a function of the number of successive laser pulses in the irradiation of the $C_{10}H_{22}/C_6H_5CH_3$ matrix at the indicated F_{LASER}.

$$\tau = \frac{1}{\alpha c} \cong \frac{1}{(3700 cm^{-1})(1500 m/sec)} < \tau_{pulse} \sim 30 \text{ ns},$$

where α is the absorption coefficient and c is the speed of sound in the material. As a result, the pressure "built-up" is too low to be of any effect.

Alternatively, the explosive-boiling model has been advanced on the basis of the Molecular Dynamics (MD) simulations relying on a "breathing sphere" representation of the particles of the substrate [[53], [54]] and thermodynamic considerations [[55], [56]]. The MD simulations predict the operation of different material ejection mechanisms below vs. above a well-defined fluence (threshold). At low F_{LASER}, desorption is found to be molecular and consistent with surface vaporization. In contrast, above a specific fluence, the simulations indicate that massive ejection of material occurs largely in the form of clusters. According to thermodynamic considerations, normal boiling is too slow to be of importance on nanoseconds-microseconds timescales. As a result, with increasing F_{LASER}, the system is overheated to higher and higher temperatures. At high enough temperatures violent boiling can occur, with the system ejected into a mixture of gas and droplets ("explosive boiling/ phase explosion"). However, adoption of this model in MAPLE studies has been rather limited.

According to the MD simulations, the onset of ablation is reflected by the ejection of material largely in the form of clusters. This criterion is different from the one established above in the experiments. However, further evidence indicates that they are closely correlated. Specifically, MD simulations have been performed on some of the systems presented here [[25]]. Due to computational limitations, the values adopted for some parameters differ from the experimental ones, thus resulting in some discrepancies. Nevertheless, there is overall good agreement with the experimental results.

Most importantly, the simulations indicate that the strongly bound dopants are ejected exclusively within clusters of the matrix, whereas the weakly bound ones are ejected mainly as monomers. Clusters have been commonly observed in ablation studies of a wide range of materials. Cluster observation has been documented in the laser irradiation of a number of cryogenic/ frozen compounds [[57]-[60]], and in particular, in the irradiation of frozen aqueous solutions of salts (evidently, because of the higher stability of the ejected clusters in the case of incorporation of ions). We have discussed in detail elsewhere [[23]] the several factors that may affect cluster detection in laser ablation studies.

The important point, however, is that the clusters are ejected directly from the substrate and they are not mainly formed by secondary collisions in the plume (as suggested in most studies). Of course, depending on their internal energy and the number of collisions they undergo in the plume, the final cluster size distribution is modified - a number of clusters are disintegrated. This indication will be further justified within the "explosive boiling" model, to be discussed in the next section. Thus, the two criteria advanced by these experimental studies and the MD simulations appear, though still not proven experimentally, to be intimately interrelated.

VI. Explosive Boiling

Besides the premise of the solid melting upon irradiation, the second fundamental premise of the "explosive boiling" model is the formation of bubbles within the (superheated) liquid.

At ~ 30 mJ/cm^2, melting of the matrix takes place. Thus, at higher F_{LASER}, the melt, under the effective zero external pressure, represents a metastable liquid that may undergo explosive boiling. Specifically, the equilibrium (saturation) pressure of the heated melt is higher than the external pressure. Thus, it is thermodynamically metastable (Figure 10) (its chemical potential μ_L is higher than that of the vapor μ_V) [[54], [61]-[65]]. Therefore, bubble nucleation/ formation is expected.

Bubble formation/ growth may be detected by optical techniques. Briefly, a sharp characteristic decrease of the intensity of the transmitted/ reflected probe beam is observed at ~60-200 ns after the UV laser pulse (Figure 11). The decrease gets more pronounced with increasing F_{LASER}, reaching maximum close to the ablation threshold. At $F_{LASER}>F_{thr}$ (100 mJ/cm^2), this peak is followed by a broad decrease (~ 10s μs) of the transmitted/ reflected probe beam. This second decrease is evidently due to the scattering of the probe beam by the ejected plume. On the other hand, the sharp decrease at ~ 100 ns shows close similarities (in time and shape/ time-decay) with the optical transients that have been observed for bubble growth in the case of superheating of liquids adjacent to laser heated surfaces [[66]-[70]].

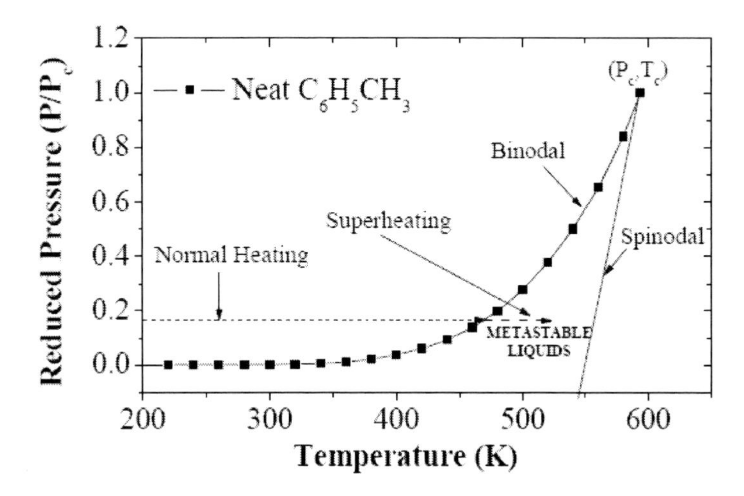

Figure 10. The binodal and spinodal curves and the region of metastability for toluene. C is the critical point, and T_C=593 ^{0}K and P_C=41 bar are respectively the temperature and the pressure of the compound at the critical point.

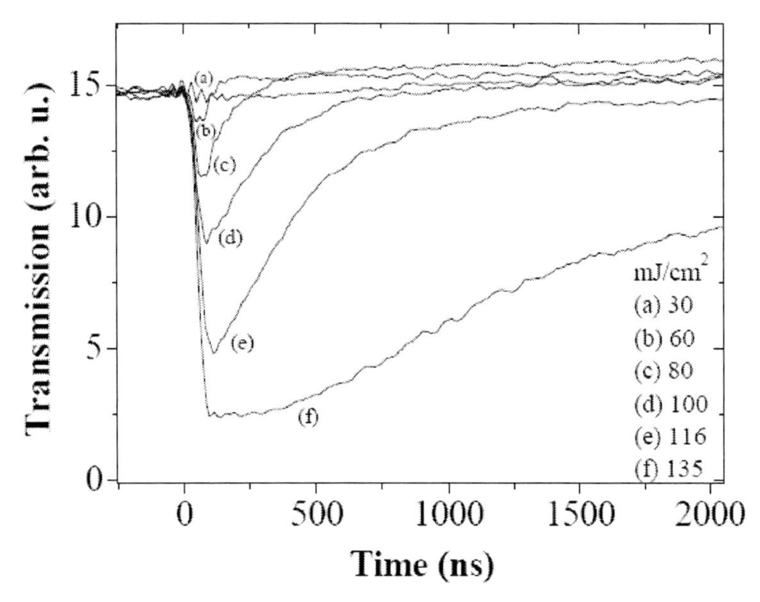

Figure 11. Time-resolved transmission at λ=633 nm, recorded upon irradiation of condensed $C_6H_5CH_3$ films with one UV pulse (λ=248 nm) at the indicated fluences.

Boiling requires bubble formation, which, however, is limited by the work necessary for the formation of a new interface within the liquid (i.e., the surface tension σ) [[61]]. The free energy for bubble formation is:

$$\Delta G = 4\pi R^2 \sigma - \frac{4}{3}\pi R^3 (P_V - P_L) + \frac{4}{3}\pi R^3 \frac{P_V}{k_B T}(\mu_V - \mu_L) \qquad (4)$$

where R is the bubble radius, σ the surface tension, k_B the Boltzmann's constant, T the saturation temperature of liquid and P_V, P_L are respectively the pressure inside the bubble and the ambient pressure of the liquid. In the above equation, the first term represents the energy necessary for the liquid-vapor interface formation, the second the work directed against the pressure forces and the third the "driving force" of bubble formation. For small R, the surface term dominates and so ΔG>0; only for sufficiently large R, ΔG<0 as necessary to lead to bubble growth. The radius for this change is specified by the condition of "mechanical" equilibrium of the bubble

$$\left(P_V - P_L = \frac{2\sigma}{R}\right)$$

and of "thermodynamic" equilibrium

$$\left(\mu_L(P_L) = \mu_V(P_{sat})\right),$$

where P_{sat} is the saturation pressure. In this case

$$\Delta G_{cr} = \frac{16\pi\sigma^3}{3(P_V - P_L)^2}$$

and the rate at which homogeneous bubbles of critical size are generated (J_{cr}) is given by:

$$J_{cr} = J_0 \exp\left(-\Delta G_{cr}/k_B T\right) = J_0 \exp\left(-\frac{16\pi\sigma^3}{3k_B T(P_V - P_L)^2}\right) \quad (5)$$

and

$$J_0 = N_L\left(\frac{3\sigma}{\pi m}\right)^{1/2}$$

where N_L is the number of liquid molecules per unit volume and m is the molar mass. Since both σ and (P_V - P_L) factors depend sensitively on temperature, critical bubble formation depends crucially on the maximum attained value and temporal evolution of the film temperature.

However, the surface film temperature drops rapidly after the end of the laser pulse as a result of evaporative cooling

$$\left(\frac{dT}{dt} \sim -\frac{P_V}{\sqrt{2\pi MRT}} e^{-\frac{\Delta H_{evap}}{RT}}\right)$$

where ΔH_{evap} is the evaporation enthalpy of the compound, M the molar mass of the compound and R the gas constant.

$$\frac{P_V}{\sqrt{2\pi MRT}}$$

represents the rate of material desorption (Figure 12). For low overheating, the reduction in the free energy upon phase change is insufficient to compensate for the surface tension limitation and, thus, bubble growth eventually halts (~ 100 ns). However, with increasing F_{LASER}/ temperatures, due to the sharp decrease of σ and the increase of $(P_V\text{-}P_L)^2$

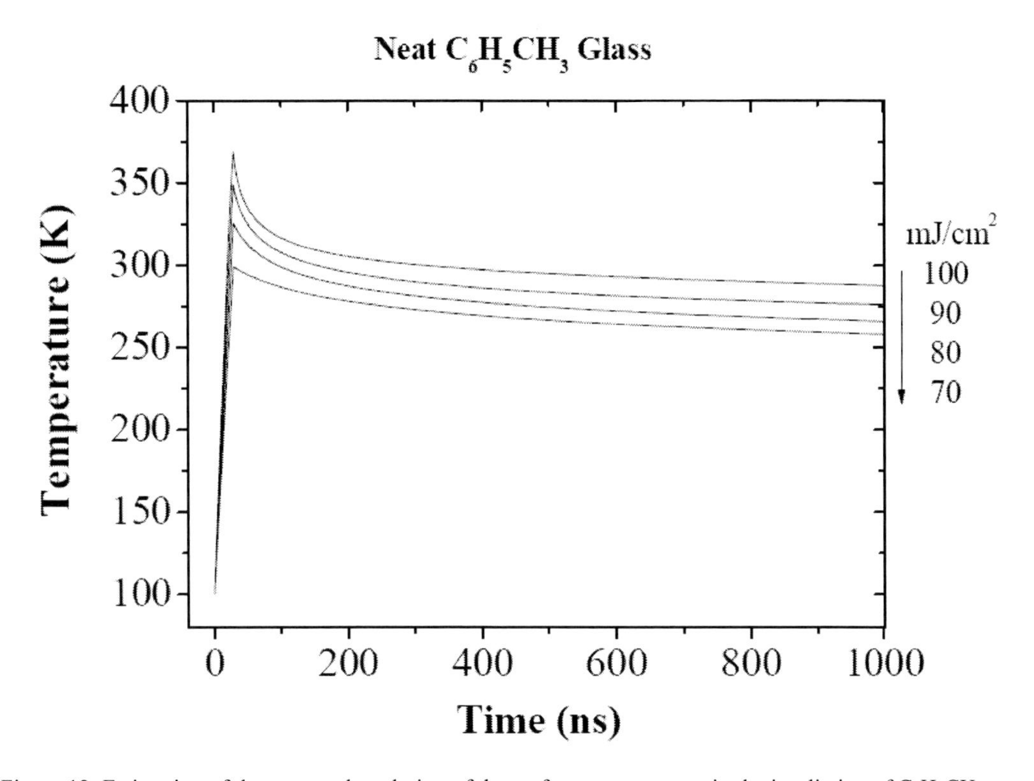

Figure 12. Estimation of the temporal evolution of the surface temperatures in the irradiation of $C_6H_5CH_3$ matrix at fluences at which desorption becomes important. The figure demonstrates the rapid drop of the surface temperatures which accounts for the limitation of the bubble formation and growth to ~100 ns.

factors, J increases sharply. At a sufficient degree of superheating, the number of interconnected bubbles and high pressure exerted by them result in the violent (supersonic beam-like) material ejection. Because J increases sharply exponentially, the onset for material ejection exhibits a "threshold-like" behavior.

Despite the overall good agreement, various discrepancies are noted in the quantitative analysis of the dynamics of superheated liquids. Both thermodynamic analysis [[54]] and the MD simulations [[49]] associate the threshold with the spinodal limit where liquid becomes unstable and "spontaneously" (i.e., without the requirement of an energy barrier to be overcome) decomposes into a mixture of liquid/ gas. This limit, spinodal limit, is specified [[61], [63]] by the criteria

$$\left(\frac{\partial P}{\partial V}\right)_T = 0 \text{ and} \left(\frac{\partial S}{\partial T}\right)_P = 0$$

(stability criteria) and occurs at ~ $0.8T_c$, where T_c is the critical temperature of the compound. For $C_6H_5CH_3$, T_{sp}~$0.8T_c$ ~ 470 ^{0}K. However, the estimated surface temperatures are much lower e.g., for neat $C_6H_5CH_3$, the temperature at threshold is estimated to be ~ 350-380 ^{0}K. This value is certainly well above the melting point of toluene, but not quite as high as expected from the models. In addition, bubble nucleation is observed at relatively low temperatures. Very recently, Perez et al [[73]] have relied on molecular dynamics simulations to re-address the process in the laser ablation of molecular solids. They have suggested that the conventional view of explosive boiling [[54]] fails to account in detail for the induced processes. They suggest that for optical penetration depths of μm, explosive boiling will be strongly influenced by stresses (tensile waves). Thus, though the general theory predicts ablation to occur at the spinodal limit, in practice, its onset is initiated at lower temperatures. This means that thermal degradation effects are considerably less than may be expected.

Most interestingly, bubble formation and efficiency of explosive desorption are found to exhibit a very high dependence on the structure of the as-deposited solid. In particular, bubble formation is pronounced in the laser irradiation of polycrystalline samples, whereas it is highly suppressed for annealed glasses. The exact reason for this sensitivity is not understood, but a likely explanation is that bubble formation and explosive boiling are facilitated by defects, voids present in the polycrystalline samples. This enhancement of bubble formation may explain why in real-life samples, the onset of explosive boiling and of ablation occur at lower temperatures than predicted theoretically.

Interestingly, in MALDI [[17]], a pronounced dependence of the biopolymer (ion) signal on the degree of crystallinity of the matrix is observed. In fact, this high sensitivity to the matrix structure has been one of the major problems in the development and optimization of MALDI. Likely, the observed dependence of explosive boiling on the matrix structure can account for several difficulties and irreproducibilities that are often noted in MALDI and MAPLE studies.

VII. Implications for Laser Based Techniques

The suggestion of explosive boiling can account consistently for several features of laser ablation at least of photoinert systems (with ns laser pulses):

It accounts for material ejection largely in the form of clusters/ droplets as suggested by the MD simulations. Indeed, even for heating of a system up to the spinodal limit (T_{sp}) the required heat $c_p(T_{sp}-T_0)$ (per unit mass) is lower than the evaporation energy (per unit mass), that is to say, even at the spinodal limit, the available heat is not sufficient for the complete evaporation to monomers, suggesting that a significant percentage of the material is ejected in the form of droplets and clusters. The important point to underline is that as a result of material ejection largely in the form of liquid droplets, the "activation energy" will be lower than expected on the basis of the sublimation/ evaporation energy of the compound.

Second, the minimum energy (per unit volume or mass) that is required for explosive boiling [[25], [63]] relates to the cohesive energy of the substrate (i.e., the intermolecular binding energy), decreasing with decreasing binding energy. Indeed, Table 2 shows that even after accounting for the different heat capacities of the systems, the ablation thresholds differ. In fact, the temperatures attained at threshold increase with the cohesive energies of the systems (the cohesive energy changes due to the dopant incorporation and changes in the intermolecular interactions). This dependence on cohesive energy can be rationalized by the fact that a higher cohesive energy of the system results in an increase of the surface tension and a decrease of P_V, thus higher temperatures (T) are required for significant bubble growth (Eq. 4). Thus, for "volatile" solvents, i.e. solvents of low intermolecular binding energy (cohesive energy), explosive boiling can be effected at relatively low fluences (temperatures), thus ensuring minimal thermal or photochemical influence on the biopolymers. Thus, this explains the common observation in MAPLE studies that better results are obtained for "volatile" solvents.

The assumption of explosive boiling also provides a rational for the ejection of the "non-volatile" species exclusively within clusters of the matrix/ solvent. Adopting the kinetic Kagan-Domler description [[61], [64]] the rate of vaporization into the bubbles can be approximated by

$$\frac{A}{(2\pi mkT)^{1/2}}\left[P_V(T)-P_L\right],$$

where A is correlated with the fraction of dopant on the bubble "surface", $P_V(T)$ is the vapor pressure of the compound at the (laser-induced) temperature T_V and P_L is the ambient pressure (on the liquid). Therefore, the relative desorption rates of the two components (dopant, matrix) into the bubbles is specified by

$$e^{-E_{dopant}} / e^{-E_{matrix}},$$

where E represents the binding energy of the component in the condensed phase. Thus, in the case of "involatile" dopant (high $E_{binding}$), the growing bubbles are mainly/ exclusively composed of $C_6H_5CH_3$ vapor, and explains the finding of the MD simulations that the strongly bound dopants are ejected exclusively within droplets of the matrix. The implication for MAPLE is that, thermodynamically, it is much more "facile" to break the weak bonds between the outer solvent layer of the cluster surrounding the biopolymer, than breaking the large number of the rather strong (hydrogen type) bonds between the biopolymer and the solvent. Thus, explosive boiling provides a simple and physically acceptable picture of how it is possible that strongly bound biopolymers are ejected at relatively low temperatures.

Table 2. Ablation thresholds and estimated surface temperatures of the solids at thresholds for the examined systems in their irradiation at 248 nm

System[1]	Threshold Fluence[2] (mJ/cm^2)	Estimated Temperature at Threshold (K)[3]	T_{sp}[4] (K)
Neat $C_6H_5CH_3$	100 ± 10	360	473
$(CH_3)_2O$/ $C_6H_5CH_3$	110 ± 10	300	450
c-C_3H_6/ $C_6H_5CH_3$	90 ± 10	270	448
c-C_6H_{12}/ $C_6H_5CH_3$	120 ± 20	330	470
$C_{10}H_{22}$/ $C_6H_5CH_3$	180 ± 30	360	480

[1] All mixtures have a 5:1 toluene to dopant molar ratio.
[2] Thresholds determined from signal~ln(F_{LASER}/F_{thr}).
[3] The values differ somewhat depending on whether the systems are assumed to be glassy or polycrystalline, though the differences are not significant. The heat capacity of the doped systems is approximated by the molar average of the heat capacities of the dopant and $C_6H_5CH_3$.
[4] T_s: spinodal decomposition temperature, T_{sp}^{mix} are taken either from literature values or approximated as the molar-averaged values of the T_{sp} values of the neat compounds.

The results account for various observations in MAPLE and MALDI studies. First, they account for the observation that the matrix desorbates are detected at fluences much lower than the biopolymers/ proteins [[12]]. In MALDI studies, the reason for this observation has been difficult to establish because ions are usually detected. Thus, the difference between the two detection limits has been plagued by arguments about the contribution of the ionization/ detection efficiency of the biopolymers [[14]]. However, here, this issue is altogether avoided, since neutral desorbates are detected. Clearly, the results show that there is a minimum threshold for ejection of the biopolymers in the gas phase. In particular, that ejection of species strongly bound to the matrix can occur *only* in the ablative regime via explosive boiling. Therefore, MAPLE is actually a misnomer. At lower fluences, because of the thermal nature of desorption, only the relatively volatile matrix/ solvent can desorb in MALDI and MAPLE, whereas ejection of the biopolymers, due to their high average binding to the matrix, is possible only in the ablative regime.

The second important point concerns the influence of clustering on desorbate intensity measurements. Since the interaction of the biopolymer with the matrix is generally large, the biopolymer may not get rid of its solvation shell through collisions in the plume. However, clustering can significantly affect ionization processes, with species (within the clusters) with lower ionization potential being preferentially ionized [[74]]. As a result, the ion intensities recorded by electron impact or multiphoton ionization may not reflect accurately the relative concentrations of the neutral species in the plume. This factor accounts for the fact that the strongly bound dopant-to-matrix intensities ratio, as defined by mass spectrometry (Figure 7), appears to be lower than the film stoichiometry (these species have a higher ionization potential than $C_6H_5CH_3$). Thus, ejection of species within clusters may be a major analytical handicap. On the other hand, for deposition processes, it may be beneficial, as the incorporation of the biopolymers within clusters may ensure their "soft" landing on the target, as well as minimization of coagulation.

Furthermore, the described studies clearly indicate the caveats of Arrhenius-type analysis of ejection signals. Such analysis would indicate the same activation energy of ejection for the involatile species (e.g. $C_{10}H_{22}$) as for toluene. However, the value determined for the dopant differs much from the binding energy determined by thermogravimetric measurements. The activation energies specified in the ablative regime do not relate to the binding energy of the specific molecules but, as shown, to the activation energy for the "explosive boiling" process.

A related misconception is that ablation may just be defined by a high amount of material/ matrix being ejected. Clearly, for matrices of a low cohesive energy, simple thermal desorption will result in a very high desorption signal (according to $e^{-\Delta E_{binding}/kT}$). Yet, this does not imply ejection of the incorporated biopolymers. The criteria for ensuring biopolymer ejection differ, as discussed in detail before.

VIII. INFLUENCE OF PHOTOCHEMICAL PROCESSES ON PHASE TRANSFORMATIONS

One of the major reasons that have hindered the study of laser-induced phase transformations in molecular/organic systems is the complications due to the fact that, in parallel, chemical processes may occur. In fact, for a long time, the interaction of laser pulses with organics has been exclusively described within photochemical mechanisms. For photoinert systems, such as $C_6H_5CH_3$, as shown above, there is no doubt for the exclusive operation of thermal mechanism(s). However, for photolabile systems, even though the thermal component appears to be significant, dissociation and reactivity occur and their influence cannot be neglected [[14], [18]]. This reactivity can be important to consider in laser processing of organic materials (generally photosensitive in the UV) [[12]-[20]], whereas in other cases, it has been exploited for the production of new materials (e.g. [[15]-[16], [75]]). It is, therefore, useful to consider briefly how the contribution of reactivity may affect laser-induced phase transformations.

It has been commonly argued that results on photoactive compounds, such as halocarbons-CH_xCl_y are inconsistent with the thermal model. For instance, for these systems, the temperatures as estimated on the basis of the absorption coefficient of the compounds are too low at the ablation thresholds (at 248 nm) for any substantial thermal desorption/ evaporation [[76]]. However, in most studies, a multipulse protocol was used. For photochemically active compounds, irradiation results in the formation of new products e.g. chlorocarbon compounds form conjugated species, such as CH_2=CHCl; C_6H_5Cl forms $(C_6H_5)_2$, C_6H_5-C_6H_4Cl, $(C_6H_4Cl)_2$, etc. Generally, these absorb strongly in the UV. As shown above, below the ablation threshold, desorption of these species in the gas phase is relatively low due to their strong interaction with the matrix. As a result of the gradual accumulation of highly absorbing products and the consequent change in laser energy absorption, ejection efficiency at fluences below the ablation threshold increases with successive laser pulses, i.e. signal induction (incubation) occurs. This effect is most pronounced in the irradiation of solids of weakly absorbing compounds, such as of halocarbons (CH_xCl_y) and C_6H_{12} at 248 nm and 308 nm [[26], [31]], but is also observed even in the irradiation of moderately absorbing ones (e.g. C_6H_5Cl at 248 nm) [[77]] (Figure 13).

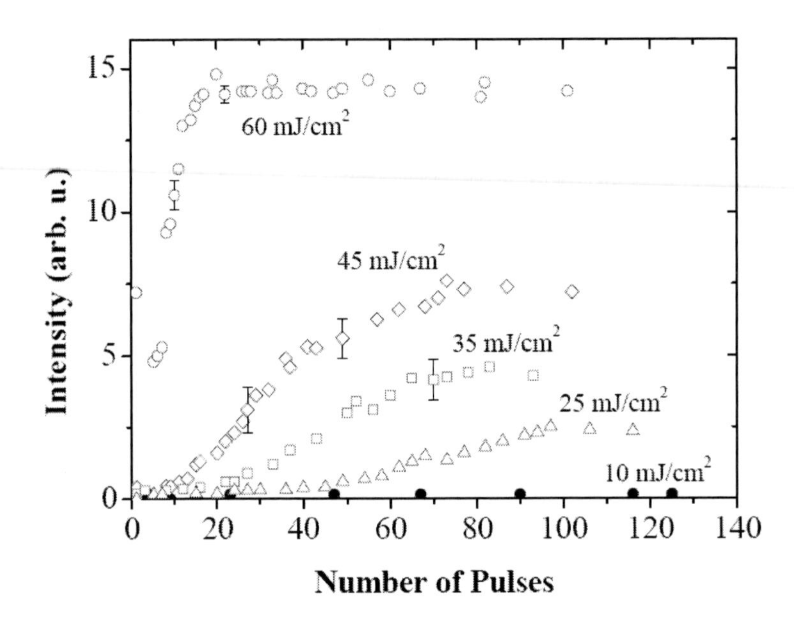

Figure 13. Pulse evolution of the desorption intensity of the parent molecule in the irradiation of condensed solids of C_6H_5Cl with successive laser pulses at different laser fluences. Reproduced with permission from ref. [[77]].

Clearly, in such systems, particular care must be paid when interpreting (analyzing) the results from multi-pulse laser irradiation experiments. First, the average "etching/ desorption yield" from multiple pulse experiment can differ significantly from the pulse-to-pulse value. In addition, it can be misleading: at moderate fluences, for the first few pulses, the process is in the sub-ablative regime, i.e. thermal desorption (in which case strongly-bound-dopants e.g. biopolymers are not ejected in the gas phase); but, after a sufficient number of pulses, ablation sets in. Thus, relying exclusively on gas-phase diagnostics for assessing the photo-induced chemical processes and effects (products) can be misleading if examination is limited at fluences below the ablation threshold. A parallel spectroscopic examination of the matrix will demonstrate a pronounced accumulation of products/species altering the absorptivity of the matrix. Most importantly, estimating the laser-induced temperatures on the basis of the absorption coefficient of the parent compound is erroneous. The discussion of mechanisms must be accordingly reconsidered, i.e. the high increase of absorptivity with subsequent pulses suggests that a thermal mechanism most likely dominates even for initially weakly absorbing systems, at variance with the arguments of Ref. [[76]].

These experimental complications aside, the question arises on how the chemical processes interfere with the thermal mechanism and the phase transformations discussed above. This issue has been considered especially for the ablation and at least there, there seems to be an evolving good understanding. Specifically, both experiments [[77]-[79]] and MD simulations [[80]-[81]] find that the threshold for ablation of condensed C_6H_5Cl is lower than that for $C_6H_5CH_3$. The two compounds are quite similar so that this difference cannot be accounted by differences in thermodynamic/physical properties. It has been indicated by the MD simulations the difference is due first to the energy released by the exothermic reactions and second by the pressure exerted by any gases produced. The first factor, i.e. the energy release contributes to attaining explosive boiling at lower fluences than estimated in the absence of reactivity. The second factor, i.e. the higher pressure, further facilitates material ejection. In all, the chemical processes can result in unexpected influence on the ''explosive boiling'' process. Such effects have not been observed in the laser-irradiation of metals and semiconductors. Thus, the study of the influence of the chemical processes may provide novel ways for further studies on phase transformations.

IX. Implications for Bubble Dynamics

Besides the importance of these studies for the elucidation of laser ablation of molecular/organic systems, it must be realized that explosive boiling is a much more general/ ubiquitous phenomenon. To this end, it is worth summarizing some issues in the theoretical description of the phenomenon, as well as its relevance to nanoscience, so as to appreciate the importance of studying the phenomenon in the laser-irradiation of cryogenic solids.

Despite extensive studies for over 100 years, especially the primary steps in bubble nucleation/formation remain unclear. There are many instances in liquids, under ambient atmosphere, at which pronounced differences are found between the predictions of the classical nucleation theory and experimental results [[61], [63], [82]]. Discrepancies have also been noted in the explosive boiling processes in the laser-irradiation of liquids and of liquids adjacent to laser-heated surfaces. However, in liquids, exclusion of dissolved gases (that may act as heterogeneous nuclei) cannot be prevented. Thus, any deviations between theory and experiment have been largely attributed to the hypothesis that long-lived cavitation nuclei, such as ultramicroscopic bubbles are present in liquids.

However, the above hypothesis suffers from the fundamental problem that bubbles with a radius smaller than r_{cr} are thermodynamically unstable and should dissolve. For resolving this difficulty, Frenkel [[83]] has resolved to the distinction between heterophases and homophases, but the physical basis for this delineation is unclear. Alternatively and equally questionably, stabilization of (nano) bubbles has been ascribed to the influence of such factors as cosmic radiation [[84]], formation of clusters of organic or ionic molecules [[85], [86]] and van der Waals stabilization [[87]]. In fact, bubble nucleation/ growth exhibits a quite complex dynamics, which cannot be accounted by classical models (e.g., “memory effects” upon neutron irradiation and in the explosive boiling of liquids adjacent to laser-heated solid surfaces [[69]]). There is increasing understanding that the above discrepancies between theory and experiment are not only due to experimental limitations, but rather to our limited understanding and specification of parameters, e.g. of the surface tension, on nanometer scale.

Besides the fundamental scientific interest, the above questions are also of direct relevance to nanoscience and nanotechnology. The formation of nanobubbles can result in undesirable effects, such as enhanced noise and even artifacts in the examination of surfaces within liquids by in situ atomic force microscopy (AFM) and other scanning microscopies [[88]]. In laser-based microstructuring and nanostructure fabrication techniques within liquids, formation of nanobubbles may compromise the focusing of the beam and the final resolution of the structures produced [[89]-[92]]. On the other hand, nanobubble formation may be used to advantage (e.g., for the directional transport of objects in microfluidics, for switching valves in microdevices, and for making optical switches, etc.). [[93]-[96]]

Despite all this importance, the study of bubble nucleation and growth phenomena in liquids is subject to major limitations, due to the presence of dissolved ambient gases. In addition, the fleeting existence of bubbles in liquids makes their study quite difficult. Generally, the time scale for most (conventional acoustic) studies has been limited to microseconds due to the instability of the transducers to generate shorter acoustic pulses of sufficient intensities to cause cavitation. By comparison, the use of cryogenic films offers the crucial advantage that the presence of dissolved gases or of other plausible sources that may act as heterogeneous nuclei can be strictly excluded. Furthermore, because of the relatively rapid cooling and subsequent solidification, bubble structures may be "arrested" and thus studied in more detail. In addition, they can be monitored in time, thus being able to establish the factors crucial for elucidating their dynamics when still in their "infancy." It is clear that bubble formation in cryogenic films can provide new insights into the technology necessary for exploiting and manipulating bubbles at the nano level.

X. Conclusion

For photo-inert compounds/ solids, irradiation results in well defined and distinct phase transformations with increasing laser fluence. Specifically, devitrification, annealing, melting and explosive boiling have been demonstrated to occur in condensed solids of toluene with increasing laser fluence. These processes exhibit various differences from the ones observed under conventional, slow heating. They can thus provide significant new insight into the thermodynamics of organics and in particular the dynamics of phase transformations in these systems.

An explosive-boiling type process seems to dominate. Several implications of explosive boiling mechanism have been examined in detail. It was shown that explosive boiling can account for most observations in the laser-material ejection (in cryogenic solids, MAPLE and MALDI) in a physically direct way. Biopolymers can be ejected in the gas phase only in the ablative regime, i.e., above the ablation threshold. At lower fluences, a thermal vaporization process operates, which can be responsible for the desorption of the solvent, but not of the polymer. Given this separation, quantitative analysis of the rates of biopolymer/ matrix activation energies of desorption can be quite misleading.

We have not discussed in this review the phase induced transformation in the irradiation with femtosecond pulses, since only preliminary studies on the ablation of cryogenic films with fs pulses have been reported. These studies indicate the high potential of femtosecond laser technology for film deposition, but in parallel they indicate that mechanisms and characteristics of material ejection differ substantially from the ones specified (described) above for nanosecond laser-induced material ejection.

Acknowledgments

We thank the UltraViolet Laser Facility operating at FORTH under the Improving Human Potential (IHP)-Access to Research Infra-structures program (contract no. HPRI-CT-1999-00074). and PENED (project no. 03ED351) administered by the Greek Ministry of

Development. We thank several previous coworkers, in particular A. Koubenakis, J. Labrakis, A. Michalakou, K. Stamataki, as well as collaboration with B. J. Garrison group (Penn State University).

REFERENCES

[1] Bäuerle D., *Laser Processing and Chemistry*; Springer-Verlag, Berlin, 2000.

[2] Stolk, P.A.; Polman, A.; Sinke, W.C. *Phys. Rev. B* 1993, *47*, 5-13.

[3] Sanders, P.G.; Aziz, M.J. *J. Appl. Phys.* 1999, *86*, 4258-4261 and references therein.

[4] Sokolowski-Tinten, K.; Solids, J.; Bialkowski, J.; Siegel, J.; Afonso, C.N.; von der Linde, D. *Phys. Rev. Lett.* 1998, *81*, 003679-003682.

[5] Boneberg, J.; Bischof, J. ; Leiderer P. *Optics Comm.* 2000, *174*, 145-149.

[6] Lee, M.; Moon, S.; Hatano, M.; Grigoropoulos, C.P. *Appl. Phys. A* 2001, *73*, 317-322.

[7] Solis, J.; Afonso, C.N. *Appl. Phys. A* 2003, *76*, 331-338 and references therein.

[8] Bounos G.; Athanassiou,, A.; Anglos D.; Georgiou S. *Chem. Phys. Lett.*, 2006, *418*, 313-318.

[9] Ediger, M.D.; Angell, C.A.; Nagel, S.R. *J. Phys. Chem.* 1996, *100*, 13200-13212.

[10] Alba, C.; Busse, L.E.; List, D.J.; Angell, C.A. *J. Chem. Phys.* 1990, *92*, 617-624.

[11] Ishii, K.; Nakayama, H.; Okamura, T.; Yamamoto, M.; Hosokawa, T. *J. Phys. Chem. B* 2003, *107*, 876-881.

[12] Miller, J. C.; Hanglund, R. F. Jr., Laser Ablation and Desorption; Experimental

[13] *Methods in the Physical Sciences*; Academic Press: San Diego, CA, 1998.

[14] Bityurin N.; Luk'yanchuk B. S.; Hong M. H.; Chong C.T. *Chem.Rev. (Washington DC)* 2003, *103*, 519-552.

[15] Shafeev, G.A.; Simakin, A.V.; Lyalin, A.A.; Obraztsova, E.D.; Frolov, V.D.; *Appl. Surf. Sci.* 1999 *139*: 461-464.

[16] Lyalin, A.A.; Simakin, A.V.; Bobyrev, V.A.; Lubnin, E.N.; Shafeev, G.A. *Quantum Electr.* 1999, *29*, 355-359.

[17] Dreisewerd, K. *Chem. Rev. (Washington DC)* 2003, *103*, 395-426.

[18] Vogel, A. ; Venugopalan, V. *Chem. Rev. (Washington DC)* 2003, *103*, 577-644.

[19] Georgiou, S.; Zafiropulos, V.; Anglos, D.; Balas, C.; Tornari, V.; Fotakis, C. *Appl. Surf. Sci.* 1998, *127-129*, 738-745.

[20] Chrisey, D. B.; Hubler, G. K. *Pulsed Laser Deposition of thin films*; Wiley-Interscience: New York, NY, 1994.

[21] Chrisey, D. B.; Piqué, A.; McGill, R. A.; Horwitz, J. S.; Ringeisen, B. R.; Bubb, D. M.; Wu, P. K. *Chem. Rev. (Washington DC)* 2003, *103*, 553-576.

[22] Georgiou, S.; Mastoraki, E.; Raptakis, E.; Xenidi, Z. *Laser Chem.* 1993, *13*, 113-119.

[23] Georgiou, S.; Koubenakis, A. *Chem. Rev.* 2003, *103*, 349-394.

[24] Koubenakis, A.; Elimioti, T.; Georgiou, S. *Appl. Phys. A* 1999, *69*, S637-S641.

[25] Yingling, Y. G.; Zhigilei, L. V.; Garrison, B. J.; Koubenakis, A.; Labrakis, J.; Georgiou, S. *Appl. Phys Lett.* 2001, *78*, 1631-1633.

[26] Koubenakis, A.; Labrakis, J.; Georgiou, S. *Chem. Phys. Lett.* 2001, *346*, 54-60.

[27] Birks, J.B. *Photophysics of Aromatic Molecules*; Wiley-Interscience, London, 1970.

[28] *NIST Chemistry WebBook*; NIST Standard Reference Database National Institute of Standards and Technology, Gaithersburg MD, 20899 (http://webbook.nist.gov)

[29] *CRC Handbook of Chemistry and Physics*; CRC Press, 75th ed., Boca Raton, FL, 1995.

[30] Goodwin, R.D. *J. Phys. Chem. Ref. Data* 1989, *18*, 1565-1636.
[31] Kokkinaki, O.; Koubenakis, A.; Michalakou, A.; Labrakis, J.; Georgiou S. in press.
[32] Kearns, K. L.; Swallen, S. F.; Ediger, M. D.; Wu, T.; Yu, L. *J. Chem. Phys.* 2007, *127*, 154702-154709.
[33] Song, Y.; Garder, Conrad, P. H.; Bradshaw, A. M.; White, J. M. *Surf. Sci.* 1991, *248*, L279-L284.
[34] Brunco, D. P.; Thompson, M. O.; Otis, C. E.; Goodwin, P. M. *J. Appl. Phys.* 1992, *72*, 4344-4350.
[35] Bounos, G.; Kolloch, A.; Stergiannakos, T.; Varatsikou, E.; Georgiou, S. *J. Appl. Phys.* 2005, *98*, 084317-084326.
[36] Oxtoby, D. W. *J. Phys.: Condens. Matter* 1992, *4*, 7627-7650.
[37] Kashchiev D. *Nucleation: Basic Theory with Applications*; Butterwoth Heinemann, Oxford, 2000.
[38] Hatase, M.; Hanaya, M.; Hikima, T.; Oguni, M. *J. Non-Crystal. Solids* 2002, *307-310*, 257-263.
[39] Vreeswijk, J.C.A.; Gossink, R.G. ; Stevels, J.M. *J. Non-Crystalline Solids* 1974, *16*, 15-26.
[40] Filipovich, V.; Fokin, V.; Yuritsin, N.; Kalinina, A. *Thermochim. Acta* 1996, *280-281*, 205-222.
[41] Sirota, E. B.; Wu, X. Z.; Ocko, B. M.; Deutsch M. *Phys. Rev. Lett.* 1997, *79*, 531-531.
[42] Zhigilei, L.V.; Leveugle, E.; Garrison, B.J.; Yingling, Y.G.; Zeifman, M.I. *Chem. Rev. (Washington DC)* 2003, *103*, 321-348.
[43] Focsa, C.; Mihesan, C.; Ziskind, M.; Chazallon, B.; Therssen, E.; Desgroux, P.; Destombes, J.L. *J. Physics –Condens. Matter* 2006, *18*, S1357-S1387.
[44] Braun, R.; Hess, P. *J. Chem. Phys.* 1993, 99, 8330-8340.
[45] Mihesan, C.; Ziskind, M.; Chazallon, B.; Therssen, E.; Desgroux, P.; Focsa, C. *Surf. Sci.* 2005, *593*, 221-228.
[46] Krasnopoler, A.; George, S.M. *J. Phys. Chem. B* 1998, *102*, 788-794.
[47] Tro, N. J.; Nishimura, A. M.; Haynes D. R.; George, S. M. *Surf. Sci.* 1989, *207*, L961-L970.
[48] Little, M. W.; Laboy, J.; Murray, K. K. *J. Phys. Chem. C* 2007, *111*, 1412-1416.
[49] Arnold N.; Bityurin N. *Appl. Phys. A*, 1999, *68,* 615-623.
[50] Tsuboi, Y.; Hatanaka, K.; Fukumura, H.; Masuhara, H. *J. Phys. Chem.* 1994, *98*, 11237-11241.
[51] Tsuboi, Y.; Hatanaka, K.; Fukumura, H.; Masuhara, H. *J. Phys. Chem. A* 1998, *102*, 1661-1665 and references therein.
[52] Rohlfing, A.; Menzel, C. ; Kukreja, L. M.; Hillenkamp, F.; Dreisewerd, K. *J. Phys. Chem. B* 2003, *107*, 12275-12286.
[53] Zhigilei, L. V.; Kodali, P. B. S.; Garrison, B. J. *J. Phys. Chem. B* 1997, *101*, 2028-2037.
[54] Zhigilci, L. V.; Garrison, B. J. *J. Appl. Phys.* 2000, *88*, 1281-1298.
[55] Kelly, R.; Miotello, A. *Phys. Rev. E* 1999, *60*, 2616-2625.
[56] Miotello, A.; Kelly, R. *Appl. Phys. Lett.* 1995, *67*, 3535-3537 and references therein.
[57] Georgiou, S.; Koubenakis, A.; Syrrou, M.; Kontoleta, P. *Chem. Phys. Lett.* 1997, *270*, 491-499.
[58] Han, C. C.; Han, Y.- L. W. ; Chen, Y. C. *Int. J. Mass Spectrom.* 1999, *189*, 157-171.
[59] Fosca, C.; Destombes, J. L. *Chem. Phys. Lett.* 2001, *347*, 390-396.

[60] Georgiou, S.; Koubenakis, A.; Kontoleta, P.; Syrrou, M. *Chem. Phys. Lett.* 1996, *260*, 166-172.
[61] Debenedetti, P. *Metastable Liquids: Concepts and Principles*; Princeton University Press: Princeton, NJ, 1996.
[62] Martynyuk, M. M. *Sov. Phys. Technol. Phys.* 1974, *19*, 793-796.
[63] Skripov, Y. P. *Metastable Liquids*; John Wiley & Sons Inc., New York, NY, 1974.
[64] Blander, M.; Katz, J. L. *AlChE Journal* 1975, *21*, 833-848.
[65] Avedisian, C. T. *J. Phys. Chem. Ref. Data* 1985, *14*, 695-729.
[66] Tam, A. C.; Leung, W. P.; Zapka, W.; Ziemlich, W. *J. Appl. Phys* . 1992, *71*, 3515-3523.
[67] Leung, P. T.; Do, N.; Klees, L.; Leung, W. P.; Tong, F.; Lam, L.; Zapka, W.; Tam, A. C. *J. Appl. Phys.* 1992, *72*, 2256-2263.
[68] Do, N.; Klees, L.; Tam, A. C.; Leung, P. T.; Leung, W. P. *J. Appl. Phys.* 1993, *74*, 1534-1538.
[69] Yavas, O.; Leiderer, P. ; Park, H. K.; Grigoropoulos, C. P.; Poon, C. C.; Leung, W. P.; Do, N.; Tam, A. C. *Phys. Rev. Lett.* 1993, *70*, 1830-1833.
[70] Park, H. K.; Grigoropoulos, C. P. ; Poon, C. C. ; Tam, A. C. *Appl. Phys. Lett.* 1996, *68*, 596-598.
[71] Yavas, O.; Leiderer, P.; Park, H. K.; Grigoropoulos, C. P.; Poon, C. C.; Tam, A. C. *Phys. Rev. Lett.* 1994, *72*, 2021-2024.
[72] Kokkinaki, O.; Michalakou, A.; Georgiou, S. submitted.
[73] Perez, D.; Lewis, L. J. ; Lorazo, P. ; Meunier, M. *Appl. Phys. Lett.* 2006, *89*, 141907-141910.
[74] Mark, T. D.; Foltin, M.; Kolibar, M.; Lezius, M.; Schreiber, P. *Phys. Scr.* 1994, *53*, 43-52.
[75] Simakin A.V.; Shafeev G.A.; Loubnin E.N. *Appl. Surf. Sci.* 2000, *154*, 405-410
[76] Mercado, A. L.; Allmond, C. E.; Hoekstra, J. G.; Fitz-Gerald, J. M. *Appl. Phys. A* 2005, *81*, 591-599.
[77] Georgiou, S.; Koubenakis, A.; Labrakis, J.; Lassithiotaki, M. *J. Chem. Phys.* 1998, *109*, 8591-8600.
[78] Koubenakis, A.; Labrakis, J.; Georgiou, S. *J. Chem. Soc. Faraday Trans.* 1998, *94*, 3427-3432.
[79] Georgiou, S.; Koubenakis, A.; Labrakis, J.; Lassithiotaki, M. *Appl. Surf. Sci.* 1998, *127-129*, 122-127.
[80] Yingling Y. G.; Garrison, B. J. *Appl. Surf. Sci.* 2007, *253*, 6377-6383.
[81] Yingling Y. G.; Garrison, B. J. *Nucl. Instrum. Methods B* 2003, *202*, 188-194.
[82] Crum, L. A. *Nature* 1979, *278*, 148-149.
[83] Frenkel, J. *Kinetic Theory of Liquids*; Oxford Press, Dover, NY, 1955.
[84] Sette, D.; Wanderlingh, F. *Phys. Rev.* 1962, *125*, 409-417.
[85] Fox, F. E.; Kertzfeld, K. F. *J. Acoust. Soc. Am.* 1954, *26*, 984-989.
[86] Bunkin, N. F.; Bunkin, F. V. *Sov. Phys. JETP* 1992, *74*, 271-273.
[87] Wentzell, R.A. *Phys. Rev. Lett.* 1986, *56*, 732-733.
[88] Holmberg, M.; Kühle, A.; Garnaes, J.; MØrch, K. A.; Boisen, A. *Langmuir* 2003, *19*, 10510- 10513.
[89] Kazakevich, P.V. ; Simakin, A.V.; Voronov, V.V.; Shafeev, G.A. *Appl. Surf. Sci.* 2006, *252*, 4373-4380.

[90] Simakin, A.V.; Voronov, V.V.; Kirichenko, N.A.; Shafeev G.A. *Appl. Phys. A* 2004, *79*, 1127-1132.

[91] Lau Truong, S.; Levi, G.; Bozon-Verduraz, F.; Petrovskaya, A.V.; Simakin, A.V.; Shafeev, G.A. *Appl. Phys. A* 2007, *89*, 373-376.

[92] Drakakis, T. S.; Papadakis, G.; Sambani, K.; Filippidis, G.; Georgiou, S.; Gizeli, E.; Fotakis, C.; Farsari, M. *Appl. Phys. Lett.* 2006, *89*, 144108-144110.

[93] Marmottant, P.; Versluis, M.; de Jong, N.; Hilgenfeldt, S.; Lohse, D. *Exp. Fluids* 2006, *41*, 147-153.

[94] Okamoto, T.; Suzuki, T.; Yamamoto, N. *Nat. Biotechnol.* 2000, *18*, 438-441.

[95] Singh, S.; Houston, J.; van Swol, F.; Brinker, C. J. *Nature* 2006, *442*, 526-526.

[96] Garstecki, P.; Fuerstman, M. J.; Whitesides, G. M. *Phys. Rev. Lett.* 2005, *94*, 234502.1-234502.4.

In: Phase Transitions Induced by Short Laser Pulses ISBN: 978-1-60741-590-9
Editor: Georgy A. Shafeev

Chapter 4

LASER-INDUCED PHOTOACOUSTIC AND VAPORIZATION PRESSURE SIGNALS IN ABSORBING CONDENSED MATTER: NEW RESULTS

Alexandr A. Samokhin* and Nikolay N. Il'ichev
A.M.Prokhorov General Physics Institute of the Russian Academy of Sciences, 38 Vavilov St., 119991, Moscow, Russian Federation

ABSTRACT

Pressure pulses generated during laser action in absorbing condensed matter due to photoacoustic, surface and bulk vaporization effects are discussed. New experimental results for water exposed to erbium laser pulses with 150-200 nanosecond length and wavelength 2.94μm are presented. At laser fluencies E>0.6 J/cm^2 short (subnanosecond) pressure peaks above smooth pressure signal are observed which can be interpreted as a manifestation of bulk vaporization (explosive boiling) process. At lower fluencies photoacoustic and surface evaporation pressure signals are investigated in the case when laser intensity was modulated with period 5 ns. It was found that amplitude of high frequency part of the pressure signals shows one or two minima during laser pulse. Such behavior is possibly due to destructive interference (mutual compensation) of photoacoustic and surface evaporation pressure signals.

INTRODUCTION

Laser irradiation of absorbing condensed matter gives rise to pressure pulses which propagate away from absorption zone and can be detected, e.g., with acoustic transducers or another type of sensor. Form and amplitude of the pulses give information about processes involved as well as thermodynamic and kinetic properties of irradiated matter. The pressure pulses generated in absorbing condensed matter under various conditions of laser irradiation

* asam40@mail.ru

are studied in many papers and used in many practical applications (see, e.g., [1-23] and references therein.)

At low laser intensity the pressure signals in irradiated absorbing matter are due to its density change (without phase transitions) induced by absorbed laser energy which causes thermal or others perturbations in the sample. In the case of direct light heating this is a well-known photoacoustic effect first described in XIX century by Bell, Tyndall and Roentgen. Since lasers came into being in the last century this effect is widely used, in particular, as a basic mechanism for photoacoustic spectroscopy [4]. In the case of nonthermal (nonequilibrium) excitation of the solid electronic subsystem, e.g., when silicon is exposed to laser radiation with a wavelength of 1,06 μm, generated pressure pulses have inverse polarity in comparison with thermal excitation [1,7].

At higher intensities in irradiated condensed matter various phase transitions occur. The first order phase transitions such as melting or vaporization give its contribution to density changes which result in modification of laser generated pressure pulses in comparison with low intensity photoacoustic signals [7]. Laser generation of pressure pulses in condensed matter is also affected by plasma formation in ablation plume above irradiated surface but this process is not considered here.

In what follows we recall some properties of photoacoustic and vaporization mechanisms of pressure pulse generation and report on new qualitative experimental results in this field obtained for water exposed to nanosecond erbium laser pulses with wavelength 2.94 μm.

PHOTOACOUSTIC PRESSURE SIGNALS

Let us consider an absorbing condensed matter in the half space $z > 0$ with absorption coefficient α , density ρ , thermal conductivity $\chi_t = c_p \rho\chi$, where c_p and χ are specific heat at constant pressure and thermal diffusivity, respectively. Photoacoustic effect in this one dimensional case is described by the following set of linearized hydrodynamic equations

$$\frac{\partial \rho}{\partial t} + \rho_0 \frac{\partial v}{\partial z} = 0 \tag{1}$$

$$\frac{\partial p}{\partial z} + \rho_0 \frac{\partial v}{\partial t} = 0 \tag{2}$$

$$\rho_0 T_0 \frac{\partial S}{\partial t} = \frac{\partial}{\partial z}\left(\chi_t \frac{\partial T}{\partial z}\right) + \alpha I = 0 \tag{3}$$

where P, ρ ,T, S, and v are, respectively, perturbations of the pressure, density, temperature, entropy, and velocity relative to their unperturbed values P_0, ρ_0 ,T_0, S_0, and v=0. Intensity of absorbed radiation in a matter $I(z,t)$ varies as $I(t,z) = I_0(t)\exp(-\alpha z)$. From eqs. (1) - (3) it immediately follows that

$$\frac{1}{v_s^2}\frac{\partial^2 P}{\partial t^2}-\frac{\partial^2 P}{\partial z^2}=\frac{\beta}{c_p}\left(\frac{\partial}{\partial z}\left[\chi_t\frac{\partial}{\partial z}\left(\frac{\partial T}{\partial t}\right)\right]+\alpha\frac{\partial I}{\partial t}\right) \tag{4}$$

where v_s is the sound speed $v_s^{\ 2}=\left(\frac{\partial P}{\partial \rho}\right)_S$, and β is the thermal expansion coefficient of a material. In deriving formula (4) it was taken into account that, according to the equation of state, $\rho=\rho(P,S)$ and

$$\left(\frac{\partial\rho}{\partial S}\right)_P=\left(\frac{\partial\rho}{\partial T}\right)_P\left(\frac{\partial T}{\partial S}\right)_P=-\rho_0\beta\left(\frac{\partial T}{\partial S}\right)_P=-\rho_0\beta\frac{T_0}{c_p} \tag{5}$$

If the heated substrate depth $z_h=\max(1/\alpha,\sqrt{\chi t})$, defined by the absorption length $1/\alpha$ or thermal effect length $\sqrt{\chi t}$ for the laser pulse time t, is small in comparison with the characteristic wavelength $z_v=v_s t$ of the excited acoustic pulse, the first term on the left-hand side of Eq. (4) can be neglected. It means, in particular, that laser pulse is sufficiently long. In this approximation, the pressure $P(z^*,t)$ in the range $z_h<z^*<z_v$ is given, after straightforward integration, by

$$P(z^*,t)=P(0,t)+\int_0^{z^*}dz'\int_{z'}^{z^*}\left(\frac{\partial^2 P}{\partial z^2}\right)dz=P(0,t)+\frac{\beta}{c_p}\left\{\chi_t\left(\frac{\partial T}{\partial t}\right)_{z=0}+\frac{1}{\alpha}\frac{\partial I_0}{\partial t}\right\} \tag{6}$$

since the absorbed intensity and temperature perturbations can be considered equal to zero in this range, i.e., $P(z^*,t)=P(t)$ is almost independent of z^*.

The quantity $P(t)$ is a boundary condition for the acoustic wave propagating to the substrate depth. If the irradiation surface is not free, the value of $P(0,t)$ should be determined by solving the complete problem in left and right half spaces with the condition of equal pressures and velocities at the interface, as it was done, e.g., in ref. [20] for the case of transparent matter at $z<0$ in contact with absorbing matter at $z>0$.

On the free surface, $P(0,t)=0$ and formula (6) is closed as a part of the considered acoustical problem because time derivative of surface temperature can be found from the heat conduction equation which follows from eq. (3) at P=const

$$\rho_0 c_p\frac{\partial T}{\partial t}=\frac{\partial}{\partial z}\left(\chi_t\frac{\partial T}{\partial z}\right)+\alpha I \tag{7}$$

The term with surface temperature time derivative in eq. (6) is important in the case of strong absorption $\alpha\sqrt{\chi t}>1$ where absorption length is small compared with thermal effect length during laser pulse time t. This is a case of surface absorption which is characteristic, e.g., for metals. In the opposite case of bulk absorption $\alpha\sqrt{\chi t}<1$ the term with intensity time derivative gives main contribution in eq. (6) so that one has no needs to solve eq. (7). For water $\alpha\sqrt{\chi t} < 1$ if t<10 μs because χ~10^{-3} cm^2/s and α~10^4 cm^{-1} at erbium laser wavelength 2.94 μm. In both cases of surface and bulk absorption eq. (6) gives bipolar pressure signals with positive and negative parts.

Bipolar form of photoacoustic pressure signals follows from eq. (4) also for short laser pulses when $v_S t$ is smaller than thermal perturbation length during laser action. In this case the right hand side of eq. (4) can be considered as a source of initial perturbation which evolves afterwards according to homogeneous wave equation.

If $\alpha\sqrt{\chi t} > 1$ then the initial pressure distribution is determined mainly by the last term in right hand side of eq. (4) which is proportional to exp(-αz). Such positive initial pressure pulse propagates into condensed matter followed by its negative counterpart which arises due to boundary condition P=0 at the free surface. It is clear that time evolution of this pressure pulse gives direct information about absorption coefficient α if value of sound velocity v_s is known [12].

At sufficiently high laser intensity negative part of pressure pulse can cause disruption in subsurface layer of irradiated matter [9, 23]. This effect of "cold" ablation is equivalent to spallation effect induced by inward reflection of strong shock wave from the free surface of condensed matter [13].

Surface and Bulk Vaporization (Explosive Boiling)

Increasing of laser intensity modifies shapes of generated pressure pulses as compared with those determined by eq. (6). The modification is, in particular, due to surface and bulk vaporization processes.

First of these processes, in the simplest approximation, can be described in the framework of heat conduction equation with moving boundary (the so-called Stefan problem). In reference frame connected with moving vaporization front at z=0 this equation has a form

$$\rho c_p\left(\frac{\partial T}{\partial t} - V\frac{\partial T}{\partial z}\right) = \frac{\partial}{\partial z}\left(\chi_t \frac{\partial T}{\partial z}\right) + \alpha I \qquad (8)$$

with boundary condition at the evaporating surface z=0

$$\chi_t \frac{\partial T}{\partial z} = \rho L V \qquad (9)$$

where V is front velocity in laboratory reference frame and L is latent heat of vaporization. The problem (8)-(9) is nonlinear even if all involved thermophysical parameters are constant because front velocity V depends on surface temperature $T(0,t)=T_S$. In more general case v as well as surface pressure $P(0,t)$ depends also on gas dynamics vapor parameters so that the problem (8)-(9) should be considered simultaneously with vapor movement problem (see, e.g. [24]).

In contrast to eq. (7) the problem (8)-(9) at I_0=const has a steady-state solution for temperature distribution $T_{ss}(z)$ in the interval between the surface value T_s and unperturbed value $T(\infty)=T_\infty$ at the depth of irradiated matter (see, e.g. [2])

$$T_{ss} = T_\infty + \Delta T \quad A\exp(-\alpha z) + B\exp\left[- \quad V/\chi \quad z\right]$$

$$A = \frac{V(c_p\Delta T + L)}{c_p\Delta T(V - \alpha\chi)}, \quad B = 1 - A, \quad \Delta T = T_s - T_\infty \tag{10}$$

$$I_0 = \rho V(L + c_p\Delta T) \tag{11}$$

Steady-state energy balance eq. (11) determines (implicitly) value of T_s at given value of absorbed laser intensity I_0. Dependencies of V and P_0 on T_s and other parameters can be found with the help of various numerical or approximate analytical methods [24]. During vaporization into vacuum $P_0=0.56\alpha_{ac}P_s(T_s)$, where P_s is saturated pressure and α_{ac} is condensation (mass accommodation) coefficient ($\alpha_{ac}<1$).

Temperature distribution $T(z)$ has a maximum value $T_m>T_s$, the difference T_m-T_s=ΔT_m being dependent on parameter $y=\alpha\chi/V$. If $y>>1$ (surface absorption) the difference is small, $\Delta T_m=L/yc_p$, while at $y<1$ its value $\Delta T_m=L/c_p$ is quite large. Making use of parameters which are approximately pertinent to water ($L/c_p\Delta T$=4 $\Delta T/T_\infty$=0.3, $\alpha=10^4$ cm^{-1}, $\chi=10^{-3}$ cm^2/s, T_∞=300 ^{o}K) one obtains from eq. (10) the curves shown in figure 1 for y=1/2 and y=2. If one takes into account a possible dependence of ΔT on vaporization velocity, *i.d.*, on y, then ΔT (y=1/2)>ΔT (y=2). Transformation of temperature distribution when ΔT is increased by factor 1,5 is illustrated in figure 1 by the difference between curve 2 ($\Delta T/T_\infty$=0.3) and 2' ($\Delta T/T_\infty$=0.45).

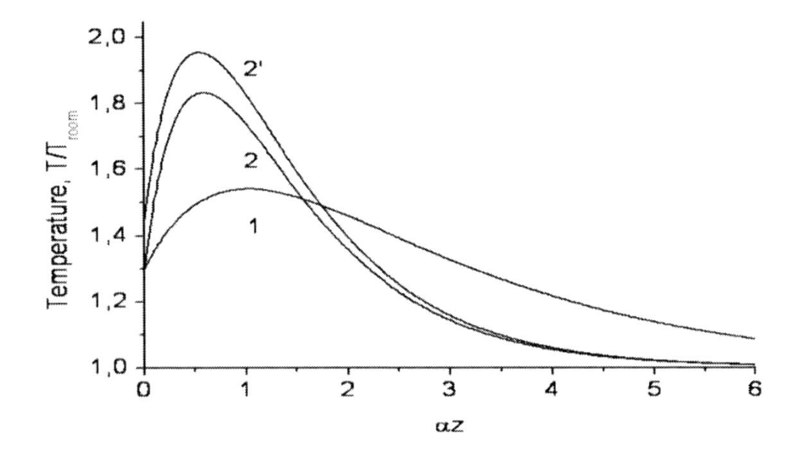

Figure 1. Steady-state temperature distributions in evaporating liquid with different values of parameter y=2 (curve *1*) and y=0.5 (curve *2*). Curve *2'* differs from curve *2* due to increased value of surface temperature: T'_s-T'_∞ =1.5(T_s-T_∞).

Such dependence of temperature distribution on parameter y means that in the case of bulk absorption T_m usually exceeds the spinodal line or critical temperature T_c of liquid-vapor phase transition (for water T_c=647 °K) and the steady-state vaporization regime (10)-(11) cannot be realized in experiment.

Instead of that, one should expect that bulk vaporization (explosive boiling) begins when T_m approaches superheating limit determined by developing homogeneous nucleation or spinodal decomposition processes. Explosive boiling results in a short (nanosecond or shorter) pressure jump whose amplitude is limited with saturation pressure value $P_s(T_m)$ and can be considerably higher than surface pressure P_0.

Rise-time of explosive boiling pressure pulse corresponds to formation time of "flat" vapor cavity in the plane where T_m approaches superheating limit. After cavity formation explosive boiling pressure diminishes due to cavity expansion and evaporating cooling of its boundaries.

The thin liquid layer between free liquid surface and adjacent to it a cavity boundary is thus ablated just as it happens in the case of mentioned above "cold" ablation induced by negative pressure pulse. In both of these cases positive pressure jumps occur while relevant kinetic processes are quite different because "hot" and "cold" ablations (or spallations) correspond to different points on the same spinodal line.

For nanosecond laser pulses initial thickness of ablated water layer can be considerably smaller than absorption depth $1/\alpha$. For this reason it is interesting to note here the paper [25], where the phase-transition dynamics of transparent liquid (isopropanol) films with thicknesses of the order of 100 nm, deposited on silicon wafer heated with a second harmonic of a Nd:YAG nanosecond laser pulse was investigated. Due to the heat transfer and subsequent explosive boiling of the superheated liquid layer adjacent to the interface, the top part of the film was mainly ejected as an intact liquid layer. The phase transition and the ejection process were monitored by reflectometry with a time resolution of about 200 ps and a spatial sensitivity on the nanometer-scale in the direction perpendicular to the wafer surface. The generated pressure was not monitored directly but calculated using a model based on the simple assumption that the pressure evolution in the vapor cavity is a polytropic process. It should be mentioned, however, that this assumption does not take into account the pressure dependence of the film velocity as well as initial value of this film velocity due to heat expansion of the silicon wafer. In any case, further detailed experimental and theoretical studies of explosive boiling under various exposure conditions are needed for better understanding of this process.

In contrast to spallation induced by negative pressure which develops after laser action, explosive boiling process occurs during laser action and it can be repeated many times if laser pulse is long enough.

Such regime of explosive boiling with repeated short pressure jumps was discussed since long ago (see, e.g., [2]). However, until recently it was not observed despite many experimental investigations of laser vaporization process. Short (subnanosecond) unipolar pressure pulses found in experiment [18] we relate namely to such explosive boiling. Some new results on explosive boiling are presented in the next section.

The pressure pulse monopolarity is one of the distinctive features of explosive boiling. Meanwhile, we also observed short pressure pulses with pronounced negative spikes at their trailing edges. In this connection, to determine the origin of this negative spike, we additionally studied temporal characteristics of the acoustic sensor. The technique and results of this study are given in Appendix. There is shown that the negative spike can be explained by design features of the used sensor, which appear at short (< 0.5 ns) pressure pulses.

It should be mentioned that for very long (millisecond) laser pulses the one-dimensional picture described above needs a modification to allow for such hydrodynamic processes as deep channel formation and cavitation effects which can result in pressure jumps considerably delayed with respect to the laser pulse beginning [17]. Experimental results described below are obtained in the conditions which in this sense correspond to a one-dimensional picture.

EXPERIMENTAL SETUP

Radiation exposing the water surface was generated by a laser with an active element (AE) based on an erbium-doped (50%) YAG crystal. The laser was passively Q-switched by Fe^{2+} -doped ZnSe crystal with an initial transmittance of 86%. ZnSe crystals were doped with Fe^{2+} ions by diffusion under conditions of thermodynamic equilibrium of solid ZnSe-solid Fe-(S_{ZnSe}-S_{Fe}-V) vapor phases [26].

The laser cavity was 75 cm long and formed by two mirrors, one of which had a reflectance of 100% and a curvature radius of 1 m. At a distance of 5 cm from the mirror, there is a ZnSe:Fe^{2+} crystal ~2ram thick, whose input face was positioned at the Brewster angle to incident radiation. Radiation reflected from the front face of the ZnSe:Fe^{2+} crystal and attenuated by filters was directed to a D-125 photodetector triggering a LeCroy 44Xi oscilloscope (passband is 400 MHz). An acoustic sensor signal was sent to the second channel of the oscilloscope.

A sapphire stack with a reflectance of ~70% was an output mirror. The lasing energy measured by a Laser power/energy monitor NOVA II with a L30A-SH-VI head was 5.6±0.25 mJ, a half-height pulse duration was ~200 ns, and the transverse distribution was close to the TEM_{00} mode.

The temporal shape of the radiation pulse was measured using an LFD-2a germanium photodiode with a resolution of ~ 1 ns and low sensitivity at a wavelength of 2.94 μm. The electrical signal induced in it under exposure to laser radiation was caused by two-photon absorption in germanium and, most likely, was nonlinear in measured radiation intensity. Figure 1 shows the typical example of the observed signal. We can see that the signal intensity is characterized by rather deep modulation which is usually associated with beats of adjacent longitudinal modes. The beat period coincides with the cavity round trip time of photons. The lasing spectrum narrowness is caused by the use of the stack as an output mirror.

Laser radiation was focused on the water surface by a lens with a focal length of ~ 20 cm. The irradiation spot diameter *d* varied from 0.3 to 1 mm, depending on exposure conditions. A ShAPR-13M lithium-niobate broadband piezoelectric sensor 22 mm in diameter and 8 mm thick with coaxial extraction of electrical pulses induced by acoustic perturbations was placed under a water layer 0.2 - 0.3 mm thick. Such sensors were used previously, in particular, in [18, 27]. The sensor resolution was controlled by the frequency characteristic with an upper bound no worse than 300 MHz. Absolute calibration of the sensor was not performed for this experimental geometry. Particular studies of temporal characteristics of the sensor were carried out; the results are given in Appendix.

RESULTS AND DISCUSSION

Figure 2(a) shows the pressure signal for modulated laser pulse intensity and low fluencies when there is no vaporization effect. The slow and fast components of this signal are shown in figure 2 (b). Such behavior of the fast component is in agreement with eq. (6) if the modulated laser intensity can be written as

$$I = I_0 \; 1 + a\sin(\omega t) \; , \quad \omega = \frac{2\pi}{\tau_m},$$

where I_0 is the smooth envelope of the laser pulse.

According to eq. (6) the high-frequency signal component in figure 2 approximately reproduces the shape of the smooth component I_0 of the laser pulse, and the modulation depth of the acoustic signal increases in comparison with laser intensity modulation depth a in proportion to the ratio of the laser pulse duration τ_p to the modulation period τ_m. In our case, τ_p/τ_m~40. It should be noted here that the observed relative value of the slow and fast signal amplitudes is affected by different acoustic extinction of these signals.

A remarkable feature of the curves in figure 2 (b) is a noticeable asymmetry between positive and negative parts of the smooth bipolar signals in contrast to the symmetric pattern for the high-frequency component. Such a difference can be caused by the dependence of acoustic diffraction distortion of the signal on its characteristic frequency.

Indeed, the characteristic diffraction length l of the acoustic signal, $l \sim d^2/\lambda s$ where $\lambda_s \sim v_s \tau_p$ and v_s are the characteristic sound wavelength and the speed of sound, respectively. Under given conditions, at squared diameter d^2~0.01 cm^2 and τ_p=200 ns, the value L<0.1 cm is smaller than the acoustic sensor thickness, but significantly exceeds it at τ_m=5 ns that is shorter than τ_p. In other words, the effect of acoustic diffraction distortions should be weak for a high-frequency signal component. Exactly this is observed in figure 2, where diffraction distortions of the bipolar signal [8] are noticeable only for the smooth (long-wavelength) signal component.

An increase in the irradiation energy significantly changes the high-frequency component of the measured signal, as is seen in figure 3. We believe that such a signal behavior is caused by the exhibition of the pressure generation mechanism due to surface evaporation which leads, in particular, to mutual suppression of high-frequency components of photoacoustic and evaporation pressures.

To our knowledge, such an effect was not previously observed, although the change in the photo-acoustic signal due to the appearance of the evaporation peak on its background was observed previously many times before (see, e.g., [3,7,15,18]). Such an evaporation peak is clearly seen in figure 3(b) in the behavior of the smooth signal component; now the amplitude of its positive half-wave substantially exceeds the amplitude of the subsequent negative half-wave.

To implement the effect of mutual compensation of the high-frequency photo-acoustic and evaporation signals, these signals should be out-of-phase, i.e., the mutual compensation depth depends on the closeness of the phase shift to π. Such a phase shift can result from the following reasons.

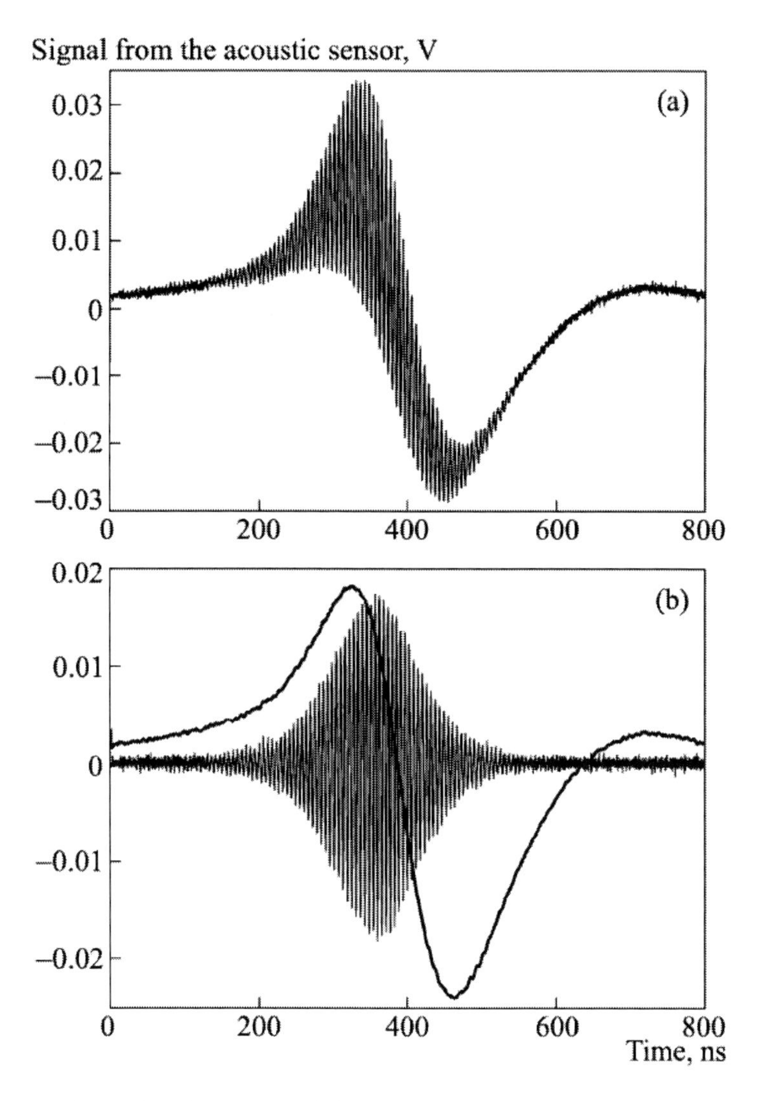

Figure 2. Photoacoustic signal in the presence of high-frequency modulation of the laser pulse intensity: (a) general view of the signal and (b) signal expansion in the sum of fast and slow components.

Formula (6) shows that the photoacoustic signal proportional to the time derivative of the laser intensity is phase-shifted by $\pi/2$ with respect to the modulated intensity part. If we further assume that the evaporation signal is proportional to the temperature change whose derivative (according to Eq. (7)) depends on the laser intensity, the phase shift for this signal will also be equal to $\pi/2$ in magnitude, but with an opposite sign with respect to the photoacoustic signal. It should be mentioned that in this case Eq. (8) coincides approximately with Eq. (7).

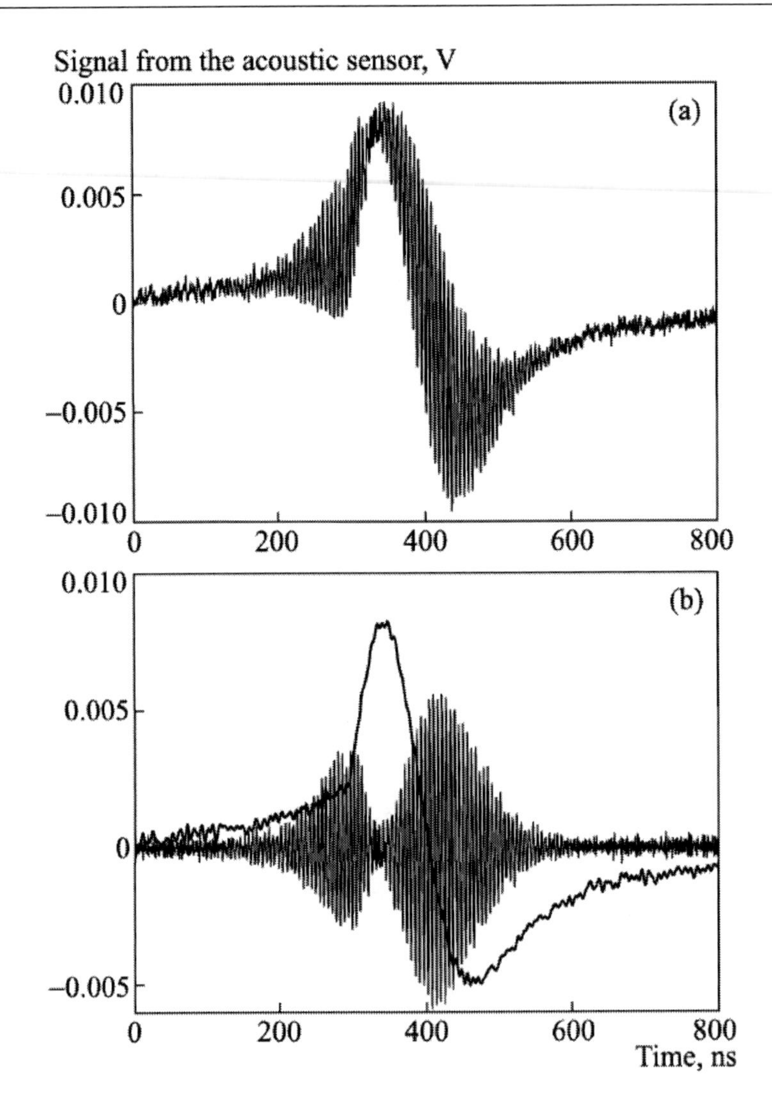

Figure 3. Photo-acoustic and evaporation signals in the presence of high-frequency modulation of the laser pulse intensity: (a) general view of the signal and (b) signal expansion in the sum of fast and slow components.

As a result, the total relative phase shift is π, which just allows mutual compensation of the photoacoustic and evaporative signals, when their amplitudes become equal upon an increase in the irradiation intensity. In figure 3, such compensation occurs exactly near the maximum of the smooth evaporation signal. As the laser pulse intensity and corresponding evaporation signal further increase, two minima can be observed in the behavior of the high-frequency component of the sum signal, which is shown in figure 4.

We recall that the evaporation signal increases much more rapidly with the intensity than the photoacoustic signal due to the strong temperature dependence of the saturated vapor pressure, which is most pronounced under non-stationary evaporation conditions. Such behavior of photoacoustic and evaporation pressure signals, mentioned above, is shown also in figure 5 (a, b) at two values of unmodulated laser pulse intensity $I_b=2I_a$ for the case where these pressure signals do not differ significantly from each other.

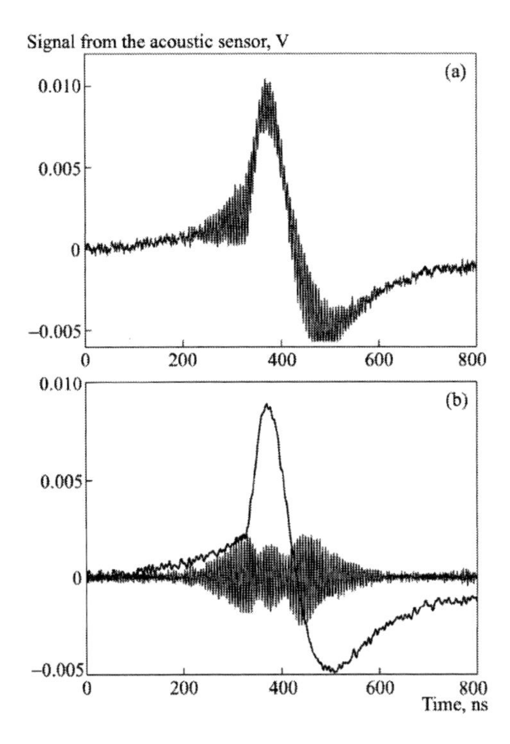

Figure 4. Photoacoustic and evaporation signals in the presence of high-frequency modulation of the laser pulse intensity: (a) general view of the signal and (b) signal expansion in the sum of fast and slow components. In comparison with figure 3, the laser radiation intensity is increased.

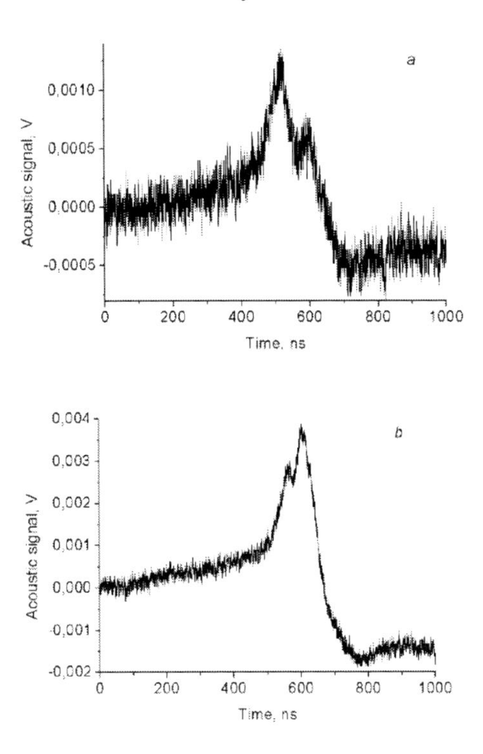

Figure 5. Photoacoustic and surface evaporation pressure signals at different laser intensities $I_b \sim 2I_a$.

It is important to note that, according to the simple estimation, absolute values of modulated parts of photoacoustic and surface vaporization pressure signals become equal at rather high values surface pressure temperature derivative $\frac{dP_0}{dT} = \rho \frac{\beta\omega^2}{\alpha^2}$ as compared with saturation pressure temperature derivative $\frac{dP_s}{dT}$ for water. A reason of this discrepancy is not clear now.

The evaporation signal is controlled by non-equilibrium gas-kinetic processes near the surface and surface evaporation kinetics which depends strongly, in particular, on the mass accommodation coefficient α_{ac} of vapor molecules to the liquid surface and other kinetic parameters. Analysis of such processes is beyond the scope of this paper. A limited accuracy of the simple estimation results probably from thermal nonlinearities and non-one-dimensional effects caused by laser intensity variation over the irradiation spot, the known uncertainty of mass accommodation coefficient [28,29], and a possible decrease in a under action of intense laser radiation [6, 18], which can affect threshold laser fluency for explosive boiling .

As it was already mentioned in the previous section, at sufficiently high laser intensity short monopolar pressure peaks appeared over smooth surface vaporization pressure signal [18]. This effect we interpret as a manifestation of explosive boiling process. Similar peaks at the top of surface vaporization pressure signal are visible in figure 6. First of these two peaks has a negative part which we attribute to the properties of acoustic sensor discussed in the previous section and Appendix.

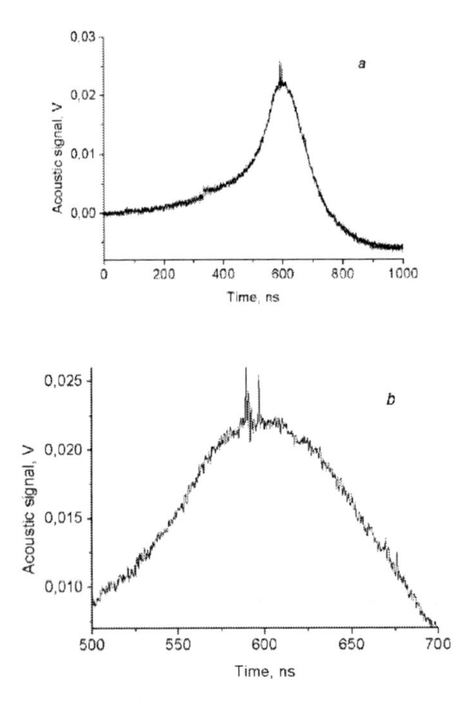

Figure 6. Surface evaporation and explosive boiling pressure signals near the threshold for the explosive boiling. Cases a) and b) differ only in time scales.

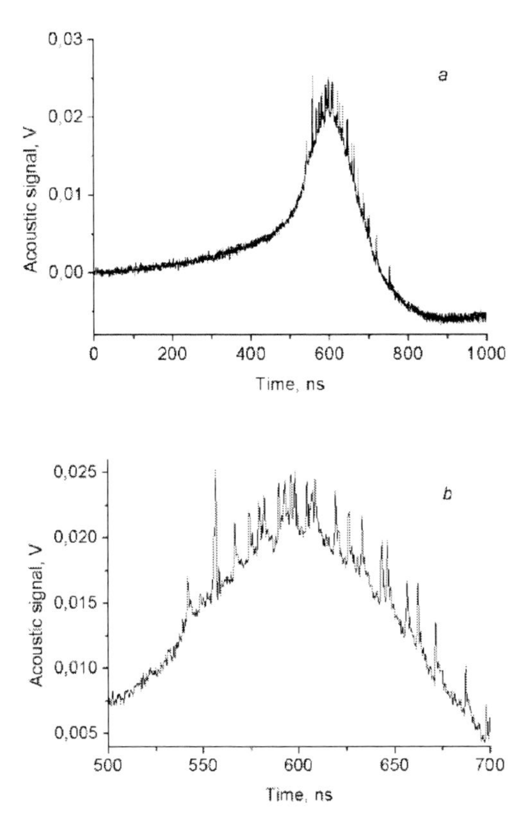

Figure 7. Surface evaporation and explosive boiling pressure signals in the case of multiply explosive boiling peaks. Cases a) and b) differ only in time scales.

In addition to the results [18] we observed here multiply pressure peaks as it is shown in figure 7. Note that these peaks are more frequent at the top of the smooth pressure pulse which approximately corresponds to the time when laser intensity reaches its maximum. This result can be considered as another piece of evidence for the explosive boiling process which can continue during laser pulse in some intensity range. It should be mentioned that relative amplitude values of explosive boiling pressure peaks and surface evaporation pressure in figures 6, 7 are affected by different acoustic attenuation for short and long pressure pulses.

CONCLUSION

In this chapter we recalled some properties of pressure signals generated in absorbing liquid exposed to laser action and reported on new qualitative experimental results in this field, obtained for water irradiated with 200 ns erbium laser pulses at wavelength 2.94 μm.

At lower laser intensities generated pressure pulses are due to photoacoustic and surface vaporization mechanisms. In the case where laser intensity is harmonically modulated with period 5 ns the new effect was found which can be explained as mutual compensation (destructive interference) of photoacoustic and surface vaporization pressure signals. Further investigation of this effect can be of interest, in particular, for determination of mass accommodation coefficient which value is important for correct description of water vaporization and condensation processes [28, 29].

At higher laser intensities short (subnanosecond) pressure peaks were observed above smooth surface vaporization pressure signal. This effect can be considered as a manifestation of bulk vaporization (explosive boiling) process [2, 18]. It was shown here that the number of the short peaks observed earlier [18] can grow significantly at some laser fluencies. We suppose that in the case of realization of one dimensional experimental conditions (constant laser intensity distribution across irradiation spot) such peaks should disappear at sufficiently high laser intensity in supercritical ablation regime where recoil pressure exceeds the critical pressure. In such a case it will be possible to determine critical pressure values in laser ablation experiments where recoil pressure is measured with sufficient time resolution.

The work was supported by RFBR Projects 06-08-01440, 06-02-16566, 07-02-12209-ofi, and 08-02-12060-ofi.

APPENDIX

Properties of an Acoustic Sensor

A.I. Study of the Sensor in the Passive Mode

The temporal characteristics of the acoustic sensor controlling the degree of electrical signal distortions during recording of short pressure pulses were studied in two modes: a passive one using an external source of electrical signals and an active mode when electrical signals resulted from pressure pulses. Figure Al shows the circuit for studying the acoustic sensor in the passive mode. An electric pulse ~ 1 ns long from a G5-78 generator was applied to a DPO 7254 oscilloscope input (passband is 2.5 GHz) through coaxial cable A 2 m long and a branch. The latter was also connected to cable B ~ 1 m long. The free cable end was connected to the acoustic sensor or remained opened. A generator pulse was initially recorded by the oscilloscope; then, in a time equal to double passage over the cable, the signal reflected from its end was recorded. Figure A2 shows the survey oscillogram with these signals. Curves *1* and *2* correspond to open end of cable B and connected acoustic sensor, respectively. The first pulse in figure A2 is the signal, when the generator pulse immediately arrives at the oscilloscope input.

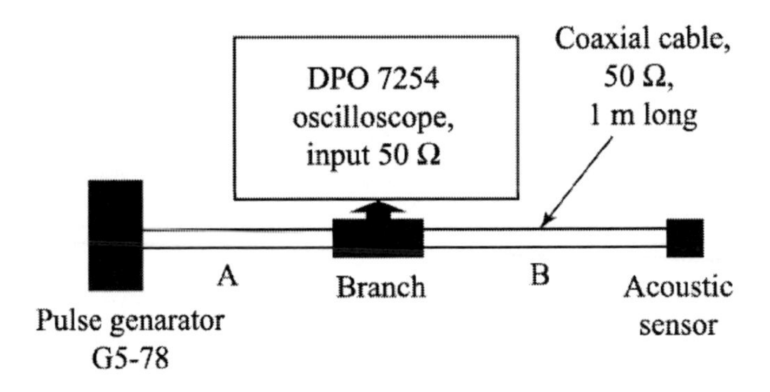

Figure Al. Electrical circuit for studying the sensor in the passive mode.

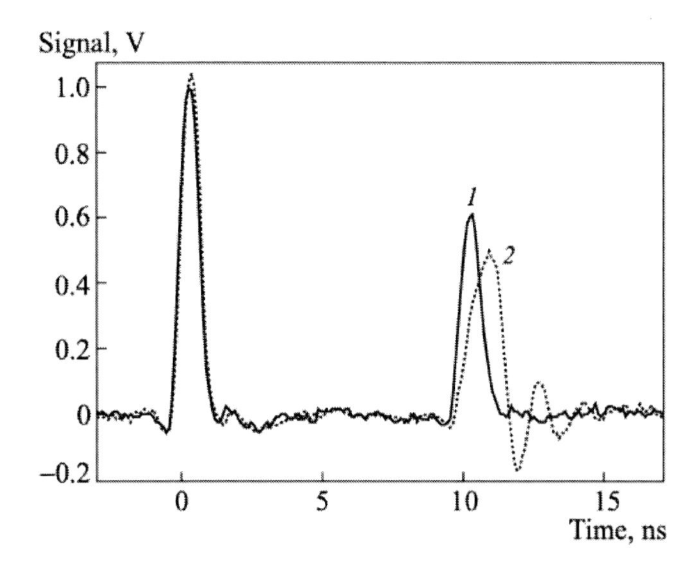

Figure A2. Oscillogram of pulses, when the cable B end (see figure A1) is open (curve *1*) and when this end is connected to the acoustic sensor (curve *2*). The first pulse is that from the generator.

Figure A3 shows the oscillograms of reflected pulses with a time resolution better than in figure A2. We can see that the pulse reflected from the acoustic sensor is broadened in comparison with the pulse reflected from the open cable end. Moreover, there is a negative spike at the trailing edge of the pulse, which is caused by inductivity of connections to the sensor due to its design features.

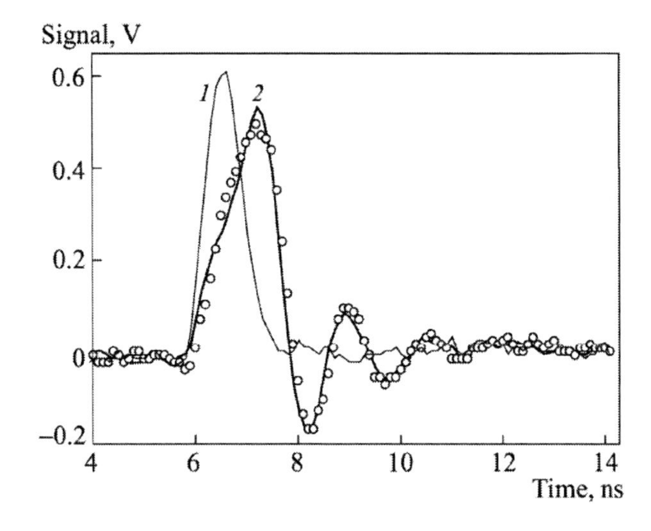

Figure A3. Oscillograms of signals: signal reflected from the cable open end (curve *1*) and that when the acoustic sensor is connected (circles). The calculated shape of pulses reflected from the acoustic sensor is also shown (curve *2*).

Figure A4 shows the equivalent electrical circuit of the sensor, used in calculating the sensor characteristics in the passive mode. Here *C1* is the capacitance caused by electrical connections and *C2* is the acoustic sensor capacitance. The parameters of the sensor equivalent circuit in the passive mode can be determined fitting the parameters (inductance and capacitances) so that the parameters of the calculated pulse would coincide with the parameters of the experimentally observed pulse.

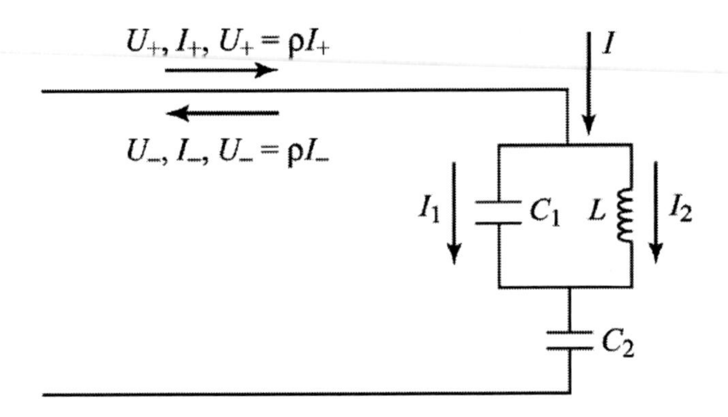

Figure A4. Equivalent electrical circuit of the sensor.

The equations for currents and voltages presented in the circuit of figure A4 are written as

$$\dot{U}_{C1} = \frac{1}{C1}\left(2\frac{U_+}{\rho} - I_2 - \frac{1}{\rho}\ U_{C1} + U_{C2}\right)$$
$$\dot{U}_{C2} = \frac{1}{C2}\left(2\frac{U_+}{\rho} - \frac{1}{\rho}\ U_{C1} + U_{C2}\right) \quad , \qquad \text{(A.1)}$$
$$\dot{I}_2 = \frac{1}{L} U_{C1}$$

where U_{C1}, U_{C2} voltages on capacitors *C1* and *C2*, respectively; I_+, I_-, I, I_1, I_2 I_+, are currents in corresponding circuit parts as is shown in figure A4, and p is the cable wave impedance. When a voltage arises on capacitor *C2* due to the piezoelectric effect, Eqs. (A. 1) take the form

$$\dot{U}_{C1} = \frac{1}{C1}\left(-I_2 - \frac{1}{\rho}\ U_{C1} + U_{C2}\right)$$
$$\dot{U}_{C2} = \frac{1}{C2}\left(\frac{U_{ac}}{\rho} - \frac{1}{\rho}\ U_{C1} + U_{C2}\right) \qquad \text{(A.2)}$$
$$\dot{I}_2 = \frac{1}{L} U_{C1}$$

Let us derive the second equation in set (A.2). Let us consider the signal formation on capacitor C2 when an acoustic signal passes through it. In the plane geometry of the acoustic sensor, the electric induction associated with the polarization P_{ac} in the dielectric, which occurs due to the piezoelectric effect, is given by $D = \varepsilon E + 4\pi P_{ac}$, where e is the permittivity of the sensor material and $div\vec{D} = 0$, i.e., D is independent of the coordinate x. As is known, the polarization P_{ac}, is proportional to the pressure σ in the acoustic wave, $P_{ac} = a\sigma(x - v_s t)$, where v_s is the speed of sound in the sensor material and a = const. Integrating the expression for the electric induction over the interelectrode distance, we obtain

$$lD = \varepsilon \int_0^l E dx + 4\pi a \int_0^l \sigma(x - v_s t) dx. \quad \text{(A.3)}$$

From this it follows

$$U_{C2} = -\int_0^l E dx = -\frac{Dl}{\varepsilon} + \frac{4\pi a}{\varepsilon}\int_0^l \sigma(x - v_s t) dx. \quad \text{(A.4)}$$

Differentiating (A.4) with respect to time, we obtain

$$\dot{U}_{C2} = -\frac{4\pi Il}{s\varepsilon} + \frac{4\pi a}{\varepsilon}\frac{\partial}{\partial t}\int_0^l \sigma(x - v_s t) dx, \quad \text{(A.5)}$$

where I is the current, s is the electrode area, and l is the interelectrode distance in the acoustic sensor. As a result, the dependence of the capacitor $C2$ voltage is written as

$$\dot{U}_{C2} = \frac{1}{C2} \quad -I + A \quad \sigma(l - v_s t) - \sigma(0 - v_s t) \quad , \quad \text{(A.6)}$$

where A=const. Then we assume that the sensor is thick; therewith, in the coaxial design, the pressure signal on one of sensor sides can be neglected. Introducing the notation

$$\frac{U(t)_{ac}}{\rho} = A\sigma(x = 0, t),$$

substituting it into (A.6), and taking into account that

$$I = \frac{U_{C1} + U_{C2}}{\rho},$$

we obtain the second equation in set (A.2).

Figure A3 (curve *2)* shows the shape of the signal reflected from the acoustic sensor. As an input signal, we take the signal shown in figure A3 (curve *1*). The calculation was performed using set (A.I). The experimental signal is best described at the following circuit parameters:

C1=0.8 pF, C2=2.8 pF, L=17.5 nH.

The cable wave impedance is ρ=50 Ω. We note that the capacitance and inductance of connections were estimated by the expressions given in [30, pp. 35, 68] as C1 = 0.9 pF and L=24 nH, which is in satisfactory agreement with the above values. We can see in figure A3 that the calculated signal describes well the shape of the pulse reflected from the sensor, including the negative spike on its trailing edge.

A.2. Study of the Sensor in the Mode of Short Acoustic Pulse Recording

In our experiments with water, short (shorter than 1 ns) pressure peaks were observed, which sometimes contained a spike of negative polarity at the trailing edge in some cases. Figure A5 shows the acoustic response pulse when water was irradiated with laser pulses having a wavelength of 2.94 μm. A short (shorter than 1 ns) pressure spike is observed against the background of the rather long pulse caused by surface evaporation. It is of interest to use the above sensor parameters (inductance and capacitance) to calculate the initial acoustic signal U_{ac} which, after substituting into set (A.2), leads to the observed signal.

The calculated results and experimental data are shown in figure A5(b). It was found that the initial acoustic signal U_{ac} (see set (A.2)) shaped as a Gaussian pulse (curve 1, figure A5(b)) with a half maximum duration of 0.37 ns and an amplitude exceeding the recorded one by 30% shows satisfactory agreement with the observed electrical signal. Thus, the observed negative spikes at the trailing edge of the recorded short acoustic signal can occur at acoustic signals shorter than 0.5 ns. The calculation also shows that the electrical signal reproduces the shape of acoustic signals longer than 1.5 ns.

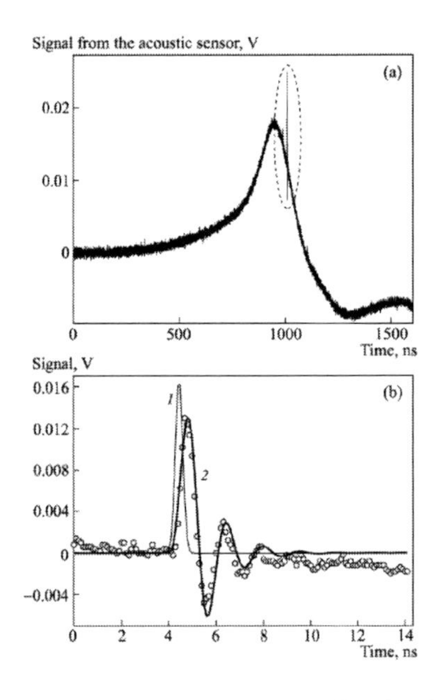

Figure A5. (a) Signal from the acoustic sensor when water is irradiated with laser (λ=2.94 μm). The dashed curve indicates the pressure spike, (b) Time dependence of the short acoustic signal generated in water irradiated with laser (λ=2.94 μm) (circles). The calculation results: the acoustic signal U_{ac} (curve *1*) and the recorded signal shape (curve *2*).

REFERENCES

[1] Gauster, W. B.; Habing, D. H. *Phys. Rev. Lett.* 1967, vol. 18, 1058.
[2] Samokhin, A. A. *Quantum Electron.* 1974, vol.9, 2056 [in Russian].
[3] Sigrist, M. V.; Kneubuhl, F. K. *JASA.* 1978 , vol.64,1652.
[4] Rosencwaig, A. *Photoacoustics and Photoacoustic Spectroscopy* R.E. Krieger Publishing Company, Malabar, Florida, 1980.
[5] Emmony, D.C. *Infrared Phys.* 1985 vol. 25, 133.
[6] Vodop'yanov, K. L.; Kulevskii, L. A.;. Mikhalevich, V. G.; Rodin, A. M. *Sov. Phys. JETP*, 1986 , vol.64, 67.
[7] Samokhin, A. A. in *Effect of Laser Radiation on Absorbing Condensed Matter;* Fedorov, V. B. ; Ed.; Proceedings of the Institute of General Physics Academy of Sciences of the USSR. Series Editor A.M.Prokhorov, Nova Science Publ., Commack, New York,1990, vol. 13, p. 1.
[8] Gusev, V. E.; Karabutov, A. A. *Laser Optoacoustics; Nauka,* Moscow, 1991 [in Russian].
[9] Paltauf, G.;Scmidt-Kloiber, H. *Appl. Phys. A.* 1996, vol. 62, 303.
[10] Kim, D. ; Grigoropoulos, C. P. *Appl. Surf. Sci.* 1998, vol.127-129, 53.
[11] Etcheverry, J. I. ; Mesaros, M. *Phys. Rev. B.* 1999, vol. 60, 94.
[12] Karabutov, A. A.; Pelivanov, I.M.; Podymova, N.B.; Skipetrov, S.E. *Quantum Electron.* 1999, vol. 29, 215.
[13] Krasyuk, I.K.; Pashinin, P.P.; Semenov, A.Yu.; Fortov, V.E. *Quantum Electron.* 2003 , vol. 33, 593.
[14] Sankin, G. N.; Simmons, W. N. ; Zhu, S. I. ; Zhong, P. *Phys. Rev. Lett.* 2005, vol. 95, 034501.
[15] Andreev, S. N.; Firsov, K, N.; Kazantsev, S. Yu.; Kononov, I. G. ; Samokhin, A. A. *Laser Phys.* 2007, vol.17, 1.
[16] Kudryashov, S.I. ; Lyon, K. ; Allen, S. D. *Phys. Rev. E.* 2007, vol.75, 036313.
[17] Vovchenko, V. I. ; Klimentov, S.M.; Pivovarov, P.A.; Samokhin, *A. A. Bulletin Lebedev Phys. Institute.*2007, vol. 34, # 11.
[18] Samokhin, A. A. ; Vovchenko, V. I. ; Il'ichev, N.N.;Shapkin, P. V. *Quantum Electron.* 2007, vol. 37, 1141.
[19] Mazhukin, V. I.; Nikiforova, N. M. ; Samokhin, A. A. *Phys. Wave Phenom.* 2007, vol.15, 81.
[20] Andreev, S. N. ; Orlov, S.V.; Samokhin, A. A. *Phys. Wave Phenom.* 2007, vol.15, 67.
[21] Vogel, A. ; Linz, N.; Freidan, S. ; Paltauf, G. *Phys. Rev. Lett.* 2008, vol.100, 038102.
[22] Samokhin, A. A. ; Vovchenko, V. I. ; Il'ichev, N.N. *Phys. Wave Phenom.* 2008, vol.16. 275.
[23] Zhakhovskii, V.; Inogamov, N.; Nishihara, K. *J. Phys. Conf. Ser.* 2008, vol. 112, 042080.
[24] Kartashov, I.N.; Samokhin, A. A. ; Smurov, I.Yu. *J. Phys. D. Appl. Phys.* 2005, vol. 38, 3703.
[25] Lang. F.; Leiderer, P. *New J. Phys.* 2006, vol. 8, 14.
[26] Il'ichev, N. N. ; Shapkin, P. V. ; Kulevsky, L. A. ; Gulyamova, E. S. ; Nasibov, A.S. *Laser Phys.* 2007, vol.17, 130.
[27] Karabutov, A. A. ; Kubyskin, A. P. ; Panchenko, V. Ya.; Podymova, N. B. *Quantum Electron.* 1995, vol. 25, 789.

[28] Laaksonen, A.; Vesala, T; Kulmala, M. ; Winkler, P. M.; Wagner, P. E. *Atmos. Chem. Phys.* 2005, vol. 5,461.
[29] Zientara, M. ; Jakubczyk, D. ; Derkachov, G.;. Kolwas, K.; Kolwas, M. *J. Phys. Appl. Phys.* 2005,vol. 38,1978.
[30] Glikman, I. Ya. ; S. Rusin, Yu. Calculation of Characteristics of Circuit Elements of Radio Electronic Devices; Sov. Radio, Moscow, 1976 [in Russian].

In: Phase Transitions Induced by Short Laser Pulses
Editor: Georgy A. Shafeev
ISBN: 978-1-60741-590-9

Chapter 5

NANOSTRUCTURES FORMATION UNDER LASER ABLATION OF SOLIDS IN LIQUIDS

E. Stratakis*[*1] and G.A. Shafeev[2]

[1] Institute of Electronic Structure and Laser, Foundation for Research & Technology—Hellas, (IESL-FORTH), P.O. Box 1527, Heraklion 711 10, Greece
[2] Wave Research Center of A.M. Prokhorov General Physics Institute of the Russian Academy of Sciences, 38, Vavilov street, 119991, Moscow, Russian Federation

ABSTRACT

Laser initiation of phase transitions at the solid-liquid interface results in the formation of self-organized nanostructures (NS) on the solid surface. Recent experimental results on the properties of such NS are described in this chapter. Formation of NS is assigned to the instability of evaporation of the liquid that surrounds the irradiated target. The morphology of NS generated on various metallic (Ag, Au, Ta, Ni, Ti, etc.) as well as non-metallic bulk solids (Si) is studied as the function of experimental parameters, such as laser pulse duration and target material. It is demonstrated that average lateral size of NS depends on the laser pulse duration and is typically of order of 100 nm. Distribution function of lateral sizes of NS usually exhibits two maxima and is transformed to a single maximum function upon decreasing laser fluence towards melting threshold of the solid. Initial (nano)relief on the target surface facilitates the formation of NS under laser exposure. Formation of NS alters optical properties of the solid due to plasmon resonance of free electrons in NS. Preliminary results on the formation of NS on the pre-patterned substrates in a wide range of laser pulse durations from femto- to picoseconds are presented. Possible applications of NS in Surface Enhanced Raman Scattering (SERS) as well as for nano-patterning of medical implants are also discussed.

* stratak@iesl.forth.gr

INTRODUCTION

Phase transitions at the solid-liquid interface, exposed to short laser pulses at an energy density sufficient to melt the solid, can be traced owing to modifications of the solid surface induced by the laser pulses. In case of a metallic target, the laser radiation is absorbed by free electrons, and the lattice temperature starts to increase due to the electron-phonon relaxation process. If the absorbed energy is sufficiently high, the metal target melts, so that a layer of the liquid that surrounds it is heated up due to heat transfer from the metal. As a result of the high pressure of the adjacent medium that contacts the melt, the latter can be modified. In particular, the liquid vapors that surround the molten layer induce in it viscous flows giving rise to the formation of various structures.

The characteristic thickness of the modified layer of the solid target strongly depends on the melt thickness, and therefore, on both the laser fluence and its duration. The thickness of the molten layer h_m can be estimated by the heat diffusion length during the laser pulse as follows: $h_m \sim (at_p)^{1/2}$, where t_p stands for pulse duration, and a stands for heat diffusion coefficient of the solid. This estimation is valid only for laser fluence close to the melting threshold of the solid. If the duration of the laser pulse is less than the time of electron-phonon relaxation, then the heating of the lattice occurs within the depth of the absorption of laser radiation. For typical metals the mean free path of excited electrons during the relaxation process is too short, and the melt thickness does not exceed a fraction of a micrometer even for nanosecond (ns) laser pulses. As a result this layer of material may be re-distributed into one or another kind of nano-structures (NS) due to the recoil pressure of the liquid medium adjacent to it.

Melting of the target surface is a necessary condition for structure formation; however the adjacent medium is responsible for the actual type of structures. The morphology of the structures that are left in the target surface after melt solidification can provide valuable information about the phase transitions taking place at the solid-liquid interface. In a sense, the observed NS are the "fingerprints" of nano-scale inhomogeneities of the medium in its supercritical state that surrounds the target.

MORPHOLOGY OF NS UNDER LASER ABLATION WITH SHORT LASER PULSES

NS imaging using scanning probe microscopy, e.g., Atomic Force Microscopy (AFM), indicates that they are densely packed nano-cones. This is illustrated in figure 1 for the case of NS grown on an Ag target under its ablation in water with picosecond laser pulses [1]. Note that both the lateral dimensions of NS and their period are much smaller than the laser spot size on the target, which in typical experimental conditions is of order of hundreds of micrometers. The estimated density of NS in figure 1 amounts to 10^{10} cm^{-2}. Therefore, the expanding vapors of the liquid that surrounds the target are unstable. Pressure difference appears within an initially smooth vapor pocket above the molten layer of the target. This pocket is split into periodic cells, and the pressure difference within these cells pushes the melt from high to low pressure areas. Solidified NS on the target are just the imprints of those cells in the adjacent to target medium in which the phase transition takes place.

This type of NS is observed under ablation of solids with sufficiently short laser pulses. The upper limit of the laser pulse duration at which the NS are observed is around 300 – 400 picoseconds (ps). Longer pulses do not favor the formation of NS in the whole range of laser fluencies, from melting threshold to intense ablation. On the other hand, NS are readily observed under ablation with shorter laser pulses down to femtosecond (fs) ones. The laser wavelength needed for NS formation is not very important as soon as metal targets are considered. This is because of the fact that optical constants of most metals similar in the range of wavelengths where lasers usually emit.

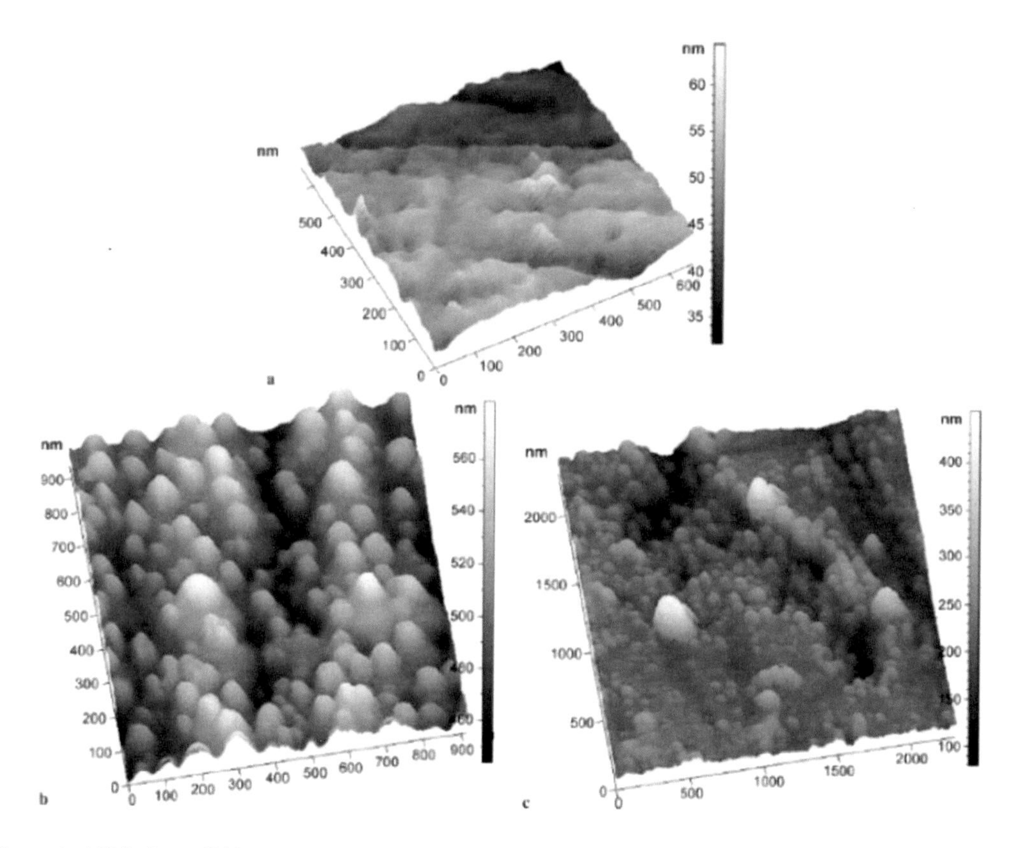

Figure 1. AFM view of NS on an Ag plate generated by its ablation in water under exposure to a 350 ps Nd:YAG laser radiation. Pristine surface (a), NS after laser ablation in water (b), concentration of NS in micro-depressions of the relief (c).

NS obtained on any other metals, such as Au, Ti, Ta, etc., look very much alike though they have different periods and lateral sizes. However, later studies showed that scanning probe microscopes are not suitable for adequate imaging of NS formed. This is due to the complex profile of NS that are realized under these conditions. In reality, NS generated via laser ablation are often just solidified drops of the melting material that is attached to the target via a thin "neck." This morphology can only be revealed with the help of high resolution scanning electron microscopy (SEM). A typical Field Emission SEM (FESEM) view of a NS on Ta target obtained via its ablation in water with a 350 ps Nd:YAG laser is presented in figure 2.

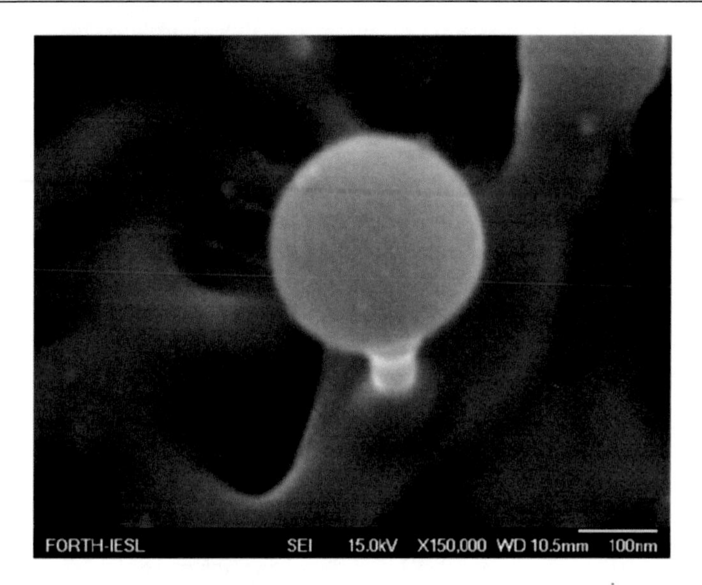

Figure 2. Enlarged view of a single NS on Ta produced by its ablation in water with 350 ps laser pulses of a Nd:YAG lasers. FE SEM, scale bar denotes 100 nm.

One can see that the shape of NS is mushroom-like, so that their lateral size is a non-monotonous function of the coordinate [2]. Figure 3 shows subsequent positions of an AFM tip along the surface containing NS presented above. The tip cannot approach the base of the NS, and the resulting profile deduced from scanning is a cone-like NS. Therefore, scanning electron microscopy provides more adequate information on the NS morphology. However, scanning probe microscopy is quite convenient for characterization of lateral dimensions of NS and their period.

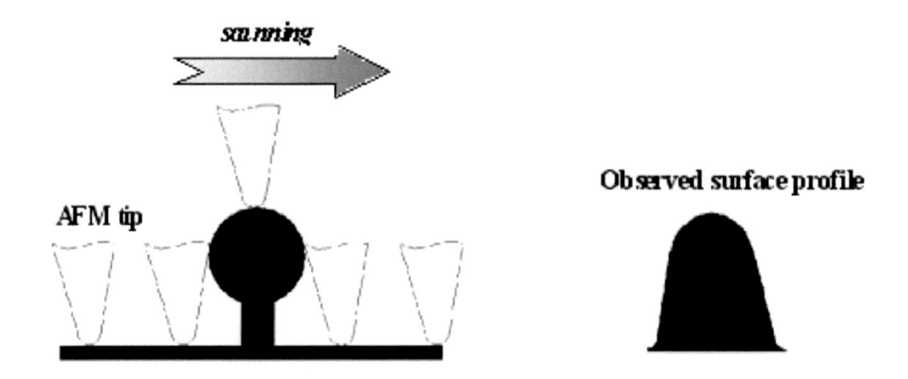

Figure 3. Imaging of NS with the help of a probe microscope. Sequence of tip positions upon scanning the NS (left). Registered profile of the NS (right).

This also concerns any type of probe microscopes, AFM or Scanning tunnel microscope (STM).

Figure 4, a-c shows the FE SEM view of NS on Ag Au, and Zn targets ablated in water with ps laser radiation. Most of NS have a mushroom shape similar to that observed for Ta NS.

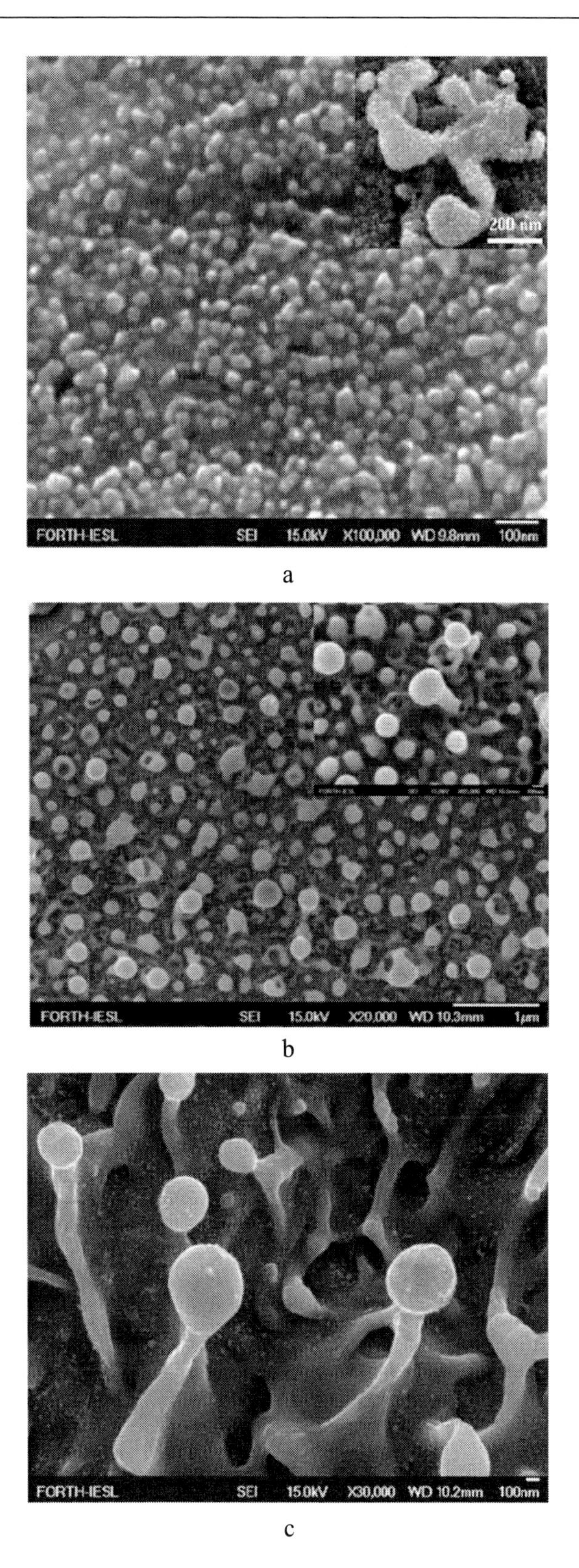

Figure 4. Continues on next page.

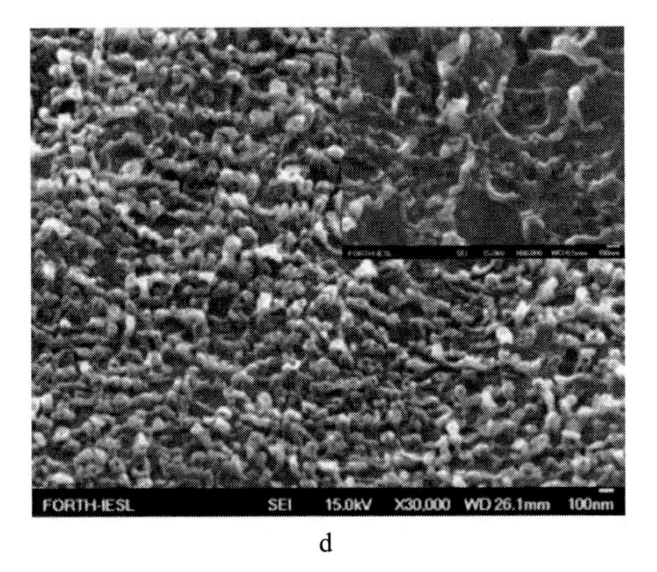

d

Figure 4. FE SEM view of NS on Ag (a) and Au (b) targets produced by their ablation in water with radiation of a 5 ps KrF laser, wavelength of 248 nm. Insets show the view at different scale. Zn NS obtained by ablation of bulk Zn in ethanol with a 150 ps laser radiation, wavelength of 1064 μm (c). NS on bulk Al ablated with 100 fs pulses of a Ti:sapphire laser in ethanol (d).

Note that for both metals the temperature of the adjacent to the surrounding liquid layer is around 1000 K. As soon as the shock wave propagates toward the free surface of the liquid, the pressure in it remains high, but when it reaches the liquid surface, the pressure above the target abruptly drops. Usually the shock wave in liquids propagates in it with the speed of sound (1497 m/s). This means that, for 1 mm thick liquid layer, the time of propagation of the shock wave is in order of microseconds, which is much longer than both the pulse duration and electron-phonon relaxation time.

Figure 5. Possible types of NS morphology. The last right scheme corresponds to detachment of the molten drop that becomes a nanoparticles suspended in the surrounding liquid.

As a result, all kinds of structures shown in figure 5 can be formed and this is corroborated by the NS images presented above. On the other hand, NS that were detached from the melt do not always take on a spherical shape. For instance, in case of Al ablation with femtosecond laser pulses either in water or ethanol, the nanoparticles found in liquid have a shape of a drop with "tails" as shown in the following figure. It is pertinent to note that this kind of NPs morphology was not observed so far on any other metals under their laser ablation in liquids [3]. This unique morphology can be due to the interplay of the melt viscosity and cooling rate of ejected nanodrops.

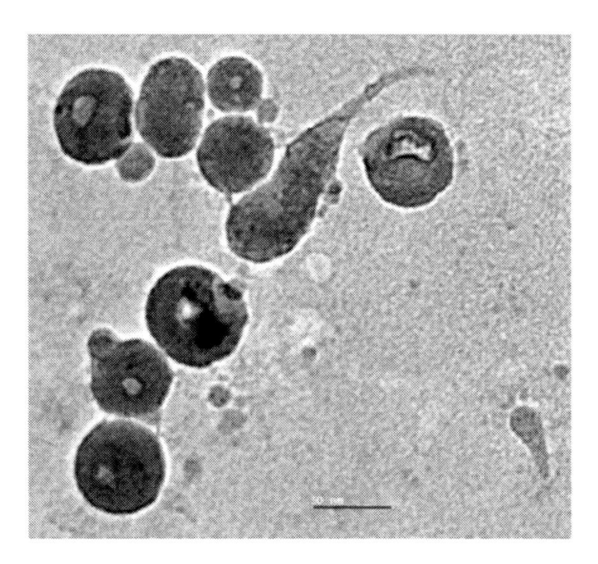

Figure 6. Transmission Electron Microscope view of nanoparticles of aluminum generated by ablation of a bulk Al target in ethanol using 100 fs Ti:sapphire laser pulses at a wavelength of 800 nm. Scale bar denotes 50 nm.

Estimation of Pressure of Surrounding Medium

The spherical structure presented in figure 2 has been formed by the melt propulsion under the action of the pressure difference within the surrounding liquid. This is supported by the observation that each NS is surrounded by a system of pits that are almost situated around the NS, shown in figure 2. Thus, the melt is pushed from the pits towards the center, where the NS is formed [2].

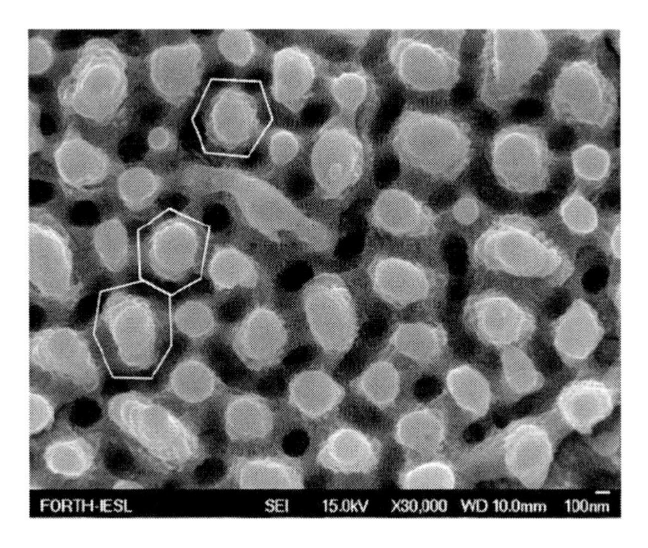

Figure 7. Top FE SEM view of NS on Ta generated by its ablation in water using 5 ps, 248 nm laser radiation. The white polygonal figures are just guides for an eye.

In this context, the formation of NS can be interpreted as the result of the work of pressure. This work is spent for the increase of the interface length against the surface tension of the melt. In view of this, the elementary work dA needed to increase a circular interface length of $2\pi R$ by dR can be expressed as follows:

$$dA = \sigma 2\pi R dR\,,$$

where σ stands for the coefficient of surface tension of the melt. The total work, A, for the generation of a spherical NS of radius R can be found as follows:

$$A = 2\int_0^R \sigma 2\pi R dR = 2\pi\sigma R^2$$

On the other hand, this work is performed by a force equal to $S\Delta p$, where S is the surface of the NS and Δp is the pressure difference. As this force acts on a distance $l=2R$ its corresponding work is:

$$A = Fl = \pi R^2 \Delta p 2R$$

Equalizing the two above expressions one finds the following estimation for the pressure difference:

$$\Delta p = \frac{\sigma}{R}$$

Substituting the parameters for NS on Ta shown above, with R = 100 nm, one obtains Δp = 400 atm. This is the order of magnitude of the pressure difference existing above the melt surface, applied on a very short distance, of the order of NS period. It is important that the estimated value for Δp exceeds the critical pressure for H_2O. Furthermore, the temperature of the liquid layer adjacent to the melt is close to the melting temperature of the target, equal to 3000° C in case of Ta. As a result, the medium that surrounds the target (H_2O) is in its supercritical state.

Many compounds, among them is water, in supercritical state are known for their enhanced chemical activity. Although H_2O is neutral to many solids in normal conditions, so that their solubility in it is negligible, it is a good solvent to many metals and dielectrics in its supercritical state. In actual experimental conditions, dissolution of the target in supercritical H_2O should take place. The dissolved target material sediments upon the pressure and temperature drop that follow the laser pulse, thus the formation of NS might be due to condensation of this material from the super-saturated solution. On the other hand, dissolution of solids in supercritical H_2O is accompanied by chemical interaction causing the formation of corresponding hydroxides. In this case, however, the re-condensed material should not be as compact as it is actually shown by the corresponding SEM images of the Ta target. The effect of supercritical dissolution of Ta on the morphology of NS has been tested studying its ablation in aqueous solutions of the base KOH.

Figures 8 and 9 show the size distribution of NS on Ta generated by its ablation in water with 1.06 μm laser radiation at different pulse durations. It is clear that the lateral size of NS on Ta increases with the increase of the pulse duration. This fact is also corroborated by comparison of lateral size of Ag NS generated with 350 (Figure 1) and 5 ps (Figure 2, a) laser radiation.

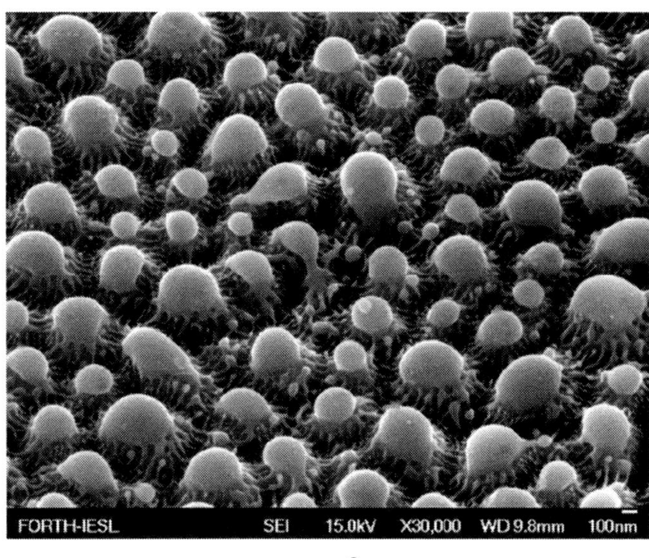

a

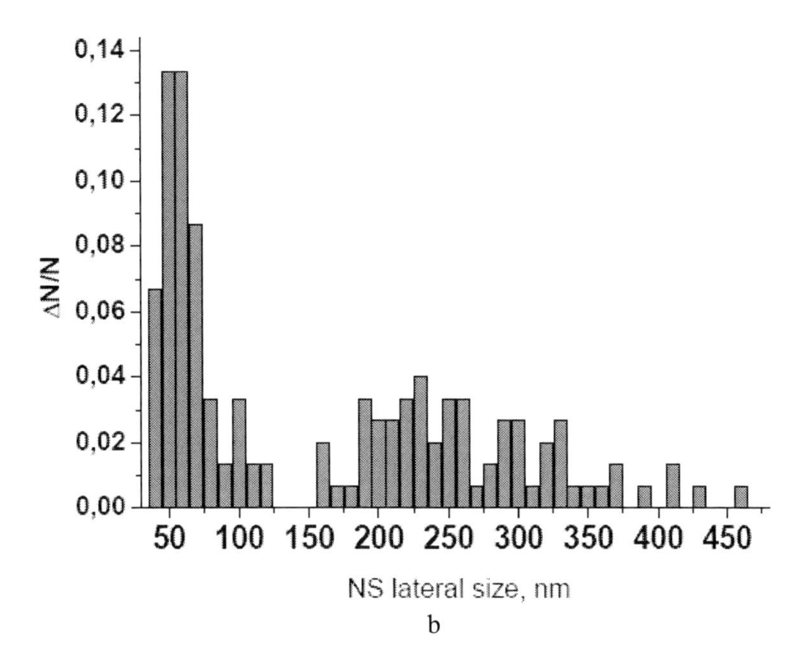

b

Figure 8. FE SEM view of NS on Ta produced by its ablation in water with a 5 ps laser pulses at wavelength of 248 nm (a). Distribution of lateral size of NS on Ta in this conditions (b).

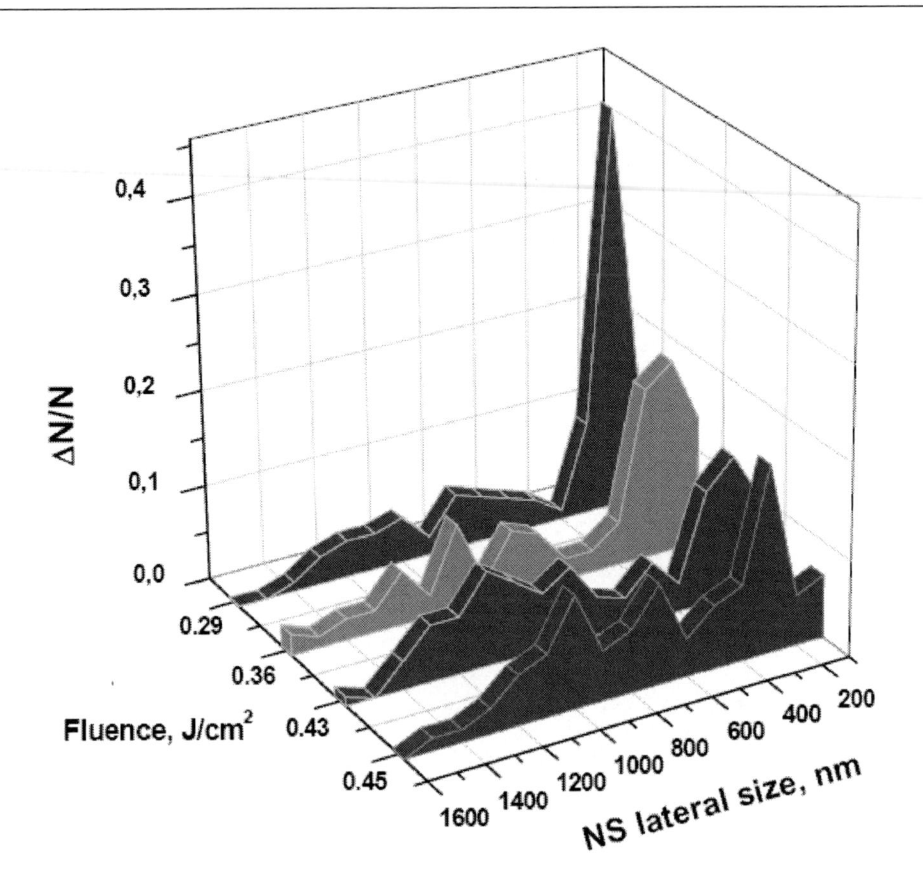

Figure 9. Distribution of the lateral size of NS on Ta with laser fluence as a parameter. Ablation in water, Nd:YAG laser, pulse duration of 350 ps at wavelength of 1.06 μm.

Furthermore, the lateral size of NS decreases with the decrease of laser fluence towards the melting threshold. It is pertinent to note that in all cases the size distribution function has two maxima, the first one at small sizes (50-200 nm) and the second in the vicinity of the laser wavelength, either at 250 or 1000 nm (see figures 8 and 9). This means that the interference of the laser radiation with Surface Electromagnetic Wave (SEW) does influence the formation of NS, though in case of ps ablation in liquids regular periodic ripples are rarely observed. In contrast, in the case of fs laser exposure in liquids, periodic ripples dominate over self-organized NS. This is observed in almost all solids, e.g., for Ge and Ni, and for Ta ablation in water it is illustrated in figure 10. It is evident that periodic ripples are the main feature of the ablated surface, whereas mushroom-like NS, typical of longer laser pulses, are situated mostly on top of them. The formation of ripples is most probably due to the same reason as for "usual" mushroom-like NS. Namely, melting of the surface occurs in the maxima of the interference pattern, and then the melt is pushed out by the recoil pressure of vapors of the surrounding liquid. Evolution of the bimodal distribution function of laser-induced structures is modeled in details in Chapter 6.

Simultaneous observation of periodic ripples and spherical NS on the same target indicates different mechanisms of their formation.

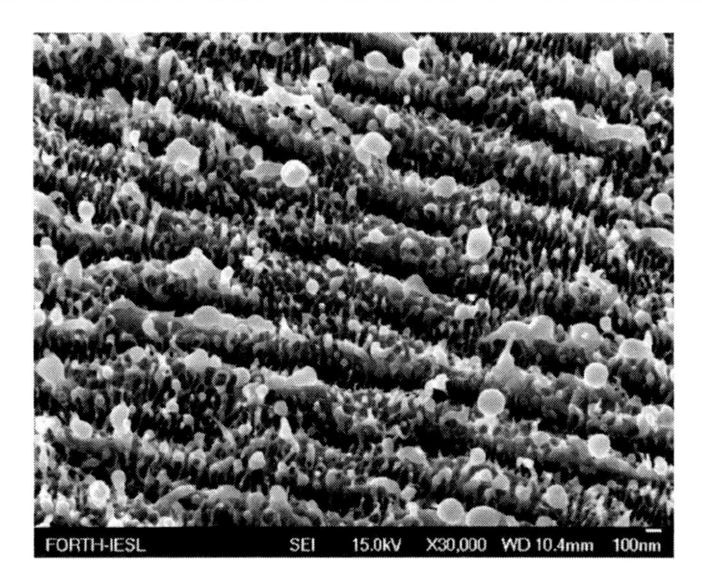

Figure 10. FE SEM view of a Ta surface ablated in water with 100 fs radiation of a Ti:sapphire laser at wavelength of 800 nm.

GROWTH OF NS ON PRE-PATTERNED SUBSTRATES

Any real surface is not perfectly smooth and is characterized by a certain degree of surface roughness. The generation of NS is sensitive to its initial value. The first realization of NS on Ag under its ablation in liquids showed that some initial surface roughness is required for their formation [1]. It is observed that the initial mean roughness should be at least 50 nm, and no NS formation occurred on optically polished targets. In the latter case, a deep crater (a groove when the laser beam scans the substrate) appears, upon increasing the laser fluence. Furthermore it was found that NS are predominantly located close to the edges or protrusions of the relief. This was attributed to a weaker thermal contact of these features with the substrate, so they are the first to melt under the laser pulse. The above findings were later confirmed for other metals as well.

A distinct property of small surface protrusions whose size lies in the nanometer range is the depression of their melting temperature compared to the bulk target material. As a result, nano-sized protrusions may melt while the rest of the target remains solid at certain laser fluence. The melting temperature drop is related to the higher fraction of surface atoms of NS, which are disordered due to the large number of dandling bonds. Theoretically, it is expressed by the following relation [4]

$$T_m = T_m^{(\infty)} exp\left(-\frac{4\delta}{\delta+2R}\right),$$

where $T_m^{(\infty)}$ stands for the melting temperature of a bulk solid and δ is the Tolman constant. The physical meaning of this constant is the thickness of a superficial layer with distorted lattice; $\delta = 6d$, where d is the lattice parameter of a bulk solid. R is the radius of a nanoparticle, which in our case is that of a local nano-protrusion. The absolute value of depression of the melting temperature depends on the nature of the target material, the maximal depression is observed for Au.

The influence of the initial target morphology on the properties of NS was studied using specially designed Ni targets pre-patterned by electron beam lithography. A layer of a photoresist (PMMA) was deposited by spin-coating on the surface of a polished Ni foil and then it was exposed to a focused electron beam. This layer was then developed in such a way that the exposed areas were dissolved. Finally, galvanic Ni deposition was carried out and the non-exposed areas of the photoresist were removed. The resulting surface comprised periodic grooves and its profile is presented in figure 11.

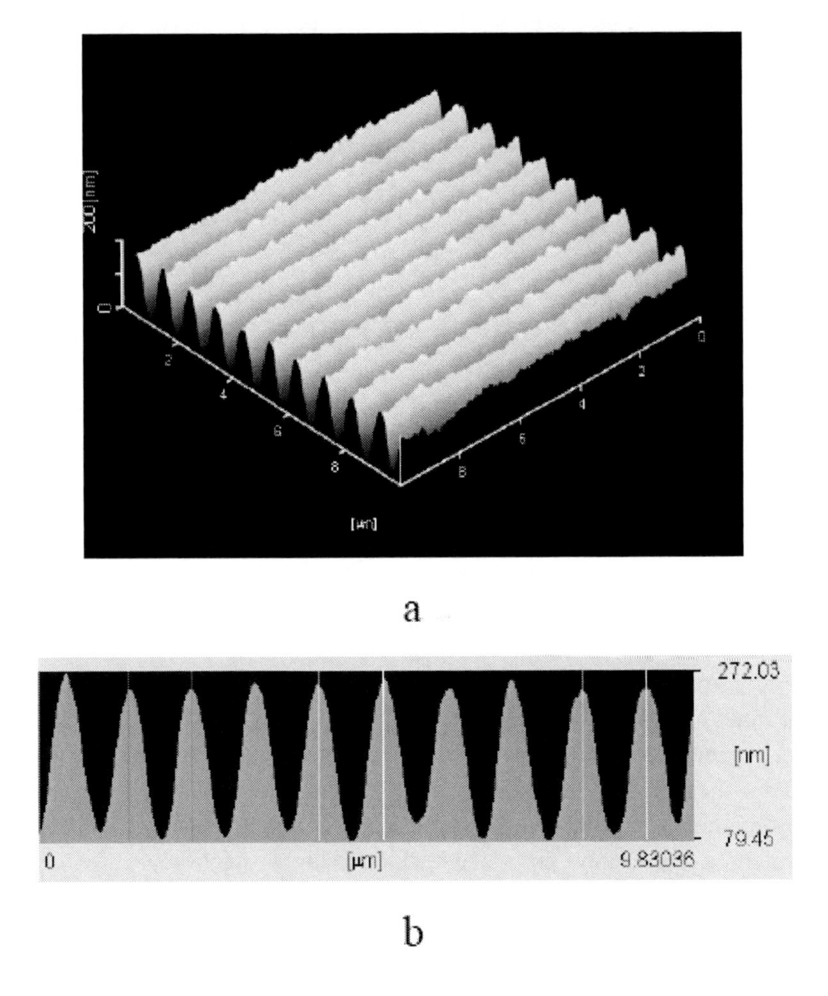

Figure 11. AFM view of a Ni target pre-patterned via electron beam lithography. 3D view (a), cross section view (b).

The effect of target pre-structuring on NS formation strongly depends on the laser pulse duration. In case of fs pulses, the influence of initial roughness on the growth of NS is illustrated in figure 12.

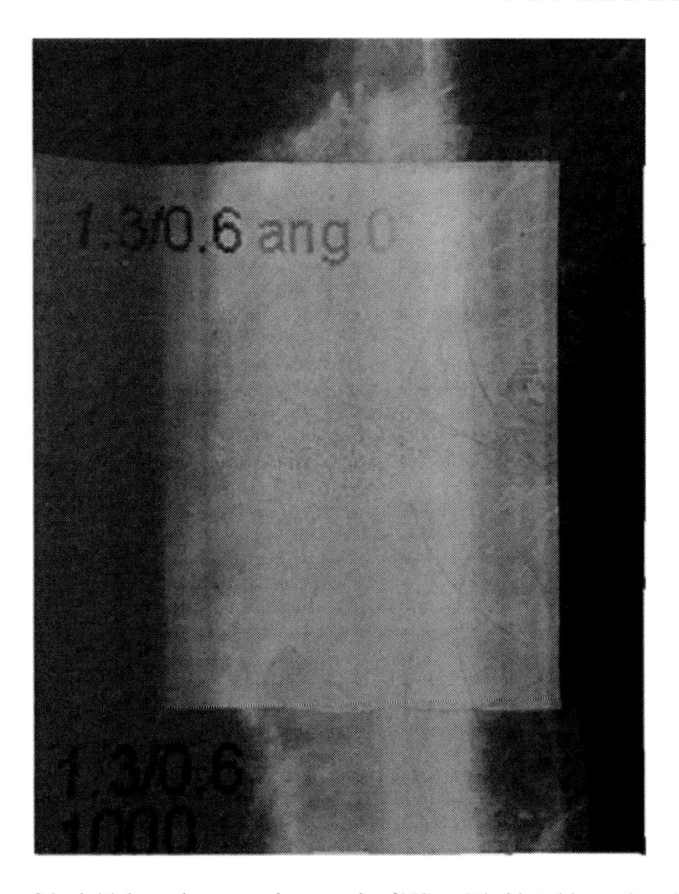

Figure 12. Influence of the initial roughness on the growth of NS on Ni ablated into ethanol using a 50 fs, 800 nm Ti:sapphire laser. The side of the pre-patterned square is 2 mm long.

The grey square shown in the figure is the pre-patterned Ni surface, whereas black areas correspond to pristine Ni. The laser beam scans the surface from top to bottom. The beam intensity profile is slightly asymmetric, and only its central part produces NS on the pristine Ni surface being visible as bright areas. Then the beam crosses the pre-patterned area of the substrate (clear square) while the area where NS are formed rapidly expands almost over the whole square. Later when the beam leaves the pre-patterned square, the area with NS remains less wide than on the square itself. This means that the threshold laser fluence needed to produce NS is much smaller on the pre-patterned areas than on the smooth one. As soon as NS are created in the surface, they are self-sustained and can be generated at much lower fluence. This result is consistent with previous observations for generation of NS on Ag [1].

A detailed view of NS formation on pre-patterned Ni surface is presented in the following set of FE SEM images.

At low laser fluence only sporadic pits are observed on the pristine Ni surface. However as soon as the laser beam reaches the initial pattern, morphology changes become more pronounced. As can be clearly seen in the enlarged view of figure 13, the surface damage is located predominantly in the summits of the initial structures. It looks more like explosion rather than melting, calling in mind the so called phase explosion (see Chapter 1). Upon further increase of the laser fluence both smooth and pre-patterned areas of the target are covered by NS (Figure 14 left).

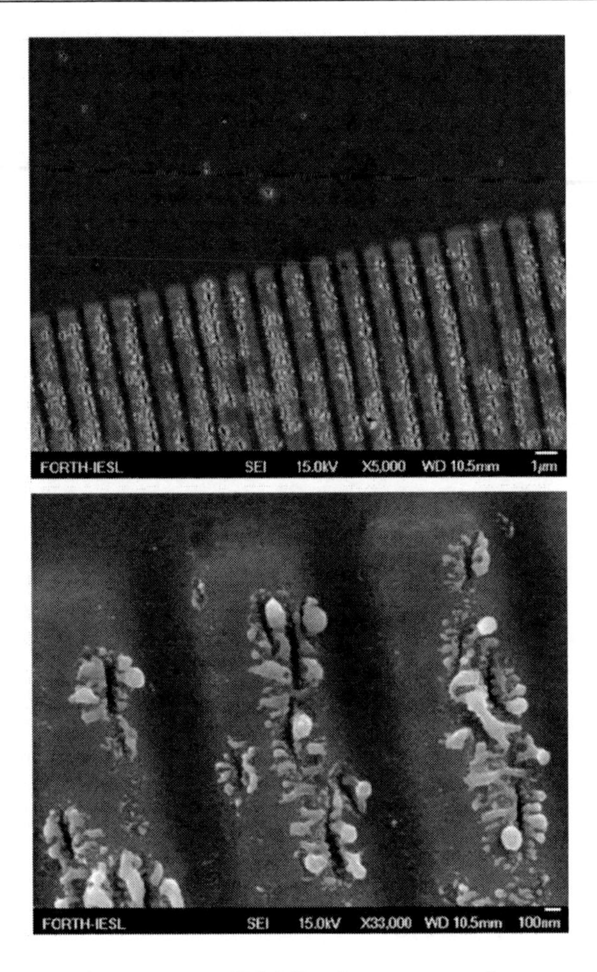

Figure 13. Formation of NS on the pre-patterned Ni foil under exposure to femtosecond Ti:sapphire laser radiation in ethanol. Low fluence of 4.5 J/cm^2. The frontier between pristine and pre-patterned areas (top), enlarged view (bottom).

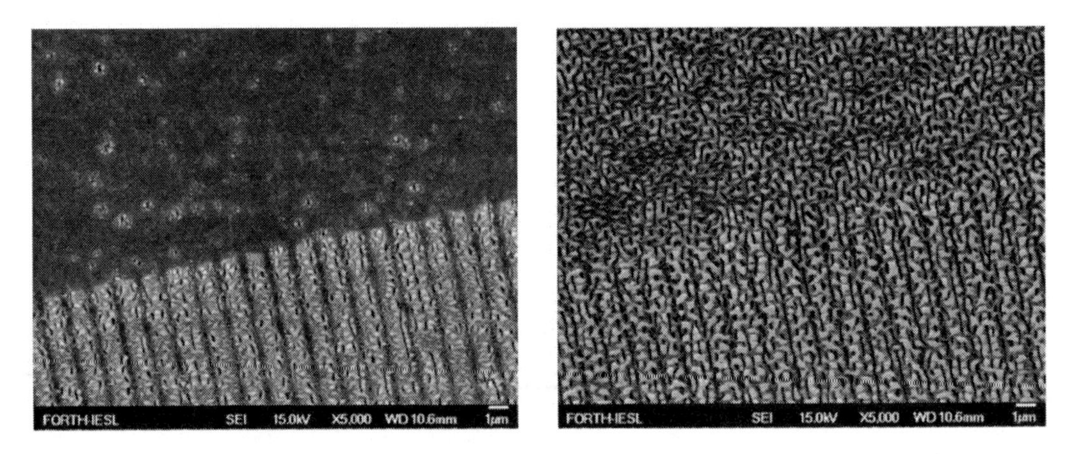

Figure 14 . Formation of NS on the pre-patterned Ni foil under exposure to femtosecond Ti:sapphire laser radiation in ethanol. The laser fluencies were 7 and 18 J/cm^2 respectively.

Finally, at much higher fluences the pre-patterned areas can be seen only due to the higher average size of NS compared to that on the pristine Ni surface (Figure 14, right).

As shown in figure 15, the effect of pre-structuring the target is less pronounced when longer laser pulses are used. The lateral size of initial structures is much less than the lateral dimensions of NS on a smooth surface, and the initial relief is just decorated with NS.

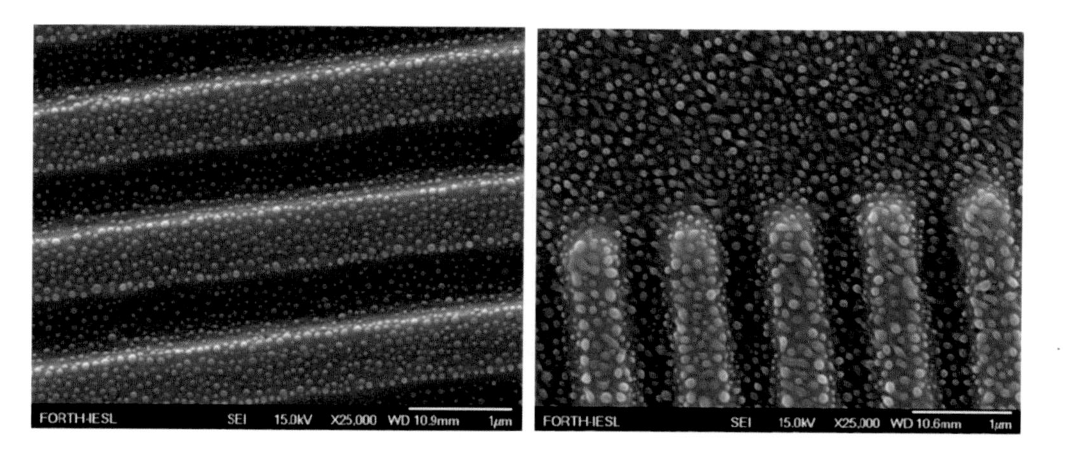

Figure 15. NS formation on pre-patterned Ni target under exposure with a 5 ps laser pulses in ethanol, wavelength of 248 nm. The laser fluences were 0.3 and 0.9 J/cm^2 respectively.

For low fluences the edges of the initial structures are decorated with large NS, while their size in the valleys is lower. It is interesting that NS are aligned along the edges of the initial stripes. This can be understood as the depression of the melting temperature on the edges of initial relief. At higher fluences, NS on top of the stripes tend to coalesce forming elongated nano-entities.

Optical Properties of NS

The visible manifestation of NS formation after laser irradiation of a metallic target is the coloration of exposed areas. A representative example is shown in figure 16, which shows the yellow colored Al surface obtained after its ablation by fs pulses in water. This coloration should be distinguished from surface oxidation, since some oxides have absorption bands in the visible. However, a thin oxide layer cannot be responsible for intense coloration of the target. The optical properties of NS are closely related to the spectral features of corresponding nanoparticles (NP). This is due to the fact that electrons are confined within NS just like in NP [5, 6].

Formation of Ag and Au NS under laser ablation of corresponding metallic targets in liquids is extensively studied, as the plasmon resonance of NS from these two metals lies in the visible. Therefore the successful formation of such NS is extremely important for optoelectronic applications. Laser exposure of Ag in liquids leads to significant modifications of the plasmon spectrum of this metal.

Figure 16. Macro view of an Al target exposed in water to radiation of 180 fs Ti:sapphire laser. The yellow circle at the center corresponds to the exposed area comprising NS. Lateral dimensions of the target are 3x3 cm^2.

Figure 17 shows the absorption spectra of an Ag surface before as well as after its exposure in water with ps laser pulses. The spectrum of the initial surface shows a peak at 315 nm corresponding to the anticipated surface plasmon oscillations of electrons in the bulk Ag (spectrum 1). This spectral feature remains in the laser-exposed surface though it is widened and shifted to higher frequencies due to the NS formation (spectrum 2) which enhance the damping of plasmonic oscillations. At the same time NS bring new absorption bands and an additional wide peak appears in the near-UV region of spectrum centered at 370-380 nm. This peak shifts to the visible (spectrum 3 in figure 17, a) after sample storage for several days in air. The wing of this peak protrudes to the visible range of spectrum resulting in yellow-gold coloration of the laser-exposed areas.

The liquid medium takes on a yellow color as well, indicating the formation of nanosized particles dispersed in it. A TEM image of these NPs is presented in figure 18. The theoretical position of the peak maximum of plasmon resonance for Ag NPs is situated exactly at 400 nm, provided that they are suspended in a liquid with refractive index $n> 1$ [7]. This peak is observed to shift towards UV, since the spectrum of the exposed surface is taken in air with $n = 1$ [8]. These spectral peaks around 400 nm indicate the formation of surface nanostructures with lateral dimensions comparable to those of Ag nanoparticles dispersed in the liquid [8]. NS of this size were indeed observed in the exposed areas of a Ag target, as shown in the previous section. Laser exposure of an Ag target in ethanol, under identical experimental conditions, leads to a similar pronounced spectral peak, though centered at 430 nm (Figure 17 b - spectrum 2).

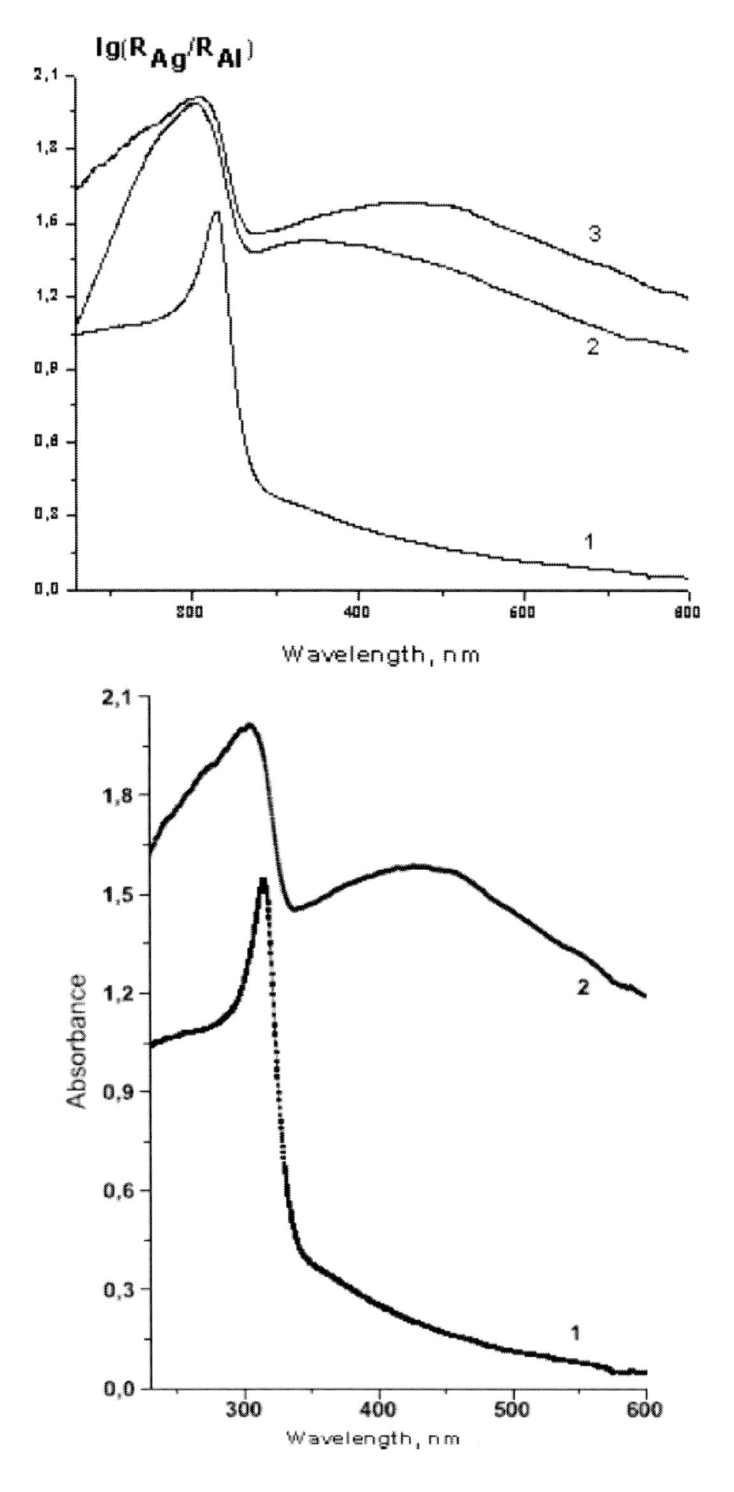

Figure 17. Modifications of reflectivity of an Ag target exposed a) in water to radiation of a Nd:YAG laser, pulse duration of 350 ps. Pristine Ag surface (1), after laser exposure (2), and after storage in air for several days (3). Reference sample is bulk Al. b) in ethanol under otherwise identical conditions.

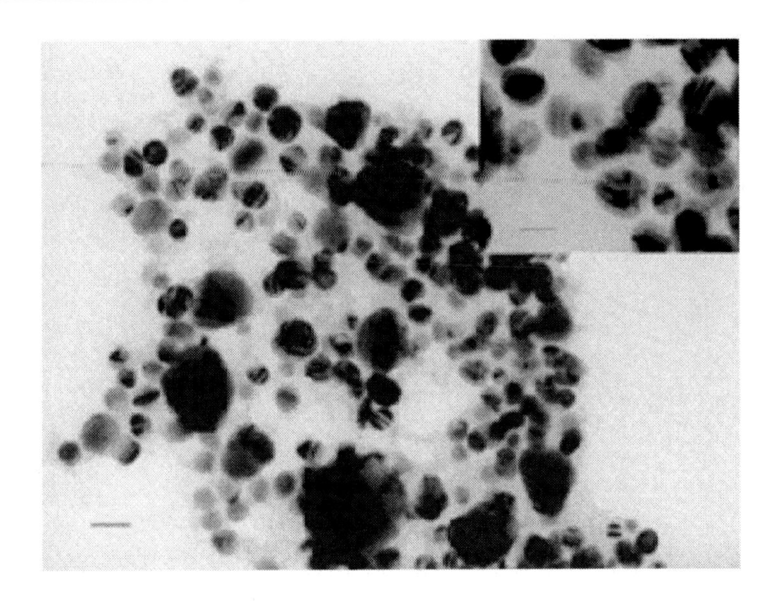

Figure 18. TEM image of Ag nanoparticles obtained by ablation of bulk Ag in water. The scale bar is 100nm. The inset shows an enlarged view of the intersection of individual nanoparticles (scale bar 30nm).

On the contrary, an Au surface with NS formed under its ablation in water looks like bulk copper. This is illustrated in figure 19 that compares absorption spectra of pristine and laser-exposed Au targets [9].

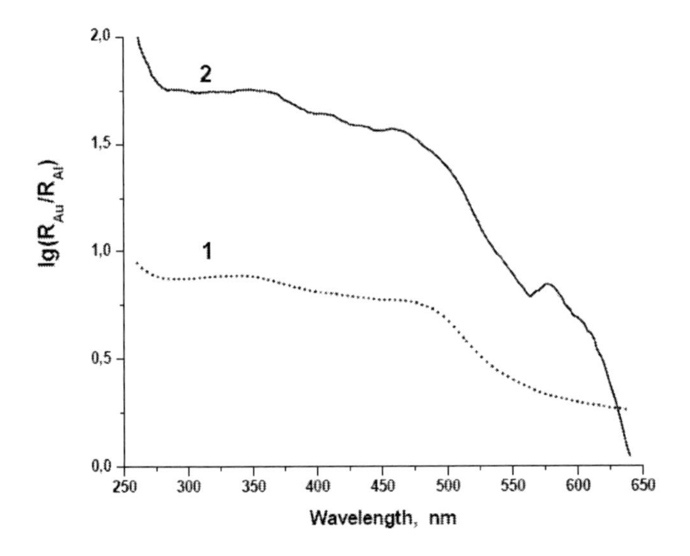

Figure 19. Absorptivity of the pristine Au target (1) and Au target with NS formed via its ablation in water with a 350 ps Nd:YAG laser (2). Spectra were taken with bulk Al as a reference.

Exposure of an Al target at short laser pulses in liquid results in the yellowish coloration of the metal, being visible at angles close to mirror reflection (Figure 16). The coloration is most pronounced in the case of exposure into ethanol, where it appears just after only a few pulses at a fluence of as low as 0.05 J/cm^2.

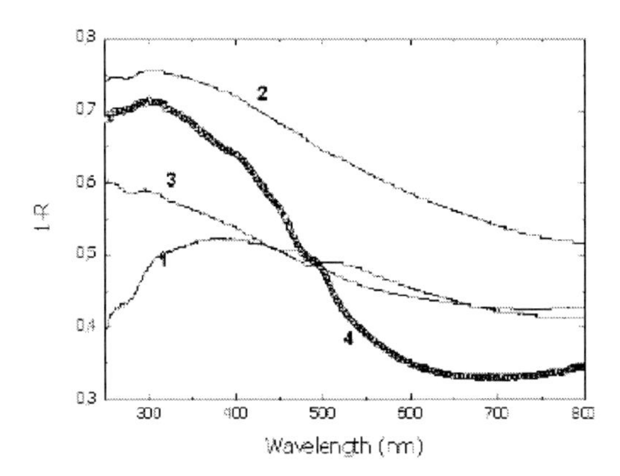

Figure 20. Absorption spectra of the initial Al surface (1) and of those exposed to the radiation of a 5 ps laser at 248 nm in water (2) and in ethanol (3). The curve 4 corresponds to exposure of Al in water to 180 fs-Ti:sapphire laser.

At lower fluences, virtually no changes of the Al surface are observed, even with an elevated number of laser shots. This colour change is permanent, for instance, the target may be wiped by a wet tissue without any change of the colour and it is observed independently of the purity of Al target with Al content ranging from 75% through 99%. Hence, this coloration should be assigned to structuring of the Al surface at the nanometre scale similarly to NS on both Ag and Au [10]. Figure 20 presents the absorption spectra of Al surfaces structured with different laser sources. In the blue and near UV region the absorption of the exposed samples exceeds that of the initial surface while in the NIR range, the samples ablated into liquid show lower absorption. In all cases, the absorption maximum is situated around 300 nm, which is justified by the yellow coloration of the corresponding exposed areas. This peak is shifted to lower wavelengths upon aging. When the Al target is exposed into liquids the latter takes on a yellow colour as well, indicating the formation of Al particles dispersed in the liquid. These NP dispersions show a characteristic absorption peak close to 300nm [11]. The lateral size of NS on Al produced by its ablation in water with 350 ps laser radiation is 200 – 300 nm. The coloration of the exposed surface in this case is not detectable at all.

Coloration of a bulk Al under ablation with a femtosecond laser has been reported recently [12]. The authors attribute the yellow coloration of Al to the formation of periodic ripples decorated with NS. The spectral reflectivity of the laser-treated sample is measured at near-normal incidence with poor resolution, so the UV features of Al NS presented in the present work were not observed. Coloration is ascribed either to different ripple periods or just to grooving the Al surface with various periods.

A comparison of laser ablation of bulk Al target in air and liquids has been reported recently [5]. It is found that coloration of Al is gained under laser action in both media. However, in case of laser exposure in air the target surface is covered by significant amount of sputtered material. NS generated in air are less compact presumably due to higher amount of oxide. The similarity of the spectrum under ablation of Al in air and in liquids suggests that the mechanism governing the NS formation is the same, independent of the medium. In all cases NS are formed due to spatial instability of the evaporation and appear as a result of action of recoil pressure onto a thin molten layer on Al. When the exposure is performed in liquids this pressure is the pressure of vapor of the surrounding liquid. In case of exposure in air this pressure is due to fast evaporation of the molten layer itself. That is why the laser fluence required to produce NS on Al in liquid is several times lower than in case of air exposure. NS in this case are formed owing to recoil pressure of metal vapors. Apparently, this process is also characterized by evaporation instability leading to formation of self-organized nano-sized areas within the evaporation cloud.

The optical absorption characteristics of a Ag surface decorated with laser-generated NS originate from the combination of plasmon resonance of electrons in bulk and nano-structured Ag. The peak at 370 nm is attributed to the generated Ag nanostructures due to the confinement of plasmon oscillations inside them [1]. This confinement occurs provided that the mean size exceeds the electron mean free path. Furthermore, the shape of the plasmon resonance peak of bulk Ag is modified by the generated nanostructures since the sharp plasmon peak at 315 nm is broadened and shifted to shorter wavelengths. The observed peak broadening evidences the enhanced scattering of electrons in the skin layer of the target, most likely due to presence of nanostructures.

The position of absorption peak of Ag nanospikes produced by ablation in ethanol can be explained by the pyrolysis of ethanol on Ag. This leads to the formation of products with high molecular mass, as already reported for the ablation of a brass target in ethanol [13]. Most probably these products are polyethylene of low molecular masses. Their deposition on Ag nanospikes also alters the refractive index of the surrounding medium and accounts for the observed red shift of the plasmon resonance.

The variation of the spectral reflectivity of the laser exposed Ag surface upon aging is due to the oxidation of nanostructures by air oxygen. Indeed, the nanospikes have a high specific surface and react easily with oxygen even at ambient temperature. The most common Ag oxide is Ag_2O which has a higher refractive index in the visible. Partial oxidation of nanospikes alters the medium around them, and the plasmon oscillations frequency shifts to the red (lower frequencies). For similar reasons, the coloration of the laser-exposed Ag plate is more pronounced when the target is immersed in a liquid. In thar case, the surrounding liquid has a higher refractive index than air, which causes the red shift of its absorption peak. The position of the peak at 370 nm for a dry Ag surface with laser-generated nanospikes agrees with the calculated position of the plasmon resonance of Ag nanoparticles in vacuum [8]. On the other hand, it is known that any surface exposed to air humidity is covered by several monolayers of water, and these layers may shift the plasmon resonance. In the absence of any adsorbates, the absorption peak of Ag nanostructures should be shifted to higher frequencies.

The optical characteristics of an Al surface after its ablation in liquids can again be explained by a plasmon resonance absorption mechanism. The theoretical position of plasmon resonance of Al NP of 10 nm in diameter in water was calculated in [8], and the maximum of absorption lies around 200 nm (more precisely – in vacuum UV region). However, it is red-shifted for NP of higher diameters. Oxidation of Al NP would also cause a red shift of their plasmon resonance, since its oxide, Al_2O_3, has higher refractive index in the UV range than water. It should be noted that oxide itself has no absorption in the range of study since its absorption only commences from 250 nm and even shorter wavelength, depending on its impurities. The presence of aluminum oxide is indicated by fluorescence measurements performed on the exposed surfaces. This oxide layer may be formed by fast oxidation of the irradiated surface upon solidification and efficiently passivates the surface of NS against further oxidation, so they are chemically stable to provide permanent coloration of Al surface. No fluorescence is detected in the case of a pristine Al target [6].

It is pertinent to note that the most probable effect caused by nano-structuring of a metallic surface is its yellow coloration. Plasmon frequencies for the majority of metallic NPs are situated in UV with the exception of only three metals, Ag, Au, and Cu. Therefore, nano-structured metallic surfaces may show enhanced absorption in the blue region of spectrum, which is closer to UV. Enhanced absorption in the blue region corresponds to yellow coloration of the surface and that is what is observed for Al and Ag. Of course, quadruple plasmon resonance of electrons in NS is usually shifted to the red region compared to the dipole one. However, the intensity of this absorption band is much weaker.

NS on Non-Metallic Targets

A recent paper [14] reported that 800nm, 100 fs laser irradiation of Si surfaces immersed in water gives rise to the formation of high-density regular arrays of nanometer-scale rods that are much smaller than the laser wavelength. The role of water environment is misunderstood by the authors; they believe that water helps faster cooling of the laser-heated areas, while it is evident that this is of least importance and may be significant only for reduction of the average temperature of the substrate.

We have conducted our own experiments on ablation of Si in ethanol using a Ti:Sapphire femtosecond laser. At fluences in the range 2-4 J/cm^2, straight ripples form with a spacing of about 120 nm (Figure 21). The long axis of the ripples is perpendicular to the laser polarization. If the ripples shown in figure 21 are irradiated for a second time in water with a polarization rotated by 90° so that it becomes parallel with the long axis of the ripples, the ripples break up into a surface that is uniformly covered with nanometer-scale rods. In general, NS generated on Si under its ablation in water with fs laser radiation show the same morphology as Ta NS obtained with fs pulses. Namely, periodic ripples coexist with spherical NS situated on top of the ripples (see figure 10). The spacing between ripples is much less than the laser wavelength. It is pertinent to note that laser ablation of materials that do not have a definite melting point results only in formation of ripples without any spherical NS.

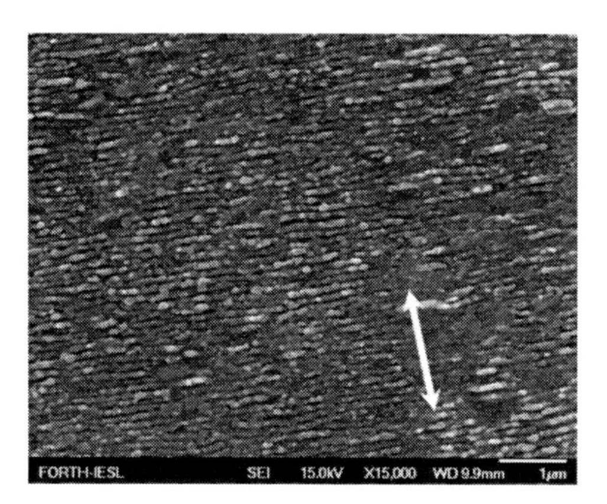

Figure 21. Single crystal Si ablated in water with femtosecond laser radiation at fluence of 2 J/cm^2. The arrow indicates the direction of laser beam polarization.

APPLICATIONS OF NS

Besides evident fundamental interest, NS that are formed under laser-induced phase transitions at the solid-liquid interface possess promising potential for different applications. Nano-sized metal features are responsible for so called Surface-Enhanced Raman Scattering (SERS) due to local amplification of the electromagnetic waves in the their vicinity. As a result, the intensity of a pumping laser beam is enhanced giving rise to a corresponding amplification of the intensity of the weak Raman-shifted scattering signal. The total enhancement factor amounts to 10^5 -10^6.

Silver and gold are the most popular metals used in SERS experiments. SERS experiments are usually carried out with freshly prepared colloidal metallic NP in absence of surface-active substances. However, the measurements can be conducted only once, since the same colloidal solution of NPs cannot be used twice due to absorbed organic molecules. Ag and Au NS generated via laser ablation at the solid-liquid interface (see figure 4, a and b), can potentially be a promising alternative for studying SERS activity since they offer the advantageous capability of performing multiple measurements on the same nano-structured substrate.

Tests were carried out on a freshly-prepared NS on bulk Ag plate with the help of its ablation in water with a 90 ps Nd:YAG laser. The probe molecule used was acridine dissolved in water.

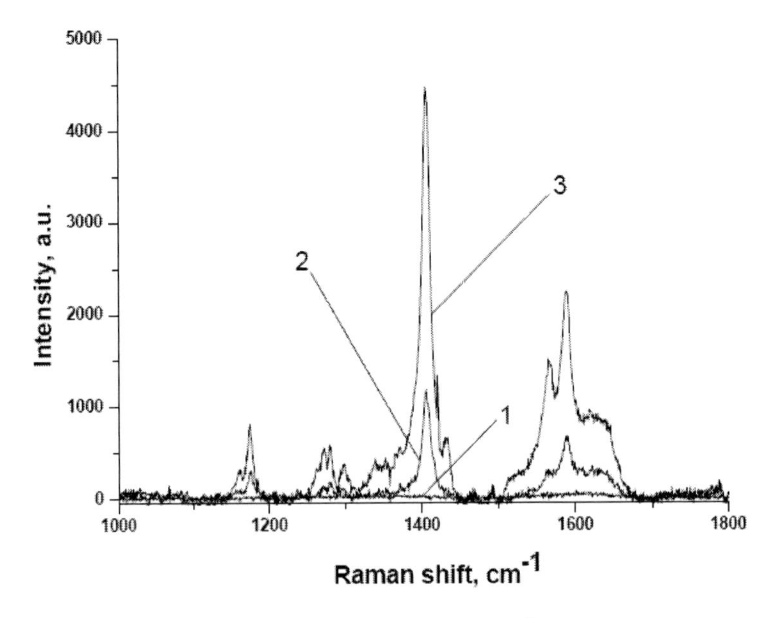

Figure 22. Raman spectra of acridine molecules (concentration of 10^{-5} M/l). Excitation by a cw Ar ion laser, wavelength of 514 nm, acquisition time of 30 s. NS on Ag were generated by exposure under water to the radiation of a 90 ps Nd:YAG laser. Initial Ag surface (1), NS generated with 2000 laser shots (2) and 6000 laser shots (3) [9].

Nanostructuring of some metals is highly desirable for various medical applications. Proliferation of biological cells proceeds more efficiently if the surface has a certain value of roughness. It was found that the optimal roughness lies in the range of several hundreds of nm. In this sense formation of NS under laser ablation of solids in liquids ideally suits the needs for fabrication of efficient medical implants. Tantalum is perfectly biocompatible with living cells owing to its high chemical stability. However, its high cost limits its use as an implant. More common biocompatible materials are Ti alloys. Figure 23 shows the example of successful nano-structuring of a medical Ti alloy by its exposure to short laser pulses in ethanol. One can see that NS on Ti alloy consist of periodic ripples decorated on top with spherical NS typical of other solids. Therefore, the initial experiments on nano-structuring of biocompatible materials were promising in this respect.

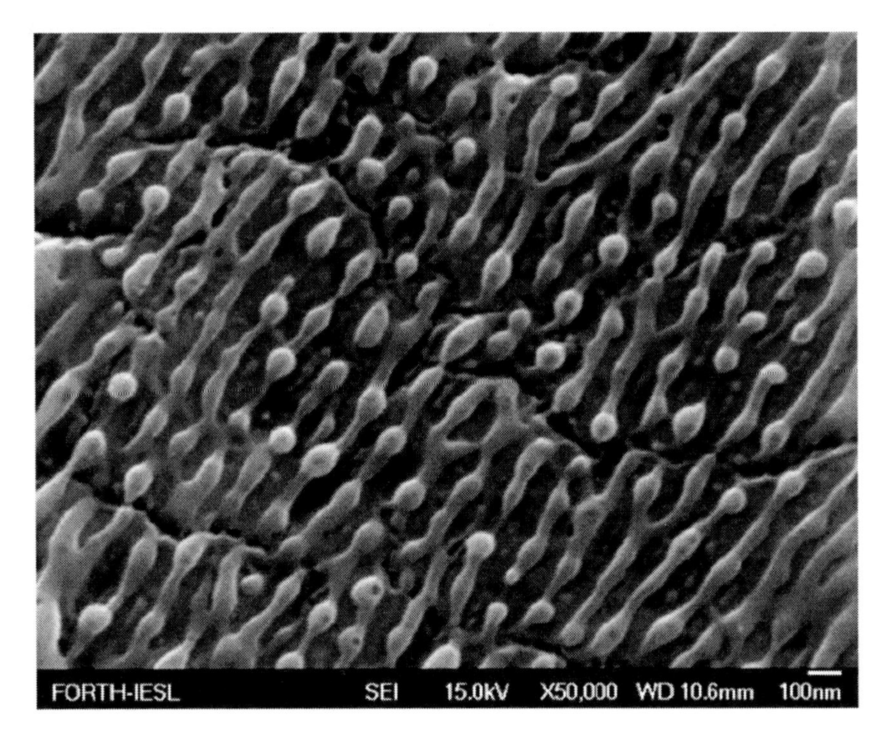

Figure 23. Nanostructuring of medical Ti alloy with the help of its ablation in ethanol with a 5 ps radiation of a KrF laser at 248 nm.

Another promising field of applications is the control over wettability of solid surfaces via laser-assisted generation of NS. This approach is very advantageous since the rate of NS generation is very high and can be carried out in the "laser writing" regime.

Conclusion

It is shown that laser-induced phase transitions at the solid-liquid interface are characterized by instability in evaporation of the liquid that surrounds the solid. Smooth temperature profile leads to formation of regions with different pressure and velocity gradients within the evaporated liquid. The melting of underlying targets serve as a recording medium to visualize these regions. This can be achieved only with sufficiently short laser pulses, since surface tension of the melt tends to dump nano-sized perturbations of its surface. Provided that the cooling rate of the target is sufficiently high, then these perturbations can be “frozen” as nanostructures. The above results show that the upper value of laser pulse width suitable for observations of NS is around 300 - 400 ps; apparently, with longer pulses, the NS are suppressed by the surface tension. The lateral size and period of NS depends on the duration of laser pulse. As a rule, the average lateral size of NS decreases with the decrease of laser fluence towards the melting threshold of the target material. In case of femtosecond laser pulses, two types of structures are observed simultaneously, namely, sub-wavelength period ripples and spherical NS. NS has been successfully realized on the following metals: Ag, Au, Al, Ta, Ti, Zn, W, and Ni. Formation of NS under laser-induced phase transitions at the solid-liquid interface is accompanied by modification of the reflectivity of the target. As a result, the surface of many metals with NS becomes colorized due to plasmon resonance of free electrons in NS. NS generated under ablation of the solid-liquid interface with short (sub-ns) laser pulses may find wide range of applications, e.g., as Raman plates and medical implants.

It is important to note that the study of phase transitions through NS left on the target surface provides complementary information about the phenomenon, since the lateral dimensions of NS are much smaller than the spatial resolution of optical and acoustical methods of characterization (see Chapters 1, 2, and 4).

ACKNOWLEDGMENTS

This work was supported in part by Russian Foundation for Basic Research, grants ## 07-02-00757, 07-02-12209, and by Scientific School 8108.2006.2. Dr. V.V. Popov is thanked for fabrication of pre-patterned Ni substrates. We are grateful to Professor C. Fotakis for helpful discussions of the results. Dr. A.V. Simakin, E.V. Barmina, P.G. Kuzmin and M. Barberoglou are thanked for their help in the experiments and data processing.

REFERENCES

[1] Zavedeev E.V.; Petrovskaya A.V.; Simakin A.V.; and Shafeev G.A.; *Quantum Electronics*, 2006 36(10) 978.

[2] Barmina E.V.; Barberoglou M.; Zorba V.; Simakin A.V.; Stratakis E.; Fotakis C.; and Shafeev G.A.; *Quantum Electronics*, 2009 39(1) 89-93.

[3] Shafeev G.A., Laser-based formation of nanoparticles, in: *Lasers in Chemistry*, Volume 2: *Influencing matter*. Edited by M. Lackner, Wiley VCH Verlag GmbH&Co, KGaA, Wienheim, ISBN: 978-3-527-31997-8, 2008, 713 – 741.

[4] Rekhviashvili S.S. and Kishtikova E.V., *Technical Physics Letters*, 2006 32(10) 50.

[5] Stratakis E.; Zorba V.; Barberoglou M.; Fotakis C.; and Shafeev G.A.; *Appl. Surf. Sci.* 2009 255 5346-5350

[6] Stratakis E.; Zorba V.; Barberoglou M.; Fotakis C.; and Shafeev G.A.; *Nanotechnology* 2009 20 105303.

[7] Shafeev G.A., Formation of nanoparticles under laser ablation of solids in liquids, in: *Nanoparticles:New Research*, editor Simone Luca Lombardi, pp. 1 -37, ISBN: 978-1-60456 – 704 – 5, 2008, Nova Science Publishers, Inc.

[8] Creighton J.A. and Eadon D.G.; *J. Chem. Soc. Faraday Trans.*, 1991 87 3881.
[9] Lau Truong S. ; Levi G.; Bozon-Verduraz F.; Petrovskaya A.V.; Simakin A.V.; and Shafeev G.A. ; *Applied Surface Science* 2007 254 1236.
[10] Lau Truong S. ; Levi G.; Bozon-Verduraz F.; Petrovskaya A.V.; Simakin A.V.; and Shafeev G.A. ; *Appl.Phys*. 2007 A89 (2) 373.
[11] Stratakis E.; Barberoglou M., Fotakis C.; Viau G.; Garcia C.; and Shafeev G.A.; Optics Express 2009 17 12650.
[12] Vorobyev. A. Y. and Guo C.; *Appl.Phys. Lett.* 2008 92 041914.
[13] Kazakevich P.V.; Voronov V.V.; Simakin A.V.; Shafeev G.A.; *Quantum Electronics*, 2004 34 951.
[14] Shen M.; Carey J. E.; Crouch C. H.; Kandyla M.; Stone H. A.; Mazur E.; *Nano Lett.*, 2008 8 (7) 2087.

In: Phase Transitions Induced by Short Laser Pulses
Editor: Georgy A. Shafeev
ISBN: 978-1-60741-590-9

Chapter 6

LASER-INDUCED SELF-ORGANIZATION OF NANO- AND MICRO-STRUCTURES OF SURFACE RELIEF VIA DEFECT-DEFORMATIONAL INSTABILITIES

Vladimir I. Emel'yanov*
Department of Physics, Moscow State University, Moscow, 119991 Russia

ABSTRACT

The Defect-Deformational (DD) approach to the problem of laser-induced formation of nano-and microstructures of surface relief in semiconductors and metals is presented. It is demonstrated that laser or ion beam-created stressed flat surface layer with point defects exhibits a threshold (in respect to the defect concentration) transition to a spatially periodic bent state with a simultaneous formation of the periodic defect piles up at the extrema of the spontaneously emerging surface relief (DD grating). The layer deformation corresponds to the displacements in a static bending quasi-Lamb wave and the deformation of the underlying elastic continuum corresponds to the displacements in the static quasi-Rayleigh wave. It is shown that not far from the instability threshold the DD selforganization is described by the closed nonlinear equation of Kuramoto-Sivashinsky type. More general analysis simultaneously involving the nonlocal character of the defect interaction with the lattice atoms and both (normal and lateral) defect-induced forces that cause the bending of the surface layer yields two maxima on the curve of the instability growth rate versus the period of the generated relief. This corresponds to the experimentally observed two scales of the surface relief modulation upon the laser and ion irradiation of semiconductors. Based on the results obtained, we propose a cooperative DD mechanism for the formation of an ensemble of the nanoparticle nucleation centers above the critical levels of the stress or the defect concentration. A new approach to the calculation of a bimodal distribution function of the nanoparticles with respect to their size is developed adequate to the DD mechanism of nucleation which expresses the distribution function through the growth rate. Nonlinear three DD gratings interactions are shown to lead to generation of second harmonic of surface relief and mixing of DD gratings wave vectors. The developed theory of DD

* e-mail: emel@em.msk.ru;

instability of the surface layer is applied for the interpretation from the unified viewpoint of experimental results obtained in studies of formation of ordered nano and microstructures on the surface of semiconductors under the action of laser pulses with different duration and fluencies. The DD structures symmetry and its evolution with increase of laser fluence and magnitude of external anisotropic stress are discussed. Similarities with formation of nanostructures under ion beam irradiation are also discussed in the framework of DD instability theory.

PACS numbers: 81.16.Rf; 68.55.Ac; 68.43.Hn; 61.80.Ba

INTRODUCTION

The formation of ordered surface nano- and microstructures on semiconductors and metals under pulsed laser irradiation has been the subject of intense interest. One may distinguish three laser-irradiation regimes that differ by laser fluencies needed for the nano- and microstructuring of semiconductor surfaces. In the first regime of a relatively low (subthreshold) fluency, the formation of surface structures results from multipulse irradiation in the absence of the melting of solid [1, 2]. In the second (single-pulse) regime, a strongly absorbing semiconductor is irradiated at medium fluencies lying in the vicinity of the melting threshold or slightly exceeding this level [3]. In the third (multipulse) regime, the pulse fluency is slightly higher than the ablation threshold and is substantially higher than the melting threshold. In the last regime, the formation of similar surface microstructures under the multipulse irradiation of semiconductors with femtosecond [4-6], picosecond [7], and nanosecond [8] laser pulses is observed.

We expose in this chapter the Defect-Deformational (DD) approach to the problem of description of laser-induced formation of nano- and microstructures of surface relief in semiconductors and metals [9-12]. It is based on the notion that laser-induced processes of the nano-and microstructures self-organization on solid surfaces are started from the generation of a surface nano- or micrometer thickness surface layer with laser-generated mobile point defects (interstitials, vacancies, electron-hole pars and doped atoms). Similar situation takes place upon ion-beam irradiation of solids. The radiation-induced surface layer that is saturated with defects has the lattice constant that differs from the lattice constant of the underlying crystal (the substrate). This leads to occurrence of a mechanical stress in the surface layer.

This defected layer can be created in different regimes of laser irradiation characterized by absorption length, pulse fluency and duration and number of pulses used. Correspondingly, different are the mechanisms of defect generation and types of point defects involved. Two thresholds manifest the change of modes and intensity of defect generation in semiconductors. The first one is the threshold fluency of plastic deformation and the second one is the threshold fluency of melting. We briefly discuss in this chapter how in conditions of multipulse irradiation above and below the melting threshold the defect-enriched layer is created, and we determine its thickness h which can lie in nanometer or micrometer range.

Due to strongly nonequilibrium conditions created by energy input (elevated temperatures, stress, recombination-enhanced mobility) defects in this layer h are highly mobile. We show that this flat tensile stressed surface layer saturated with mobile point defects becomes unstable above certain critical levels of the defect concentration and the layer exhibits a transition to a periodically bent state with the simultaneous accumulation of defects at the relief extrema (the surface DD instability). The medium displacements inside the layer and the substrate are determined as in the bending quasi-Lamb wave and the quasi-Rayleigh wave, respectively. The resulting coupled static Lamb-Rayleigh deformations in the layer and the substrate are maintained due to the self-consistent distribution of point defects that deform the elastic continuum. Such a deformed state of the layer and the substrate represents a static analog of a dynamic Lamb-Rayleigh wave that propagates in a thin surface layer whose density is higher than the substrate density [13]. In the considered case of DD instability, the defect enriched layer (the film) also has elastic characteristics different from the underlying part of the sample (the substrate). So the DD model considered in this chapter can be referred to as "the film on substrate" model.

The physical mechanism of the surface DD instability consists of the following [9]. Initially, mobile point defects are distributed uniformly along the surface. The fluctuating local increase of surface defect concentration leads to the rising of surface relief corrugation and corresponding surface strain. The defects interact, owing to the deformation potential, with this long-range surface strain field. The strain gives rise to the lateral defect flux, proportional to the defect concentration and directed opposite to the diffusion flux. At low defect concentration the latter exceeds the strain-induced flux and fluctuations of the spatial defect distribution decay so the surface remains flat.

The second stage of surface structuring begins when the concentration of mobile defects exceeds a certain critical value. At this critical point the lateral, strain-induced defect flux starts to exceed the lateral diffusion flux. Due to this DD instability develops, in which defects (interstitials, vacancies or e-h pairs) auto-localize in periodic self-consistent strain wells. In doing so, the homogeneous lateral distribution of defects makes a transition to a spatially periodic (along the surface) state. This is accompanied by the appearance of a periodic corrugation of the surface with defects piling up at extrema of the surface relief. In doing so interstitials are piled up at hillocks and vacancies at valleys of the surface relief. This periodic corrugation of the surface, coupled with the periodic piling up of defects, comprises the DD grating with wave vector $\boldsymbol{q}$ (Figure 1). When the DD instability is developed, the DD-grating amplitudes increase with time as $\exp \lambda_q t$, where λ_q is the growth rate. The value $q = q_m$ corresponding to the maximum growth rate determines the period of dominant gratings $\Lambda_m = 2\pi/q_m$ which are selected in Fourier spectrum of the surface relief. Salient feature of the surface DD instability is a proportionality of the characteristic scale Λ_m to the layer thickness h.

A superposition of the surface DD gratings with different $\boldsymbol{q}$ yields a cellular or lamellar seed DD structure on the surface (see figure 3). The characteristic scale of heterogeneities therein is determined by Λ_m and its symmetry can be changed by a selection of the $\boldsymbol{q}$ vector directions (see figure 8).

The analysis simultaneously involving the nonlocal character of the defect interaction with the crystal lattice atoms up to the fourth-order term in the expansion of the kernel of the interaction operator and both (normal and lateral) defect-induced forces that cause the layer bending yields two maxima on the curve of the instability growth rate versus relief period

Λ far enough above the instability threshold (far above the critical defect concentration or the critical stress) [12]. In contrast, the growth rate exhibits only a single maximum not far above the threshold. When the periodic surface relief is induced by energy beams, two maxima of the growth rate must give rise to two scales of the relief modulation with the large(micron)-scale modulation supplementing the small(nano)-scale modulation. The occurrence of two scales of the surface relief modulation is a characteristic feature of the nanorelief self-organization upon both laser and ion-beam irradiation of semiconductors as was pointed out in [11]. Due to the triple interactions between DD gratings further enrichment of spectrum of surface relief harmonics can occur due to generation of second harmonic and mixing of wave vectors of interacting DD gratings (see [14] and Secs.3 and 6.4).

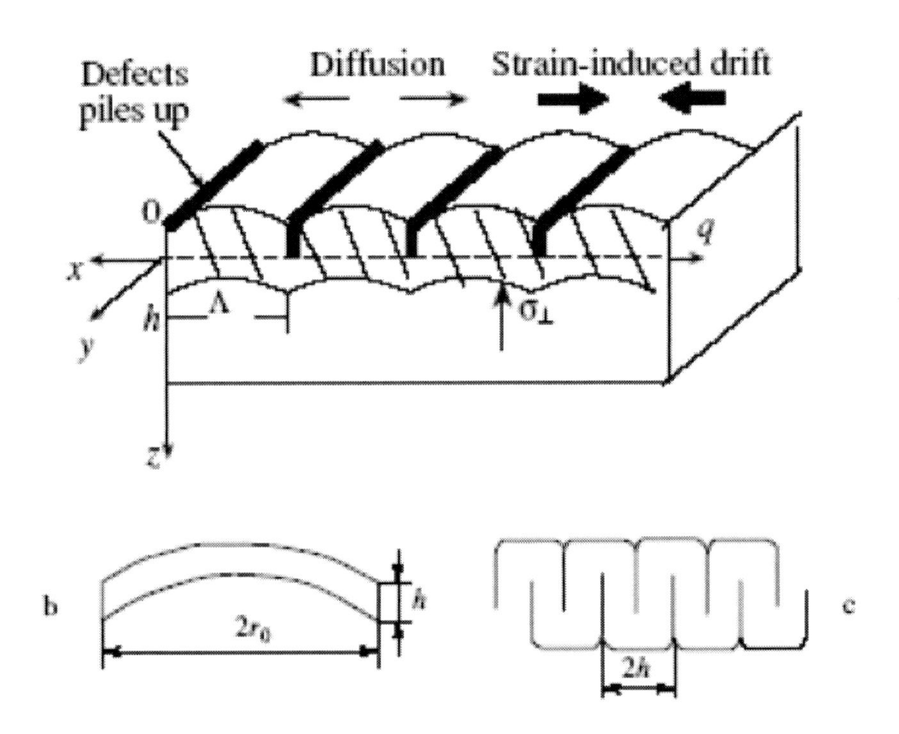

Figure 1. The periodically bent state of the surface layer h with defects formed due to DD instability (DD grating with wave vector $\boldsymbol{q}$). The diffusion tends to smooth out defects piles up, nonlinear strain-induced drift is directed oppositely to diffusion flux and leads under exceeding of critical defect concentration to auto-localization of defects in self-consistent periodic potential well. $\sigma_{\perp}$ is the normal stress exerted on the surface layer by the substrate; a film in the state corresponding to the first bending mode from the spectrum of bending modes (b), r_0 is the radius of laser beam; a film in the state corresponding to the limiting bending mode (c). The limiting mode, which yields the relief-modulation period, $\Lambda = 2h$ can be observed in experiments (see figure 7).

We develop in this chapter a new approach to the calculation of the size distribution function of the nanoparticles formed due to laser or ion-beam irradiation [12]. This approach is adequate to the cooperative DD character of the formation of the nucleation centers ensemble and makes it possible to represent the distribution function in terms of the DD instability growth rate $\lambda(\Lambda)$. Function $\lambda(\Lambda)$ exhibits one or two maxima depending on the irradiation conditions. Therefore, the calculated function $n_{\text{dot}}(\Lambda)$ also exhibits one or two maxima. Note that, under certain regimes of the pulsed laser nanostructuring of solid surfaces, the size distribution function of nanohills exhibits a transition from the unimodal to bimodal shape upon a variation in the irradiation conditions [15] (Sec.6.3).

In the case of strong and prolong enough laser or ion-beam irradiation the seeding DD structure is etched, so that the regions of the defect accumulation are etched at a rate that differs from the etching rates in other regions. The etching leads to the visualization of the seed DD structure, which imposes its periodicity and symmetry to the resulting permanent structure of the surface relief [11]. Due to this it is possible with the help of computer analysis of SEM or AFM images of irradiated surface to test the predictions of the DD instability theory against experimental data. We consider a few examples of laser-induced nano and microstructures formation and make comparison of predictions of the above DD theory and experiment. In particular, the generation of relief second harmonics and mixing of wave vectors are revealed and described (Sec.6.4).

The chapter is organized as following. In Sec.1 the starting equations of surface DD instability are formulated, in Sec.2 the basic set of coupled equations for Fourier amplitudes of surface defect concentration field is derived and analyzed in the linear approximation and existing of two scales of relief modulation is established. Sec.3 is devoted to consideration of the nonlinear regime of DD instability in which three DD gratings interactions lead to generation of surface relief harmonics and mixing of wave vectors. In Sec.4 it is shown that not far from the threshold, the DD instability is described by the closed nonlinear equation for the defect concentration, which, in the mean field approximation, is reduced to the form of well known Kuramoto-Sivashinsky equation. In Sec.5 the bimodal size distribution function of nanoparticles formed due to DD instability is obtained. In Sec.6 the developed theory of DD instability of the surface layer is applied for the interpretation from the unified viewpoint of some experimental results obtained in studies of formation of ordered nano and microstructures on the surface of semiconductors under the action of laser pulses with different duration and fluencies. Of particular interest will be the bimodal dependence of the growth rate of DD gratings on its period and corresponding two scales modulation of the surface relief, the size distribution function and its transformation to a bimodal one with change of irradiation conditions and also the generation of harmonics of surface relief due to three-wave DD interactions. These new topics are not covered in the recent review [11], in which the systematic comparison of DD theory and experiment is carried out with focus made on the universal linear dependence of the period of the surface structures on the thickness of mobile point defect-enriched subsurface layer formed due to the laser or ion beam action.

1. Equations of the DD Instability of the Stressed Surface Layer with Mobile Defects

We assume that the laser or ion beam irradiation of a crystal leads to the generation of point defects with concentration n_d (d = v and d=i for vacancies and interstitials, respectively) in the surface layer with thickness h (Sec.6.1). The plane $z = 0$ coincides with the free surface of the sample and the z-axis is directed from the surface to the bulk (Figure 1).

The spatial distribution of defect concentration in this surface defect layer is written in the form

$$n_d(x,y,z,t) \equiv N_d(x,y,t) f(z), \tag{1}$$

where $N_d(x,y,t)$ is the surface concentration of the defects and a function $f(z)$ determining the defect distribution along the normal to the layer will be defined below (see (14)).

The equations of the model are written on the assumption of the isotropic surface diffusion and drift. Surface flux of defects $\mathbf{j}_d$ consists of the diffusion- and deformation-induced components:

$$\mathbf{j}_d = -D_d \nabla N_d + N_d \frac{D_d}{k_B T} \mathbf{F}(\mathbf{r}). \tag{2}$$

Here, the lateral nonlocal force that is exerted on the defect by the deformed elastic continuum is given by

$$\mathbf{F}(\mathbf{r}) = \theta_d \nabla \left(\xi_f + l_d^2 \Delta \xi_f + L_d{}^4 \Delta^2 \xi_f \right)_{z=0}, \tag{3}$$

where $\mathbf{r} = (x,y)$, $\nabla \equiv \hat{e}_x \frac{\partial}{\partial x} + \hat{e}_y \frac{\partial}{\partial y}$, $\hat{e}_x$ and $\hat{e}_y$ are unit vectors along the x and y axes, $\theta_d = \Omega_d K$ is the deformation potential of the defect, Ω_d is a change of the volume of the medium due to the formation of one defect, K is the elasticity modulus, $\xi_f = \xi_f(x,y,z) = (\text{div}\ \mathbf{u}_f)$ is the strain in the layer, and $\mathbf{u}_f = \mathbf{u}_f\ \ x,y,z,t$ is the material displacement vector in the layer. In formula (3) we introduce parameters l_d^2 and $L_d{}^4$ which describe the nonlocal interaction of defects with the crystal lattice atoms and are assumed to be given. In the microscopic approach, they result from the expansion of the nonlocal deformation that enters the integrand along with the interaction kernel that decays at the characteristics lengths of the defect--atom interaction $\sim l_d$ and L_d. The typical values are several nanometers (see, for example, [16]).

Using expression (3) in the continuity equation for N_d, we obtain the surface-diffusion equation allowing for the strain-induced drift of the defects:

$$\frac{\partial N_d}{\partial t} = D_d \Delta N_d - \gamma_d N_d - \frac{D_d \theta_d}{k T_B} \text{div}\left[N_d \nabla \left(\xi_f + l_d^2 \Delta \xi_f + L_d{}^4 \Delta^2 \xi_f \right) \right]_{z=0} \tag{4}$$

where $\Delta = \frac{\partial^2}{\partial x^2} + \frac{\partial^2}{\partial y^2}$, D_d is the surface diffusion coefficient, and $\gamma_d = \tau_d^{-1}$, τ_d is the defect lifetime. We neglect the renormalization of the diffusion coefficient due to deformation.

To find $\xi_f = \text{div}\, \mathbf{u}_f$, we will consider the defect-enriched layer with thickness *h* as a film with density ρ and Young modulus *E*. The film is connected to the substrate (the remaining part of the crystal) with elastic parameters ρ_S and E_S. The free surface of the film is $z = 0$, and the deformation in the substrate is described using the material displacement vector $\mathbf{u}\ x, y, z, t$. We assume that the film exhibits the bending deformation and that the *z* component of the film displacement vector is $u_{fz} = \zeta\ x, y$ where ζ - is the bending coordinate of the film (the *z* displacement of the points on the median plane of the film from the equilibrium position).

The shear stresses in the film and in the substrate are equal to each other at $z = h$:

$$\mu_f(z = h)(\partial u_{fx_\alpha} / \partial z + \partial u_{fz} / \partial x_\alpha)_{z=h} = \mu_s(z = h)(\partial u_{x_\alpha} / \partial z + \partial u_z / \partial x_\alpha)_{z=h} \quad (5)$$

where $x_\alpha = \{x,y\}$, $\mu_f(z = h)$ and $\mu_s(z = h)$ are shear moduli in the film and substrate, respectively, at $z = h$.

We assume that $\mu_s(z = h) \to 0$ due to the generation of dislocations (e.g., misfit dislocations) at the film--substrate interface or strain-induced pumping of vacancies from the surface to the bulk (see sec.6.1). Then, expression (5) yields $(\partial u_{fx_\alpha} / \partial z + \partial u_{fz} / \partial x_\alpha)_{z=h} = 0$. The same zero boundary condition for the shear component of the stress tensor is satisfied at the free film surface $z = 0$. Besides, to determine the film deformation in the zero approximation in respect to the substrate reaction along the normal, we assume that normal component of the stress tensor in the film is zero at the interface:

$$\sigma_{fzz}(x, y, z = h) \equiv (\sigma_f)_\perp = 0. \quad (6)$$

At the free surface, we also have $\sigma_{fzz}(x, y, z = 0) = 0$. Note that a conventional approximation in the theory of thin-film bending [17] involves a neglect of the external force exerted along the normal to the film in the boundary condition upon derivation of the thin-film deformation under the action of this force. Thus, in the above approximation, the free-film condition $\sigma_{fxz} = \sigma_{fyz} = \sigma_{fzz} = 0$ is satisfied on both film interfaces. Then, the film strain $\xi_f = \text{div}\, \mathbf{u}_f$ is represented as [17]:

$$\xi_f = -\nu(z - \frac{h}{2})\Delta\zeta, \quad (7)$$

where $\nu = (1-2\sigma_p)/(1-\sigma_p)$, σ_p is the Poisson coefficient of the film. The linear sign-alternating deformation in the layer as a function of z (expression (7)) is characteristic of the Lamb wave in plates [13].

Substituting expression (7) in formula (4), we derive the equation

$$\frac{\partial N_d}{\partial t} = D_d \Delta N_d - \frac{N_d}{\tau_d} - \frac{\nu h D_d \theta_d}{2kT_B} \operatorname{div}\left[N_d \nabla\left(\Delta\zeta + l_d^2 \Delta^2 \zeta + L_d{}^4 \Delta^3 \zeta \right)\right]. \quad (8)$$

Aiming to find the growth rate of the DD structure growth, we restrict consideration to the initial (linear) regime of the DD instability. The following linear equation for the ζ coordinate follows from the generalization of the conventional equation for bending of a free film [17]:

$$\frac{\partial^2 \zeta}{\partial t^2} + l_0{}^2 c^2 \Delta^2 \zeta - \frac{\sigma_{||}}{\rho_f} \Delta\zeta = \frac{\sigma_\perp}{\rho_f h} - \sum_d \left\{ \frac{\theta_d}{\rho_f h} \int_0^h \frac{\partial n_d}{\partial z} dz + \frac{\nu \theta_d}{\rho_f h} \int_0^h \left(z - \frac{h}{2} \right) \Delta n_d dz \right\}, \quad (9)$$

where $c^2 = E_f / \rho_f \; 1-\sigma_f{}^2$ is the film rigidity, E_f is the Young modulus, and $l_0{}^2 = h^2/12$. The summation on the right-hand side involves vacancies (d = v) and interstitials (d = i). Note that the film bending rigidity (the coefficient of $\Delta^2\zeta$) depends on film thickness h that serves as a scale parameter specific of the DD instability.

The generalization of the conventional bending equation is as follows. On the left-hand side of expression (9), the term proportional to $\sigma_{||}$ takes into account the effect of the lateral stress in the film resulting from the misfit of the lattice parameters of the film and substrate and/or the defect generation in the surface layer. We assume that $\sigma_{||} > 0$, so that the film is under tensile stress which is assumed to be known. In the first term on the right-hand side of expression (9), $\sigma_\perp$ is the stress that is normal to the film surface and that arises due to the substrate action on the film (substrate reaction). Note that in the analysis of the film bending, expression (9) takes into account the substrate reaction, which has been above neglected in the consideration of the internal deformations in the film. Such an approach is not controversial and can be substantiated. The second term in formula (9) takes into account the defect-induced bending force that acts along the normal to the film surface due to a nonuniform distribution of the defects along the z axis. The third term on the right-hand side takes into account the defect-induced bending lateral force resulting from a nonuniform distribution of the defects along the film. Thus, Eq. (9) takes into account both forces (normal and tangential) exerted on the film by the defect subsystem.

The film bending gives rise to displacement vector $\boldsymbol{u}$ in the substrate, which obeys the equation

$$\frac{\partial^2 \boldsymbol{u}}{\partial t^2} = c_t{}^2 \Delta \boldsymbol{u} + \left(c_l{}^2 - c_t{}^2 \right) \operatorname{grad} \operatorname{div} \boldsymbol{u}, \quad (10)$$

where c_l and c_t are longitudinal and transverse velocities of sound in the substrate.

We have three boundary conditions at the film-substrate interface. The *z* displacement is continuous, so that

$$u_z(z=h)=\zeta\,. \tag{11}$$

The normal stress in the substrate determines the force exerted on the film along the *z* axis:

$$\left[\frac{\partial u_z}{\partial z}+(1-2\beta_s)\left(\frac{\partial u_x}{\partial x}+\frac{\partial u_y}{\partial y}\right)\right]_{z=h}=\frac{\sigma_\perp(x,y)}{\rho_s c_l^2}, \tag{12}$$

where $\beta_s=c_t^2/c_l^2$.

The tangential stress at the interface is zero:

$$\left[\frac{\partial u_{x_\alpha}}{\partial z}+\frac{\partial u_z}{\partial u_{x_\alpha}}\right]_{z=h}=0 \tag{13}$$

where $x_\alpha=\{x,y\}$.

We do not impose limitations on the tangential components of the displacement vector in the substrate at the interface $u_{x_\alpha}(z=h)$. This assumption and condition (13) follow from the assumption on the generation of the misfit dislocations at the interface.

The system of equations (8)-(13) represents a closed system of equations that describes the DD instability of a stressed flat thin surface layer with mobile defects. We demonstrate below that such instability is related to the defect-induced instability of the static quasi-Lamb waves in the layer coupled with the static quasi-Rayleigh waves in the underlying elastic continuum.

2. Two Maxima of the Growth Rate of the Surface DD Gratings as the Function of the Wave Number

One can show that since $h<\Lambda$, n_d is rapidly adjusted to the *z* distribution of the bending deformation and is given by antisymmetric function of *z*:

$$n_d(x,y,z,t)=\frac{2}{h}\left(\frac{h}{2}-z\right)N_d(x,y,t)\,. \tag{14}$$

Whence it follows

$$n_d(z=0) = -n_d(z=h) = N_d. \tag{15}$$

Substituting expression (14) to the right-hand side of equation (9) and calculating the integrals with regard to expressions (14) and (15) we obtain the following equation on the assumption that the deformation is instantaneously adjusted to the defect subsystem ($\frac{\partial^2 \zeta}{\partial t^2} = 0$):

$$\Delta^2 \zeta - \frac{1}{l_{||}^2}\Delta\zeta = \frac{\sigma_\perp}{h l_0^2 \rho_f c^2} + \sum_d \left(A_d - \frac{2\nu\theta_d}{\rho_f c^2 h}\Delta\right) N_d, \tag{16}$$

where $A_d = \frac{2\theta_d}{h l_0^2 \rho_f c^2}$ and the characteristic scale parameter is given by

$$l_{||} = h\left(\rho_f c^2 / 12\sigma_{||}\right)^{1/2}. \tag{17}$$

To eliminate substrate reaction $\sigma_\perp$ in the system of equations (12) and (16), we search for a solution to boundary problem (10)-(13). In expression (12), we employ the Fourier-series expansion:

$$\sigma_\perp = \sigma_\perp(\mathbf{r},t) = \sum_{\mathbf{q}} \sigma_\perp(q)\exp(i\mathbf{qr} + \lambda_q t).$$

We search for the displacement vector in the substrate that satisfies Eq. (10) as superposition of quasi-Rayleigh (static) waves that represent modifications of the surface acoustic Rayleigh waves [17]. Each displacement vector in this superposition is represented as a sum of the longitudinal and transverse components ($\boldsymbol{u} = \boldsymbol{u}_l + \boldsymbol{u}_t$) which satisfy the conditions $\mathrm{rot}(\mathbf{u}_l) = 0$ and $\mathrm{div}(\mathbf{u}_t) = 0$. For the longitudinal component, we have

$$u_{lx_\alpha} = -i\sum_{\mathbf{q}} q_{x_\alpha} R(q)\exp\left(i\boldsymbol{qr} - k_l z + \lambda_q t\right),$$

$$u_{lz} = \sum_{\mathbf{q}} k_l R(q)\exp\left(i\boldsymbol{qr} - k_l z + \lambda_q t\right). \tag{18}$$

For the transverse component, we obtain

$$u_{tz\alpha} = -i\sum_{\mathbf{q}} \frac{q_{x_\alpha}}{q} k_t Q\ t\ \exp\ iq\mathbf{r} - k_t z + \lambda_q t\ ,$$

$$u_{tz} = \sum_{\mathbf{q}} qQ\ t\ \exp\ iq\mathbf{r} - k_t z + \lambda_q t \tag{19}$$

where $k_{l,t}^2 = q^2 + \lambda_q^{\ 2}/c_{l,t}^2$ and $R(q)$ and $Q(q)$ are the fluctuation amplitudes. It is seen from expressions (18) and (19) that the frequency is $\omega_q = 0$ in the static quasi-Rayleigh wave.

The solution of the system of equations (10)-(13) in the Fourier representation yields the relationship

$$\zeta_q \left\{ \frac{2k_l k_t - k_t^2 - q^2}{k_t^{\ 2} - q^2\ k_l} (q^2 + k_t^2) - 2k_t \right\} = \frac{\sigma_\perp(q)}{\rho_s c_t^2}.$$

Using the expansion of the expression in braces in terms of the small parameter $\lambda_q^{\ 2}/c_{l,t}^2 q^2 << 1$, we find the reaction stress of the substrate caused by the film bending:

$$\sigma_\perp(q) = \zeta_q 2q(\beta_s - 1)\rho_s c_t^2. \tag{20}$$

Below, we employ the Fourier-series expansions

$$\zeta(\mathbf{r},t) = \sum_{\mathbf{q}} \zeta_{\mathbf{q}} \exp(i\mathbf{qr} + \lambda_q t), \tag{21}$$

$$N_d(\mathbf{r},t) = \sum_{\mathbf{q}} n_d(\mathbf{q}) \exp(i\mathbf{qr} + \lambda_q t) \tag{21a}$$

Formulas (21) and (21a) determine the superpositional DD structure consisting of the coupled 2D DD gratings of the surface relief and defect concentration, respectively. Each DD grating with wave vector $\boldsymbol{q}$ can be interpreted as the bending static Lamb wave with the wavelength $\Lambda = 2\pi/q$ that is maintained by the self-consistent distribution of the defects. Each quasi-Lamb wave (21) is related to the static Rayleigh wave with the same wave vector $\boldsymbol{q}$ from superposition (18) and (19). The Fourier amplitudes of each DD grating with wave vector $\boldsymbol{q}$ and the coupled quasi-Rayleigh wave increase in time with growth rate λ_q The summations in superposition (21) and (21a) as well as in superposition (18) and (19) involve both directions and magnitudes of vectors $\boldsymbol{q}$. Note that the summation with respect to the magnitude $|\mathbf{q}|=q$ is performed in the limits $q_1 \le q \le q_c$, where $q_1 = \pi/L$ is the wave number of the first bending mode (L is the lateral size of the region with mobile defects) and $q_c = \pi/h$ is the wave number of the limiting bending mode. In the last case, the periodically bent film represents an accordion-type structure with the period $\Lambda_c = 2h$ of the corresponding modulation of the surface relief (see figure 1 and figure 7).

Using expression (20) in (16), we find the linear relation between bending coordinate and defects concentrations:

$$\zeta_{\mathbf{q}} = \sum_d \eta_d(\mathbf{q}) N_d(\mathbf{q}), \tag{22}$$

where the coefficient of the DD coupling in the linear approximation is given by

$$\eta_d(\mathbf{q}) = -\frac{2\theta_d}{\rho_f c^2 h l_0^2}(1+\nu l_0^2 q^2)\left[q^4 + l_{||}^{-2} q^2 + \frac{2(1-\beta_s)\mu_s}{h l_0^2 \rho_f c^2} q\right]^{-1}. \tag{23}$$

To derive this formula, we employ the relationship $\mu_s = \rho_s c_t^2$. Thus, the initial amplitudes of the DD grating with the wave vector q are expressed through the initial fluctuation of defect concentrations: $N_d(\mathbf{q}) \equiv N_d(\mathbf{q}, t=0)$.

The term in parentheses in expression (23) takes into account the bending effect of two defect-induced forces (compare with expression (9)). In the denominator in expression (23), the first two terms take into account the effective bending rigidity of the film allowing for the lateral stress in the film (the term proportional to $l_{||}^{-2}$). The last term in the denominator, which is proportional to ($\sim \mu_s$), takes into account the reaction of the elastic substrate to the film bending. Neglecting this effect, we arrive at a simplified DD model (the model of a free film with mobile defects), in which the coefficient of the DD coupling is represented as

$$\eta_d(\mathbf{q}) = -\frac{2\theta_d}{\rho_f c^2 h l_0^2}(1+\nu l_0^2 q^2)\left[q^4 + l_{||}^{-2} q^2\right]^{-1}. \tag{24}$$

The comparison of expressions (23) and (24) yields the condition under which the substrate reaction can be neglected:

$$\sigma_{||} > 2(1-\beta_s)\mu_s / qh. \tag{24a}$$

The shear modulus μ_s at the interface is decreased due to the generation of the misfit dislocations at the interface between the surface layer and substrate and leading to transition in the regime of plastic deformation [18, 19] or due to the strain-induced pumping of vacancies from the surface to the bulk [11]. Thus, the substrate reaction can be neglected at a relatively large stress $\sigma_{||}$.

For simplicity, we only consider the contribution of the defects of one type in formula (22) and neglect defect recombination ($\gamma_d = 0$). Performing Fourier transformation of (4) with substitution of (7) and (22), (24) we obtain equation for Fourier amplitude

$$\frac{\partial N_q}{\partial t} = \lambda_q N_d(\mathbf{q}) + D_d \frac{1}{N_{cr}} \sum_{\mathbf{q}_1} (\mathbf{q}\mathbf{q}_1) \frac{1+\nu l_0^2 q_1^2}{\left[1+l_{||}^2 q_1^2\right]} N_{\mathbf{q}_1} N_{\mathbf{q}-\mathbf{q}_1} \tag{25}$$

Where in the nonlinear term we put $l_d = 0$, $L_d = 0$ and introduced the growth rate of the DD grating

$$\lambda_{\mathbf{q}} = -D_d q^2 + D_d q^2 \frac{N_{d0}}{N_{cr}} \frac{1+\nu l_0^2 q^2 \left[1 - l_d^2 q^2 + L_d^4 q^4\right]}{1+l_{||}^2 q^2}, \tag{26}$$

where the critical concentration of the defects

$$N_{cr} = \sigma_{||} \frac{k_B T}{\nu \theta_d^2}. \tag{27}$$

The first and second terms in the parentheses in the numerator of expression (26) correspond to the normal and tangential defect-induced bending forces, respectively, exerted on the film. The first term in the brackets in the numerator of expression (26) takes into account the local force exerted by the deformation of the elastic continuum on the defect, and the remaining terms proportional to l_d^2 and L_d^4 take into account the nonlocal character of this force.

For numerical investigation of dependence (26) it is convenient to introduce dimensionless variables $x = l_{||} q$ and $\lambda_x = \lambda_{\mathbf{q}} l_{||}^2 / D_d$, and parameters $a = l_0^2 / l_{||}^2$, $b = l_d^2 / l_{||}^2$, $c = L_d^4 / l_{||}^4$ and the control parameter $\varepsilon = N_{d0} / N_{cr}$. Then, we obtain:

$$\lambda_x = -x^2 + x^2 \varepsilon \; 1 + a\nu x^2 \left[1 - bx^2 + cx^4\right] / 1 + x^2 \tag{27a}$$

The dependence of the growth rate λ_x, (27a), at two values of control parameter $\varepsilon = 5$ and $\varepsilon = 10$, is shown in figure 2. We used here the following values of dimensionless parameters: $\nu = 0.2$, $a = 1.4 \cdot 10^{-2}$, $b = 1.5 \cdot 10^{-2}$ and $c = 1,8 \cdot 10^{-3}$, which correspond, for example, to the following set of physical parameters characteristic for the case of nanostructure formation: $h = 10^{-6}$ cm, $\rho_f c^2 = 7 * 10^{11}$ erg * cm^{-3}, $\sigma_{||} = 10^{10}$ erg * cm^{-3}, $l_d = 3 * 10^{-7}$ см, $L_d = 5 * 10^{-7}$ см.

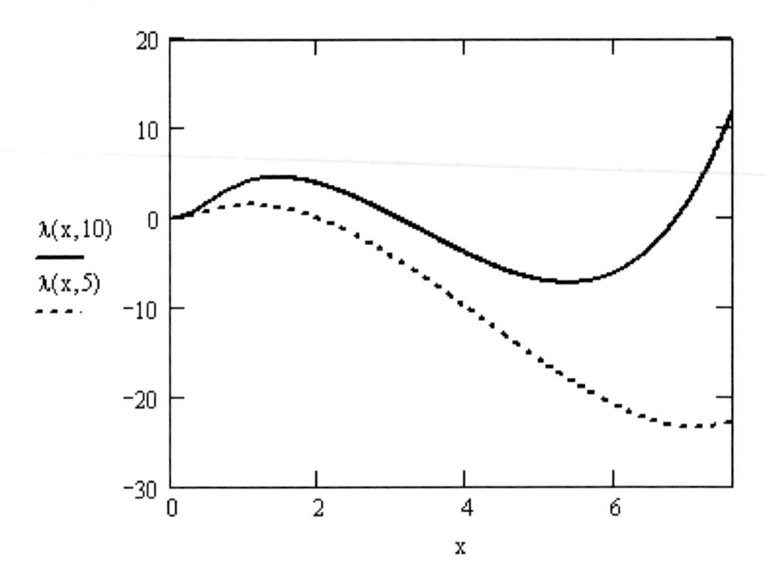

Figure 2. The dependence of the dimensionless growth rate of Fourier-amplitude of DD grating on dimensionless wave vector at two values of control parameter: ε =5(dash) and ε =10 (solid). The curves are calculated with formula (27a) at the values of parameters given in the text. The end of the *x*-axis is at the point $x_c = l_{||} q_c$ -corresponding to the wave number of the limiting bending mode $q_c = \pi/h$.

From figure 2 it is seen that the growth rate exhibits a single maximum at $q = q_m$ in the long-wavelength range at concentrations of the defects exceeding the threshold value and that an additional maximum emerges at $q = q_c$ in the short-wavelength range when the concentration is further increased. Corresponding wave vector is equal to the wave vector of limiting bending mode

$$\mathbf{q}_c = \pi/h, (\Lambda_c = 2h) \tag{28}$$

Therefore, at high enough concentrations of defects two DD gratings with $\mathbf{q} = \mathbf{q}_m$ ($\Lambda = \Lambda_m$) and $\mathbf{q} = \mathbf{q}_c$ ($\Lambda_c = 2h$) have the maximum growth rates.

The long-wavelength maximum of the growth rate at $q = q_m (\Lambda = \Lambda_m)$, can be analytically described provided that the nonlocal character of the DD interaction is neglected in expression (26) (i.e., $l_d = L_d = 0$) and the lateral bending force is also neglected ($\nu l_0^2 q^2 < 1$). Then, using expression (26), we have

$$\lambda_{\mathbf{q}} = -D_d q^2 + D_d q^2 \frac{N_{d0}}{N_{cr}} \frac{1}{1 + l_{||}^2 q^2}. \tag{29}$$

The maximum on the curve λ_q is reached at $q = q_m$, such that

$$q_m = \frac{1}{l_{||}}\left(\left(\frac{N_{d0}}{N_{cr}}\right)^{1/2} - 1\right)^{1/2}. \quad (30)$$

The corresponding period of the dominant DD grating with wave vector q_m is given by

$$\Lambda_m = 2\pi/q_m = 2\pi h\left(\frac{\rho_f c^2}{12\sigma_{||}}\right)^{1/2} \frac{1}{\left(\left(N_{d0}/N_{cr}\right)^{1/2} - 1\right)^{1/2}}. \quad (31)$$

and is proportional to layer thickness *h*. We note that Λ_m can, by order of magnitude, be larger than Λ_c.

The maximum growth rate for the grating with $q = q_m$ is

$$\lambda_m = D_d q_m^{\ 2}\left(\sqrt{N_{d0}/N_{cr}} - 1\right) = \frac{D_d\left(\sqrt{N_{d0}/N_{cr}} - 1\right)^2}{l_{||}^{\ 2}} \mathrm{Sign}\left(\sqrt{N_{d0}/N_{cr}} - 1\right). \quad (31a)$$

It follows from expressions (31) and (31a) that a real value of q_m emerges and growth rate λ_m becomes positive when the critical concentration of the defects is exceeded ($N_{d0}/N_{cr} > 1$). For T = 300 K, $\theta_d = 10^2$ eV, $\sigma_{||} = 10^{10}$ erg * cm^{-3}, $\nu = 0.5$, expression (27) yields the critical concentration $N_{cr} = 2 * 10^{16}$ cm^{-3}.

To construct the computer image of the surface with relief given by Eq. (21) we introduce the dimensionless growth rate $\tilde{\lambda}_{\tilde{q}} = \lambda_q l_{||}^{\ 2}/D_d$, getting from (29) $\tilde{\lambda}_{\tilde{q}} = -\tilde{q}^2 + G\tilde{q}^2/1+\tilde{q}^2$ with $\tilde{q} = l_{||}q$ and $G = N_{d0}/N_c$. The dimensionless *z*-coordinate of the surface is given by:

$$\tilde{Z}(\tilde{\mathbf{r}}, T) = \sum_{|\tilde{\mathbf{q}}|<\tilde{q}_0} \cos(\tilde{\mathbf{q}}\tilde{\mathbf{r}} + \Phi(\tilde{\mathbf{q}}))\exp(\tilde{\lambda}_{\tilde{\mathbf{q}}}T), \quad (32)$$

where $\tilde{\mathbf{r}} = \mathbf{r}/l_{||}$, $T = t \cdot D_d / l_{||}^2$ and we also assumed that the initial amplitude $\zeta_{\mathbf{q}} = \mathrm{const} = \zeta$ for all $\tilde{\mathbf{q}}$, so that $\tilde{Z} = Z/\zeta$, and $\tilde{q}_0$ is determined from the condition $\tilde{\lambda}_{\tilde{q}_0} = 0$; $\Phi(\tilde{\mathbf{q}})$ is a random phase uniformly distributed in the interval [0,2π]. The result of computer modeling of the 3D image of the surface relief is shown in figure 3a in comparison with the AFM image of the real irradiated surface from ref.[20] (figure 3b). The lamellar-like structure of figure 3a generated on computer with the help of Eq. (32) looks quite similar to the real one (figure 3b).

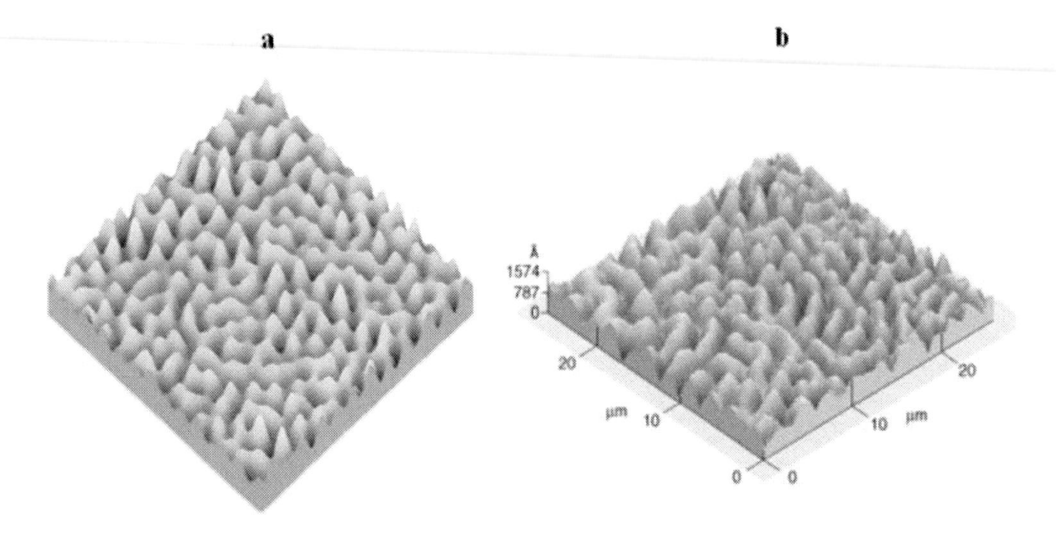

Figure 3. Comparison of computer-generated [21] (a) and experimental [20] (b) 3D images of surface relief. Computer image is constructed with the help of Eq. (32) with $G = 1.5$, $T = 100$.

In ref. [20] the relief pattern shown in figure 3b was generated on the SiO_2 layer surface for a SiO_2/Si layered structure irradiated by 23 ns excimer laser pulses ($\lambda = 248$ nm) at a normal incidence. The ripple period was observed to increase almost linearly with the laser fluency and to increase with the number of pulses. Importantly, the ripple period also has a linear dependence on the oxide layer thickness, which, as it was noted in [20], provides a new approach for controlling the pattern size of the surface microstructures. In the work [21] we developed the Defect-Deformational (DD) mechanism of ripples formation on Si surface with SiO_2 cap involving the plasma driven instability of static quasi-Lamb elastic modes similar to the one described in sections 1 and 2 that is capable of explaining and describing all three above mentioned dependencies observed in [20]. It also consistently describes the symmetry of the ripples pattern and points out the way of changing the symmetry of ripples in a directed way.

3. Three DD Gratings Interactions and Generation of Surface Relief Harmonics

3.1. Equations of Three DD Gratings Interactions

We consider now the nonlinear regime of surface relief generation. We confine to the case of three DD gratings interactions with collinear wave vectors: $\mathbf{q}_1 = \mathbf{q}_c$, $\mathbf{q}_2 = \mathbf{q}_m$,

$\mathbf{q}_3 = -(\mathbf{q}_c + \mathbf{q}_m)$, and $\mathbf{q}_1 \| \mathbf{q}_2 \| \mathbf{q}_3$ [21]. In particular, the role of $\mathbf{q}_c$ and $\mathbf{q}_m$ can play the wave vectors (28) and (30) at which the maximum of the growth rate $\lambda_{\mathbf{q}}$ is achieved.

Equations for Fourier-amplitudes of interacting DD gratings follow from (25) and have the form

$$\begin{aligned}
\partial N_{\mathbf{q}_m} / \partial t &= \lambda_{\mathbf{q}_m} N_{\mathbf{q}_m} + D_d \frac{A_m}{N_{cr}} N_{\mathbf{q}_c + \mathbf{q}_m} N_{-\mathbf{q}_c} \\
\partial N_{\mathbf{q}_c} / \partial t &= \lambda_{\mathbf{q}_c} N_{\mathbf{q}_c} + D_d \frac{A_c}{N_{cr}} N_{\mathbf{q}_c + \mathbf{q}_m} N_{-\mathbf{q}_m} \\
\partial N_{\mathbf{q}_c + \mathbf{q}_m} / \partial t &= \lambda_{\mathbf{q}_c + \mathbf{q}_m} N_{\mathbf{q}_c + \mathbf{q}_m} + D_d \frac{A_{cm}}{N_{cr}} N_{\mathbf{q}_c} N_{\mathbf{q}_m},
\end{aligned} \tag{33}$$

where the coefficient of three DD gratings interactions are

$$A_m = -(\mathbf{q}_c \mathbf{q}_m)\ 1 + \nu l_0^2 q_c^2\ \Big/ \Big[1 + l_{||}^2 q_c^2\Big] + ((\mathbf{q}_c + \mathbf{q}_m)\mathbf{q}_m)\ 1 + \nu l_0^2 (\mathbf{q}_c + \mathbf{q}_m)^2\ \Big/ \Big[1 + l_{||}^2 (\mathbf{q}_c + \mathbf{q}_m)^2\Big], \tag{34}$$

$$A_c = -(\mathbf{q}_m \mathbf{q}_c)\ 1 + \nu l_0^2 q_m^2\ \Big/ \Big[1 + l_{||}^2 q_m^2\Big] + ((\mathbf{q}_c + \mathbf{q}_m)\mathbf{q}_c)\ 1 + \nu l_0^2 (\mathbf{q}_c + \mathbf{q}_m)^2\ \Big/ \Big[1 + l_{||}^2 (\mathbf{q}_c + \mathbf{q}_m)^2\Big], \tag{35}$$

$$A_{cm} = (\mathbf{q}_m (\mathbf{q}_c + \mathbf{q}_m))\ 1 + \nu l_0^2 q_m^2\ \Big/ \Big[1 + l_{||}^2 q_m^2\Big] + \mathbf{q}_c (\mathbf{q}_c + \mathbf{q}_m))\ 1 + \nu l_0^2 \mathbf{q}_c^2\ \Big/ \Big[1 + l_{||}^2 \mathbf{q}_c^2\Big]. \tag{36}$$

Introducing the real variables $N_j = n_j \exp i\varphi_j$, we derive from (33) the system of three equations for the real amplitudes $n_j (j = c, m, cm)$ and the equation for the phase difference $\Phi = \varphi_c + \varphi_m - \varphi_{cm}$:

$$\begin{aligned}
\frac{\partial n_m}{\partial t} &= \lambda_m n_m + D_d A_m \frac{n_c n_{cm}}{N_{cr}} \cos\Phi, \\
\frac{\partial n_c}{\partial t} &= \lambda_c n_c + D_d A_c \frac{n_m n_{cm}}{N_{cr}} \cos\Phi, \\
\frac{\partial n_{cm}}{\partial t} &= \lambda_{cm} n_{cm} + D_d A_{mc} \frac{n_m n_c}{N_{cr}} \cos\Phi,
\end{aligned} \tag{37}$$

$$\frac{\partial \Phi}{\partial t} = -\frac{D_d}{N_{cr}} \left[A_m \frac{n_c n_{cm}}{n_m} + A_c \frac{n_m n_{cm}}{n_c} + A_{mc} \frac{n_m n_c}{n_{cm}} \right] \sin\Phi. \tag{38}$$

The equation (38) describes phase relaxation: $\Phi = \varphi_c + \varphi_m - \varphi_{cm} \to 0$. From comparison of equations (38) and (37) it is seen that the characteristic time of phase relaxation τ_{phase} divided by the characteristic time of redistribution of defects from one DD grating to another τ_0 is of order of $\tau_{phase}/\tau_0 \sim n_j/N_{cr} \sim n_j/N_{d0} << 1$. Because of that one can put $\Phi = 0$ in equations (37) upon consideration of nonlinear transformations of DD gratings.

3.2. Second Harmonics Generation

Let us consider one DD grating. For example, not far from the threshold of the DD instability, when the growth rate of the DD gratings has a single maximum, it may be the DD grating with wave number q_m, (30). (Similarly, one can consider the DD grating with wave number q_c, (28)). Let us show that the three gratings' interactions render possible the process of summation of two identical wave numbers: $q_{mm} = -q_m - q_m$. This process leads to generation of additive DD grating with wave number $2q_m$. This corresponds to the optical SHG in a nonlinear crystal when the condition of exact synchronism is met. The coupling coefficients (34), (36) in this case acquire the form:

$$A_m = q_m^2 \; 1 - 2l_{||}^2 q_m^2 \; / \; 1 + l_{||}^2 q_m^2 \;\; 1 + 4l_{||}^2 q_m^2 \; , \; A_{mm} = 4q_m^2 / \; 1 + l_{||}^2 q_m^2 \; .$$

For numerical analysis of the system of equations (37) we introduce dimensionless time $t' = D_d A_{cm} t$ and dimensionless amplitudes $n_j' = n_j/N_{cr}$ and put Φ=0. Then for considered case of SHG the system of equations (37) assumes the form:

$$\frac{\partial n_m'}{\partial t'} = n_m' + \frac{A_m'}{\lambda_m'} n_m' n_{mm}',$$

$$\frac{\partial n_{mm}'}{\partial t'} = \frac{\lambda_{mm}'}{\lambda_m'} n_{mm}' + \frac{A_{mm}'}{\lambda_m'} n_m' n_m' \qquad (39)$$

The numerical solution of the set of equations (39) is shown in figure 4.

It is seen that on times exceeding the characteristic time of linear growth of the amplitude of the fundamental harmonic ($\lambda_m'^{-1}$), the amplitude of the second harmonic due to SHG starts to exceed the amplitude of the fundamental one. The linear growth of the SH, for chosen value of λ_{mm}', is insignificant since at the values of parameters used $\lambda_m' / \lambda_{mm}' = 11.5$.

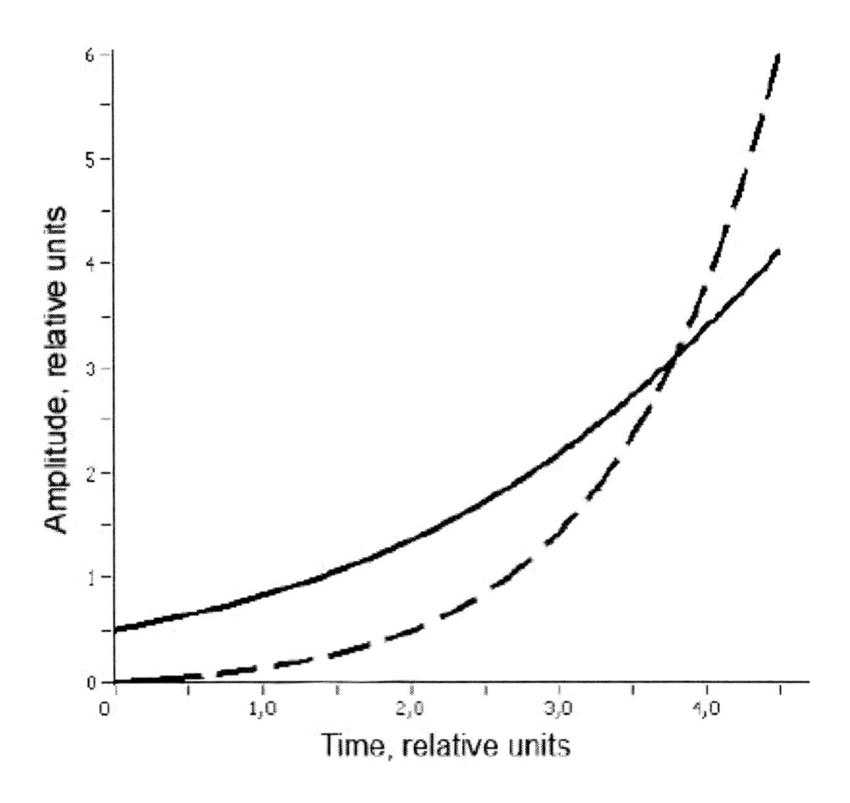

Figure 4. The dependence of the amplitudes of the SH n_{mm}' (dash) and the first harmonic n_m' (solid) of the surface relief on the dimensionless time $t' = \lambda_m t$. The numerical solution of the set of equations (39) at the initial conditions $n_m'(t=0) = 0.5$, $n_{mm}'(t=0) = 10^{-5}$. The values of growth rates λ_m' and λ_{mm}' are determined from (27a) at $a = 1.4 \cdot 10^{-2}, b = 4 \cdot 10^{-5}, c = 10^{-8}, \varepsilon = 10$.

3.3. Wave Vectors Mixing

Let initially on the surface there are two collinear DD gratings with wave numbers $q_1 = q_m$ and $q_2 \equiv q_c = 2q_m$. Consider the process of wave vectors mixing resulting in arising of the DD grating with wave number $q_3 \equiv q_{cm} = -(q_1 + q_2) = -3q_m$. In this case, the system of equations (37) in dimensionless variables at $\Phi = 0$, assumes the form

$$\frac{\partial n'_m}{\partial t'} = n'_m + \frac{A'_m}{\lambda'_m} n'_c n'_{cm},$$

$$\frac{\partial n'_c}{\partial t'} = \frac{\lambda'_c}{\lambda'_m} n'_c + \frac{A'_c}{\lambda'_m} n'_m n'_{cm}$$

$$\frac{\partial n'_{cm}}{\partial t'} = \frac{\lambda'_{cm}}{\lambda'_m} n'_{cm} + \frac{A'_{cm}}{\lambda'_m} n'_m n'_c \qquad (40)$$

Figure 5 shows the results of numerical solution of the set of equations (40) at the negative growth rate of the third harmonic: $\lambda'_{cm} / \lambda'_m = -2$ and $\lambda'_m / \lambda'_c = 11.5$ and the values of parameters given in figure caption.

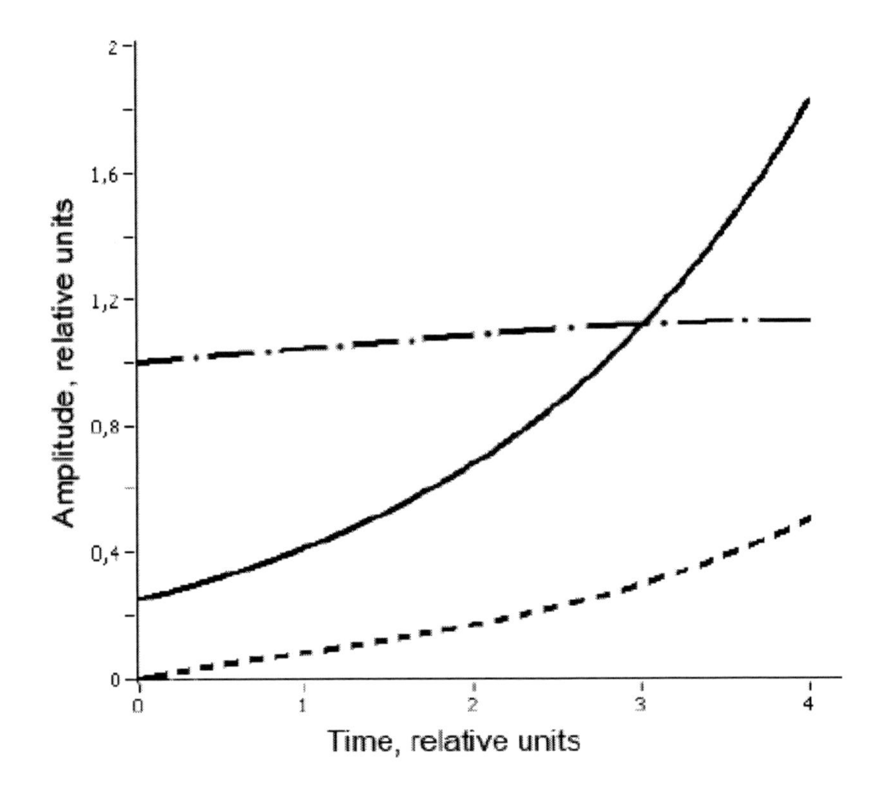

Figure 5. The dependence of the amplitude of the first harmonic (n'_m, solid), the second (n'_c, dash dot) and the third (n'_{cm}, dash) harmonics of the surface relief on the dimensionless time $t' = \lambda_m t$. The numerical solution of the set of equations (40) at the initial conditions: $n_m'(t=0) = 0.25$, $n_c'(t=0) = 1$ and $n_{cm}'(t=0) = 10^{-5}$. The values of the growth rates are calculated with the formula (27a) at $a = 1.4 \cdot 10^{-2}, b = 4 \cdot 10^{-5}, c = 10^{-8}, \varepsilon = 10$.

It is seen that even at the negative growth rate of the third harmonic its amplitude undergoes essential increase due to the nonlinear integrating interactions. The comparison with experiment is given in Sec.6.

4. KURAMOTO-SIVASHINSKY EQUATION FOR SURFACE DD INSTABILITY

4.1. Derivation of KS Equation

The mode representation expressed by the set of coupled equations for Fourier amplitudes, (25), following from the set of initial differential equations (4), (7) and (9), is the most general but not the only possible approach to the description of surface DD instability. In this section we show that, under certain limitations, it is possible to derive a closed DD nonlinear partial differential equation for defect concentration from the same set of equations [55]. This equation has the form of famous Kuramoto-Sivashinsky (KS) equation [22, 23], widely used for description of selforganization phenomena in different systems [24]. In particular, the KS equation, specific for the case of ion-beam sputtering of solids, is the main tool of description of selforganization of nanostructures during this process [25, 26].

We neglect in (16) the substrate reaction ($\sigma_{\perp} = 0$) and the longitudinal bending force (the second term in the right hand side) and transform (16) to the equation for the surface strain using the relation $\xi \equiv \xi(z=0) = (\nu h/2)\Delta\zeta$, following from (7). Then we obtain the following equation

$$\Delta\xi_f - \frac{1}{l_{||}^{2}}\xi_f = -\frac{\nu\theta_d}{l_0^2 \rho_f c^2} N_d , \tag{41}$$

The equation (41), can be solved using the theory of perturbation in respect to the small parameter $l_{||}^{2}/\Lambda_m^{2} << 1$. We represent the bending coordinate in the form $\xi_f = \xi_f^{(0)} + \xi_f^{(1)}$, where $\xi_f^{(0)}$ and $\xi_f^{\ 1}$, are, respectively, the solutions of zero and the first order and find from (41) the expression for the surface strain nonlocally dependent on the surface defect concentration

$$\xi_f(z=0) = \frac{\nu\theta_d}{\sigma_{||}} N_d + \frac{\nu\theta_d}{\sigma_{||}} l_{||}^{2} \Delta N_d , \tag{42}$$

Substituting (42) in (4), we obtain the closed nonlinear DD equation

$$\frac{\partial N_d}{\partial t} = D_d\left(1 - \frac{N_d}{N_{cr}}\right)\Delta N_d - D_d \frac{N_d}{N_{cr}} l_{||}^{2} \Delta^2 N_d - \frac{D_d}{N_{cr}} (\nabla N_d)^2 - \gamma_d N_d , \tag{43}$$

where the critical defect concentration N_{cr} is given by (27).

In derivation of equation (43) we neglected the nonlinear term $D_d l_{||}^{2} (\nabla N_d)\nabla\ \Delta N_d / N_{cr}$, since it is small (of order of $l_{||}^{2}/\Lambda_m^{2}$) compared with the retained nonlinear term $\sim (\nabla_{||} N_{d1})^2$.

The nonlinear DD equation (43) is similar to the KS equation [22, 23] with the following essential differences. The coefficient in the first term in the right hand side of (43) and one at the dispersion term ($\sim l_{||}^2$) are not constants as in the KS equation [22, 23], but depend on the dynamical variable with the former changing its sign with increase of this variable. The condition under which the DD equation, (43), was derived is the neglect of substrate reaction, which is justified if $\sigma_{||} > \mu_s \Lambda \; 1-\beta_s \; /\pi h$, where Λ is the characteristic scale of the surface DD structure(see (24a)).

To obtain the analytic solution of the DD equation (43) is not possible and we leave this problem for future numeric investigation. Here we show that the DD equation (43), in the mean field approximation, is reduced to the KS equation and we carry out comparative analysis of solution of this equation for the considered case of DD instability.

We represent the defect concentration at the surface in the form

$$N_d(x,y,t) = N_{d0} + N_{d1}(x,y,t), \tag{44}$$

where N_{d0} is spatially uniform part of defect concentration playing the role of control parameter and $N_{d1}(x,y,t)$ is spatially nonuniform part which arises spontaneously under exceeding of instability threshold. Under the condition $|N_{d1}(x,y,t)| < N_{d0}$, introducing dimensionless control parameter $\varepsilon = N_{d0}/N_{cr}$, we reduce (43) to the form

$$\frac{\partial N_{d1}}{\partial t} = -D_d(\varepsilon-1)\Delta N_{d1} - D_d \varepsilon l_{||}^2 \Delta^2 N_{d1} - D_d (\nabla N_{d1})^2 / N_{cr} - \gamma_d N_{d1} \tag{45}$$

The closed nonlinear DD equation (45) has the form of KS equation [22, 23] with the essential difference that the coefficient in diffusion term is not constant but depends critically on the control parameter. Besides, it takes into account the damping term (last term in the right-hand side of (45)), which is essential for formation of ordered hexagonal DD dot structures [56].

4.2. Formation of DD Structure Described by KS Equation in the Linear Regime

We may investigate the properties of the DD equation (45) in the linear approximation and compare results with the results of modal analysis. We use Fourier transforms

$$N_{d1}(\mathbf{r},t) = \sum_{\mathbf{q}} N_d(\mathbf{q}) \exp(i\mathbf{qr} + \lambda_q t) \tag{46}$$

$$\zeta(\mathbf{r},t) = \sum_{\mathbf{q}} \zeta_{\mathbf{q}} \exp(i\mathbf{qr} + \lambda_q t) \tag{47}$$

Then, linearizing (45), obtain the growth rate

$$\lambda_q = D_d q^2(\varepsilon - 1) - D_d \varepsilon l_{||}^2 q^4, \tag{48}$$

Growth rate λ_q reaches the maximum at $q = q_m$, with

$$q_m = \left(1 - \varepsilon^{-1}\right)^{1/2} \Big/ \sqrt{2} l_{||} \tag{49}$$

Corresponding period of dominant DD grating with wave vector q_m is

$$\Lambda_m = 2\pi/q_m = 2\pi\sqrt{2}l_{||}\Big/\left(1-\varepsilon^{-1}\right)^{1/2} = 2\pi h\left(\rho_f c^2\big/6\sigma_{||}\left(1-\varepsilon^{-1}\right)\right)^{1/2} \tag{50}$$

The period Λ_m is proportional to *h*, critically increases at $N_{d0} \to N_{cr}$ and saturates at its minimal value at $N_{d0} \to \infty$. Such behavior of Λ_m justifies introduction of small parameter $l_{||}^2/\Lambda_m^2 = (1-(N_{cr}/N_{d0})/8\pi^2$, used as the expansion parameter in the perturbation theory upon derivation of (13).

The maximum value of the growth rate of the DD grating with $q = q_m$ is

$$\lambda_m = D_d \varepsilon (1-\varepsilon)^2 \big/ 4 l_{||}^2 \tag{51}$$

As is seen from (49), upon exceeding the critical defect concentration (ε>1), q_m becomes a real that corresponds to appearance of periodic DD structure. It can be verified that expressions for λ_q, q_m, Λ_m, and λ_m, (48)-(51) follows from corresponding formulas from Sec.2 if one expands the latter in power series of small parameter $l_{||}^2/\Lambda_m^2$.

Summing up DD gratings with different ***q***, entering in (47) in a way indicated in the discussion after Eqs. (21) and (21a), we obtain the superpositional surface DD structure. The result of such summation is quite similar to the one shown in figure 3a. Thus figure 3a gives the spatial distribution of reduced defect concentration $N_{d1}/|N_d(\mathbf{q})|$ along the surface, the distribution of surface strain, (42), and surface relief (47).

It is seen that in the linear regime chaotic lamellar surface relief is formed. The relief shown in figure 3 is similar to the relief obtained with numerical solution of KS equation in [26] at small exceeding over the threshold and short duration of instability development.

The expressions (22), (24) indicate that interstitials piles up (θ_d>0) in such seeding lamellar DD structure occur at maxima of surface relief (expanded regions) while vacancy piles up (θ_d<0) occur at minima of surface relief (compressed regions). Vacancy piles up are etched with greater rate than interstitial ones. So in the case of semiconductors where both interstitials and vacancies are generated, laser or ion-beam etching must lead to formation of the ensemble of hillocks while in metals, where vacancies are defects generated ensemble of voids must be formed. This is the case upon ion-beam sputtering of the surface of GaSb, where ensemble of nanodots is formed [27], while in the case of Ag surface sputtering the ensemble of voids is formed [28]. The periodicity and the symmetry of the final structure formed by energy beam irradiation of solid follow those characteristics of the seeding cellular DD structure. The isotropic DD equation KS obtained in this chapter may be generalized to the anisotropic case to study the influence of the anisotropy of the DD system on the geometry of seeding DD structure. Recently, computer solutions of nonlinear two-dimensional DDKS equation (45) were obtained, describing formation of a variety of DD structures lamellar-like structures, chaotic dot ensembles and ordered hexagonal dot structures[56].

We note close similarities between the dot structures formed upon ion and laser beam irradiation of solids (compare figure 9 and figure 10). The DD mechanism was suggested in [11] as a universal cause of formation of surface structures in both cases. The fact of relevance of the KS equation for the DD mechanism that was established in [55] and in this chapter strongly supports this suggestion.

5. Bimodal Size Distribution OG Nanoparticles Formed by Laser Irradiation

In the case of the laser or ion-beam etching, the strain extrema with defect piles up in the seed surface DD structure, that results from the superposition of the DD gratings with different periods Λ, serve as the nucleation centers for the subsequent growth of nucleuses which leads to the formation of the nanodot ensemble.

By way of example, we consider the (100) surface that exhibits two selected mutually orthogonal directions along which the vectors of the DD gratings are oriented. The superposition of two DD gratings with wave vectors $\mathbf{q}_1$ and $\mathbf{q}_2$ such that $|\mathbf{q}_1| = |\mathbf{q}_2| \equiv q$ and $\mathbf{q}_1 \perp \mathbf{q}_2$ yields a 2D periodic square superlattice of the strain extrema (nucleation centers) with the period $\Lambda = 2\pi/q$. Alternative approach consists in using instead of (21), (21a) the expansion of variables in Fourier-series of eigen functions of Laplace operator in square region:

$$\zeta(\mathbf{r},t) = \sum_{\mathbf{q}} \zeta_{\mathbf{q}} \exp(\lambda_q t)\cos(qx)\cos(qy), \tag{52}$$

$$N_d(\mathbf{r},t) = \sum_{\mathbf{q}} n_d(\mathbf{q}) \exp(\lambda_q t)\cos(qx)\cos(qy) \tag{53}$$

If the intermode interaction is neglected in the linear regime of the DD instability, the resulting DD structure (21), (21a) (or (52), (53)) consists of a superposition of square gratings with the periods $\Lambda = 2\pi/q$ and the magnitudes of the wave vectors of these gratings are selected by the maxima of the DD instability growth rate. The gratings are independent of each other in the linear approximation.

Each surface 2D DD grating with period Λ in the superposition (21), (21a) is considered as a crystal (a periodic square superlattice) whose regular superlattice cites are occupied by the nanoparticle nucleuses (compare with atoms that occupy the regular lattice cites in a crystal). As in the crystalline lattice, the free energy of the superlattice is minimized when a fraction of regular cites in the superlattice is vacant. The surface concentration n_V (cm^{-2}) of such vacancies in the square superlattice with period Λ is given by the thermodynamic formula by analogy with a crystal:

$$n_{SV}(\Lambda) = \Lambda^{-2}\exp(-E_1(\Lambda)/k_B T),$$

where $E_1(\Lambda)$ is the bond energy of one nucleus at the regular superlattice cites (the energy needed for the formation of a vacancy). Then, the size distribution function of the nanoparticle (nanodot) nucleation centers is given by

$$n_{\rm dot}(\Lambda) = \Lambda^{-2} - n_{SV}(\Lambda) = \Lambda^{-2}\ 1-\exp(-E_1(\Lambda)/k_B T)\ . \quad (54)$$

In the adiabatic approximation with respect to the bending coordinate, we express the bond energy $E_1(\Lambda)$ (potential energy well depth) through the square modulus of Fourier-amplitude of the growing DD grating with $q = 2\pi/\Lambda$ (i.e., through the growth rate $\lambda_{\mathbf{q}} = \lambda(\Lambda)$. For this purpose, we use the following expression for the DD interaction energy stored in bending-deformed surface layer with thickness *h*:

$$W = -\int_S d\mathbf{r}\int_0^h dz\,\theta_d n_{d1}(\mathbf{r},z)\xi(\mathbf{r},z),$$

where *S* is the surface area of the layer. The Fourier-series expansion of the variables yields

$$W = -S\int_0^h dz\sum_{\mathbf{q}}\theta_d n_{dq}(z)\xi_{-q}(z) \equiv \sum_{\mathbf{q}} W_q\,. \quad (55)$$

Then, the energy of one superlattice is written as

$$W_q = -S\int_0^h dz\theta_d n_d(q,z)\xi_{-q}(z) = -S\frac{4\nu\theta_d^{\,2}}{h\rho_f c^2}\ 1+\nu l_0^{\,2}q^2\ \frac{1}{\left[q^2 + l_{||}^{\;-2}\right]}\left|N_d(q,t)\right|^2. \quad (56)$$

$$W_q = -S\int_0^h dz\theta_d n_d(q,z)\xi_{-q}(z) = -S\frac{4\nu\theta_d^2}{h\rho_f c^2}\, 1+\nu l_0^2 q^2 \,\frac{1}{\left[q^2 + l_{||}^{-2}\right]}\left|N_d(q,t)\right|^2$$

Dividing W_q by the number of the superlattice regular sites S/Λ^2, we obtain the energy of one site:

$$W_1 \equiv W_q\Lambda^2/S = -\Lambda^2\frac{4\nu\theta_d^2}{h\rho_f c^2}\, 1+\nu l_0^2 q^2 \,\frac{1}{\left[q^2 + l_{||}^{-2}\right]}\left|N_d(q,t)\right|^2 \equiv -E_{SV}(q)\,, \quad (57)$$

where $E_{SV}(q)$ is the bond energy of the nucleation center at one superlattice regular site for the superlattice with the wave number q. Quantity $E_{SV}(q)$ determines the energy of the defect formation: the absence of the nucleation center (vacancy denoted with subscript *SV*) at the regular site for the superlattice wave number q.

The quasi-equilibrium concentration of vacancies $n_{SV}(q)$ in the surface superlattice q is given by

$$n_{SV}(q) = \Lambda^{-2}\exp(-E_{SV}(q)/k_B T)\,. \quad (58)$$

Then, the concentration of the nanodot nucleuses in the superlattice q is represented as

$$n_{\text{dot}}(q) = \Lambda^{-2} - n_{SV}(q) = \Lambda^{-2}\, 1-\exp(-E_{SV}(q)/k_B T\,. \quad (59)$$

When the condition $E_{SV}(q)/k_B T < 1$ is satisfied, we employ expression (59) allowing for $\Lambda = 2\pi/q$ and formula (57) and obtain after transformation

$$n_{\text{dot}}(q,t) = \frac{h}{3N_{cr}}\frac{1+\nu l_0^2 q^2}{\left[1+l_{||}^2 q^2\right]}\left|N_d(q,t)\right|^2\,. \quad (60)$$

We put $t = t_0$ in relationship (40) and use $\left|N_d(q,t_0)\right|^2 = \left|N_d(q,0)\right|^2\exp(2\lambda_q t_0)$, where t_0 is a characteristic time of the nucleation, which depends on the type of the nanostructuring process and will be defined below (see Sec.6.1). Then we obtain the following formula that represents the sought distribution:

$$n_{\text{dot}}(q,t_0) \approx \frac{h}{3N_{cr}}\frac{1+\nu l_0^2 q^2}{\left[1+l_{||}^2 q^2\right]}\left|N_d(q,0)\right|^2\exp(2\lambda_q t_0)\,. \quad (61)$$

With taking into account that $\Lambda = 2\pi/q$ the formula determines the size distribution function of nanoparticles formed by laser irradiation: $n_{\text{dot}}(\Lambda) = n_{\text{dot}}(q, t_0)$ .Since the growth rate undergoes transition from unimodal to bimodal dependence of wave number with increase of defect concentration (figure 2), so does the size distribution function (61) (see figure 14 for experimental example of such a transition).

6. Comparison with Experimental Results

We consider below a few examples of laser-induced nano and microstructures formation and make comparison of predictions of the above DD theory and experiment. Of particular interest here will be the bimodal dependence of the growth rate of DD gratings on its period, the particle size distribution function and its transformation with change of irradiation conditions and generation of harmonics of surface relief due to three-wave DD interactions. These new topics in the DD theory are not covered in recent review [11], in which the comparison of the DD theory and experiment is carried out with focus made on linear dependence of the structure lateral size on the thickness of defect-enriched layer.

6.1. Laser-Induced Creation of the Surface Defect-Enriched Layer and the Limiting Mode of the Layer Bending Modes Spectrum

Defects which take part in the DD instability in semiconductors are interstitials, vacancies and electron-hole pairs. As it was shown above, at high enough defect concentration, the surface relief modulation formed due to laser irradiation must have two scales: Λ_m ~h, (31), and Λ_c=2h, (28), (see figure 2). The ratio of these two scales

$$\Lambda_m/\Lambda_c = \left(\rho_f c^2/12\sigma_{||}\right)^{1/2}\left(\left(N_{d0}/N_{cr}\right)^{1/2}-1\right)^{-1/2} \sim \left(\rho_f c^2/12\sigma_{||}\right)^{1/2} \sim 10 \quad (62)$$

at $N_{d0}/N_{cr} \sim 10$, $\rho_f c^2 \sim 10^{12}$ erg*cm^{-3}, $\sigma_{||} \sim 10^9$ erg*cm^{-3}.

Since the characteristic time of formation of DD gratings can be roughly estimated as $t = \Lambda^2/4\pi^2 D_d$, the ratio of corresponding times of formation is $t_m/t_c = 10^2$.

The nature of the surface layer h and the value of h depend on the material and characteristics of irradiation. Below we discuss some of the possible variants.

As far as the nature of nanoscale surface layer h in semiconductors is concerned, we note that a universal feature of semiconductors is the presence of a near surface space-charge region (SCR) with excess or deficiency of free carriers in respect to the bulk. The thickness h of this SCR is of order of 10^{-5}-10^{-6} cm [33]. Because the carriers in the SCR have a deformation potential, they deform the crystal lattice. For this reason, the SCR with thickness *h* has elastic characteristics differing from those of the bulk. We suggest considering this SCR as a film of thickness *h* saturated with mobile defects taking part in the surface DD instability described in Sec.2. The thickness of the SCR $h \sim 10^{-5}$-10^{-6} cm can serve as a universal scaling parameter determining the period of nanometer modulation of the surface relief in semiconductors subjected to the action of laser radiation. In particular, it suits the case of Ge studied in [31, 32] (sec.6.2).

Other scaling parameters of interest for semiconductors in nanoscale region are the thickness of subsurface layer perturbed by technological treatment upon producing the sample ($h \sim 10^{-6}$ cm) and the thickness of the surface layer melted by the laser pulse action h_m that depends on the laser fluency and, at low enough fluencies, may lie in the range of tens or hundreds nanometers.

In the case of irradiation of semiconductors in water with intense short-wave (UV) laser pulses, one should take into account also the possibility of photocatalytic water splitting [34] at semiconductor (or its oxide) surface with formation and penetration into the subsurface layer of hydrogen or oxygen dimmers that can play the role of mobile defects.

The thickness of defect-enriched layer in this hypothetical case could depend on the mobility of these defects. For example, in the case of femtosecond pulses, assuming the surface melt duration to be $t_m \sim 10^{-9}$ s and using as an estimate of dimmer diffusivity in silicon melt the value $D_d \sim 10^{-4}$ cm^2 * s^{-1} (as given in [35] for As in molten Si) we have the estimate of the thickness of subsurface defect-enriched layer $h \sim (D_d t_m)^{1/2} \sim 3$ nm (compare with figure 6).

We note, also, that laser hydriding of silicon and titanium was studied in ref. [36] that leads to incorporation of large amounts of hydrogen into the surface layers of the sample. The process consists of laser-irradiating the sample in hydrogen atmosphere at elevated hydrogen gas pressures (1-3 bar). About 250 pulses of a XeCl excimer laser (308 nm, 55 ns pulse duration) were used. Interesting that even before laser irradiation of silicon, a peak of hydrogen concentration was recorded on untreated surface with maximum of about 2 at. % at the depth ~ 5 nm and width of about 10 nm which results from adsorbed water vapor of the residual gas in the chamber. After laser irradiation with fluence ~6 J*cm^{-2} the maximum is only slightly shifted but its intensity rises up to at. 8 %.

In any of the above cases a surface defect-enriched layer is formed which can be considered as a surface film taking part in the DD instability. We note that the picture of surface film periodically bent due to the surface DD instability (Figure 1) finds its counterparts in experimental data on laser-induced generation of submicron structures in semiconductors and micrometer- scale structures in metals (see figure 6 and figure 7).

The thickness of the surface film in figure 6 is of order of 2-3 nm and experimental value of the structure period Λ ~300nm. With $\left| \rho_f c^2 / 12\sigma_{||} \right|^{1/2}$ ~10 (see (62)) we conclude that the theoretically estimated value Λ_m ~200 nm, (31), corresponds to the experimental one from [37].

In the case of single nanosecond laser pulse excitation of a semiconductor, the optical absorption length $h = h_{abs}$ may serve as a scaling parameter (see [29]). In metals, the thickness of molten layer h_m determines, at small enough fluency, the submicron scaling parameter: $h = h_m$. (see Sec. 6.3).

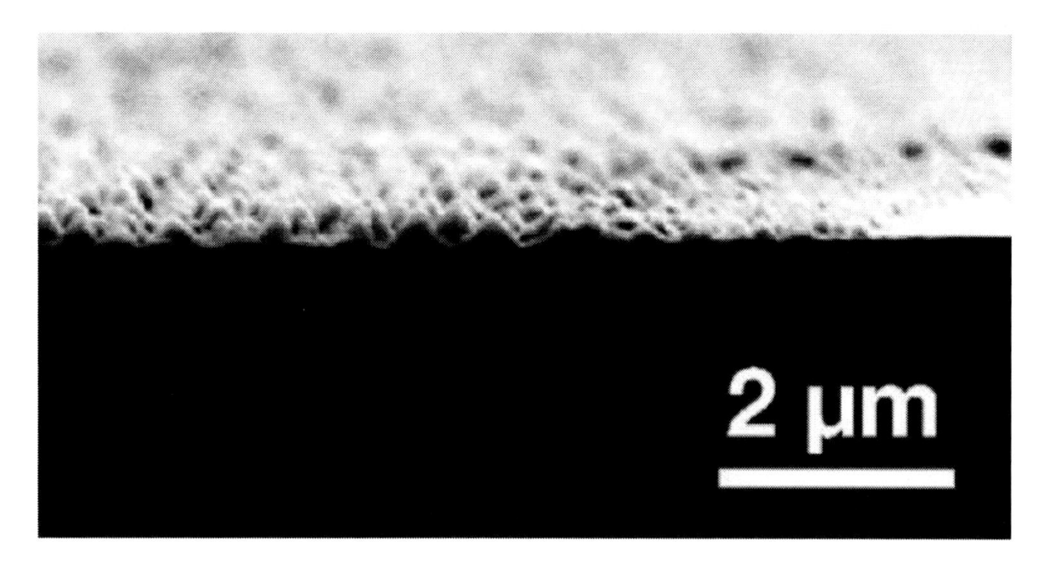

Figure 6. Scanning electron micrograph of cross-section of silicon sample with periodic surface structure with period of order of 300 nm formed by irradiation with a train of 100 fs, 400 nm, 60 μJ laser pulses in water [37]. The top layer is two-dimensionally periodically bent and can be considered as a periodically bent surface film (compare with figure 1a and figure 6).

The same parameter h may determine the larger (micrometer) lateral size $\Lambda_m \sim 10\,\Lambda_c$ of DD structures in semiconductors and metals in the case when the growth rate exhibits both maxima at q_c and q_m. The defect diffusion length in the direction normal to the surface may give another micrometer scaling parameter: $h = h_{\text{diff}}$, when pulse duration is long enough (for example, in the case of millisecond pulse (see review [9] and [11]).

In the case of multipulse laser irradiation the strain-induced drift of vacancies from the surface molten layer of thickness h_m to the bulk may essentially increase the thickness of subsurface defect-enriched layer. In this case the lateral size of DD structures may increase to tens of micrometers. One example is the surface periodic structure formed in brass in water by a train of laser pulses [30] (see figure 7). The cross section of the structure shows that it can be interpreted as a surface periodically bent film in the state close to limiting bending mode (compare figure 1c with figure 7) in which the period of the bending grating: $\Lambda = 2h$, where h is the film thickness. In [30] it was found that $\Lambda \sim r_0$, where r_0 is the radius of the irradiated laser spot. This size effect was described in [38] (see also [11]) from the viewpoint of the DD theory.

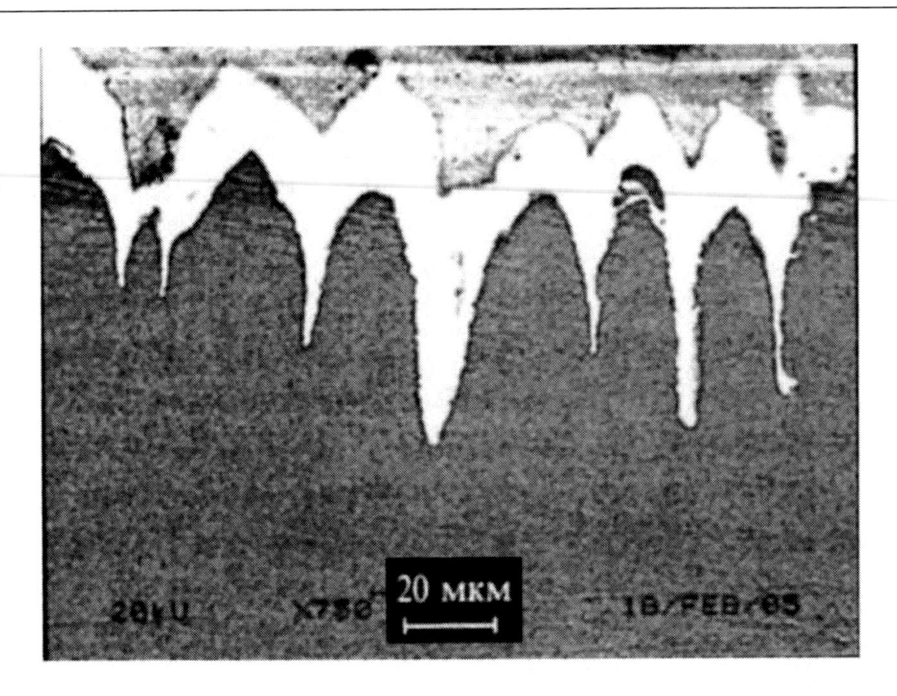

Figure 7. Scanning electron microscope image of the cross-section of a brass sample with a two-dimensional surface structure produced by irradiation in water by a train of nanosecond pulses from copper vapor laser with fluency 16 J^*cm^{-2} [30]. For clearness, colors in the photograph are reversed compared to the photograph in [30]. The bar represents 20 μm . Periodically bent regions can be considered as "a film" in a state close to the limiting bending mode in which the period of bending wave is $\Lambda = 2h$, where *h* is the film thickness (compare with figure 1c).

6.2. Two Scales of Laser-Induced Surface Relief Modulation

In experiments [1, 2], Ge was irradiated using laser pulses with $\tau_p \geq 0.4\,\mu s$, λ =0.53 μm . The irradiation was performed over several scanning regions with sizes 3×5 mm^2. The fixed levels of the pulse fluency F_i were realized in each of the scanning regions at a constant number of pulses per each point N=10^3 (number of photodeformations). Pulse fluencies F_i were different for different scanning regions varying in the vicinity of threshold $F_0 \sim 0.1 J^*cm^{-2}$ corresponding to transition to inelastic deformations. Defects of interest here are vacancies that are assumed to be generated by dislocations motion [32].

At low fluencies, the accumulation of defects under inelastic photodeformations gives rise to randomly distributed clusters (Figure 8a). Then, at $F > F_1 = (1.2 - 1.5)F_0$ the further self-organization of light-induced point defects leads to appearance of 2D periodic cluster grating (Figure 8b) formed by superposition of two 1D gratings with wavevectors directed along the sides of scanned rectangular region. At higher F , only one of these two gratings is left and the formation of the 1D periodic nanostructures is observed (Figure 8c,d).

The isotropic [31] and generalized anisotropic [32] film DD models of formation of surface nanostructures taking into account anisotropic stress in scanned region were used to qualitatively and quantitatively interpret the formation and evolution of periodic nanostructures of Ge surface relief. The theoretical predictions [31, 32] are in agreement with the experimental data [1, 2] on the period and time of nanostructures formation and on the critical defect concentration of DD instability. The DD theory [32], that takes into account two-dimensional anisotropic surface stress inside rectangular-shaped scanned region, also describes the evolution of the symmetry of surface structure with fluency increase (Figure 8).

One of the distinctive features of the process of nanostructures formation in [1, 2] is the presence at higher fluencies of large scale (~3 μm) modulation of the surface relief. It is seen in AFM images (figure 8c and figure 8c) .and also in profilogram (figure 8e).

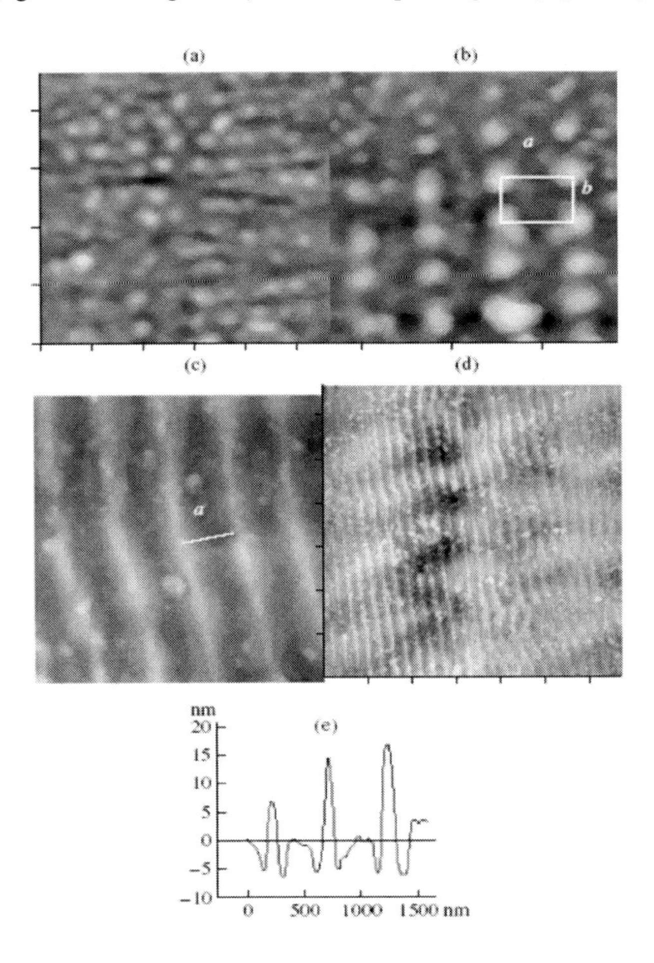

Figure 8. The hierarchy of the nanostructure formation in (111)Ge [1, 2]: J=(a) 85, (b) 110, (c) 150 mJ*cm^{-2}, and (d) J>150 mJ*cm^{-2}. Gratings periods are a=600 and b=400 nm. Figure 6e demonstrates the part of the profilogram of the surface relief $\zeta \quad x$ for the image shown in figure 6b. It is seen that the small scale (b~400 nanometers) is supplemented by large-scale (B~3 μm) modulation.

We note that the ratio of the periods of large scale modulation B and small scale modulation b (figure 8) is B/b~10 in accordance with prediction of the DD theory, (62). It enables one to relate the small scale b to Λ_c=2h and large scale B to Λ_m, (31).

The occurrence of two scales of modulation of the surface relief, fitting the DD criterion (62), was observed also in CdTe upon nanosecond single laser pulse irradiation [3]. The plasma-strain (PS) instability theory of laser-induced formation of the ensemble of nanostructures in CdTe was developed in [29], where the defects taking part in PS instability are electron-hole pairs and the scaling parameter is the optical absorption length $h = h_{abs}$. The PS theory describes well the extremum dependence of the lateral nanostructure size on the laser pulse fluency observed in experiments [3].

We note that upon ion-beam etching of semiconductors the dot nanostructure is formed which at more prolong irradiation is supplemented by micrometer scale modulation [39, 40] (see figure 9).

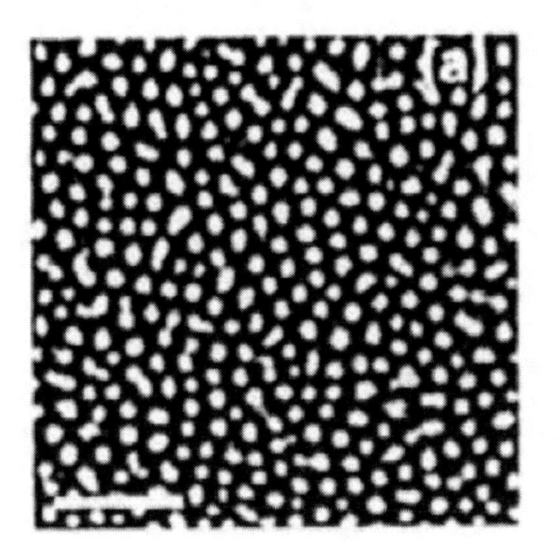

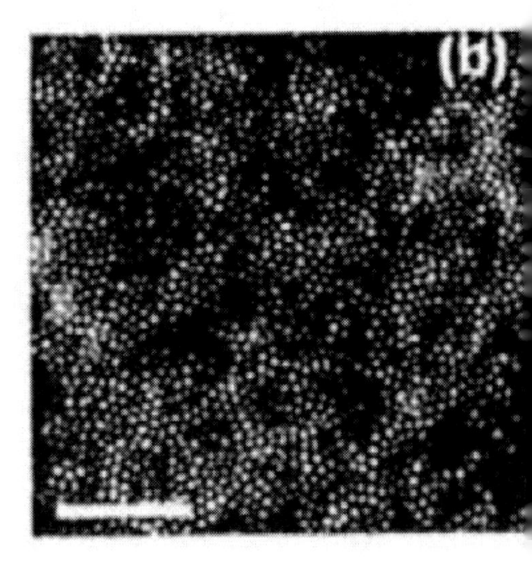

Figure 9. Nanopatterning of solid surface under ion-beam sputtering [39, 40]. Hexagonal dot structure formed on Si surface irradiated by the beam of Ar-ions with energy 1.2 keV at normal incidence during 6 min. The bar represents 277 nm (a); the large (micrometer) scale modulation of surface relief appears when the irradiation time is increased to 960 min. The bar represents 831 nm. (b).

The nano- and micrometer scale modulations of Si surface relief in figure 9 and times of their formations upon ion-beam irradiation fit the scaling relation of the DD theory (62). Figure 10 gives an example of similar periodic dot structure formation in Si under laser beam irradiation. For more detailed comparison from the unified DD point of view of formation of nanostructures upon laser and ion-beam irradiation see [11].

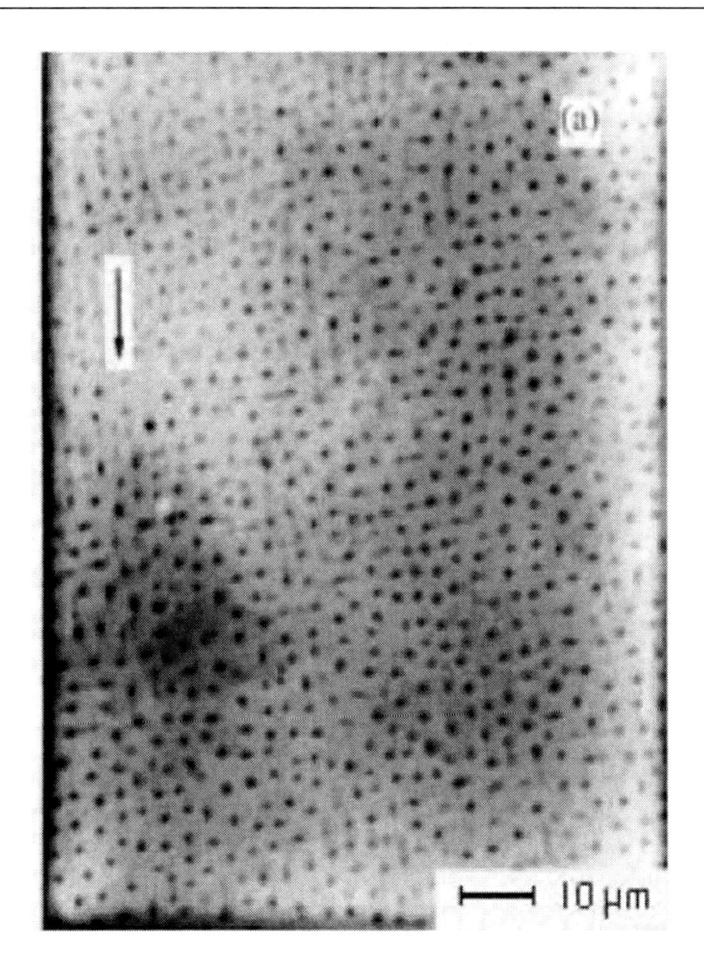

Figure 10. Photograph in optical microscope of Si(100) surface irradiated by millisecond pulse of linear polarized radiation at normal incidence [45]. The arrow shows the (100) direction. Two-dimensional crystallographically oriented dot structure is formed with period ~3.5 micron. Compare with figure 9.

6.3. Bimodal Particle Size Distribution Function

In the work [41] tantalum surface under the water layer was irradiated by a train of laser pulses. Two regimes of irradiation were used. In the first one, the KrF laser with wavelength 248 nm, pulse duration 5 ps with repetition rate 10 Hz was used. Ordered ensemble of nanoparticles was obtained with bimodal size distribution function (Figure 11).

In the second regime Nd-YAG laser with wavelength $1.06\,\mu m$, pulse duration $350\,ps$, repetition rate 300Hz with fluency variated in the range 0.29-0.45 $J*cm^{-2}$ was used. As a result of irradiation, an ensemble of nanoparticles with lateral dimensions 200-250nm was formed. The lateral size distribution function in this ensemble has characteristic shape with intense asymmetric narrow peak in the range of small nanoparticles and less intense broad peak in the range of large (micrometer size) particles. With fluency increase, the intensity of former peak is decreased and it is shifted to the range of larger sizes (Figure 12).

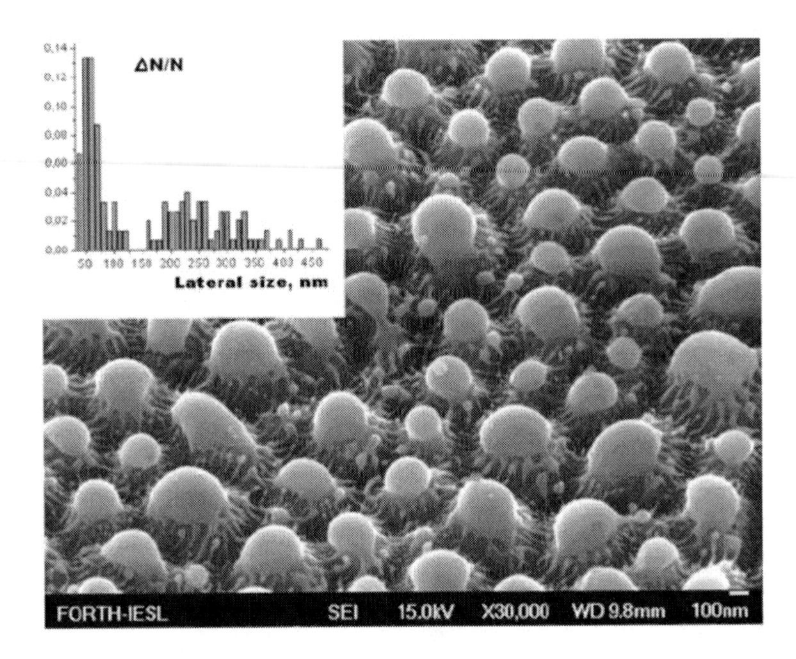

Figure 11. The tantalum surface nanostructured by irradiation under the water with a train of 5 ps laser pulses. The bar in the right bottom corner corresponds to 100 nm. Inset shows bimodal lateral size distribution function [41]. Horizontal axis shows the lateral size in nanometers (see Chapter 5).

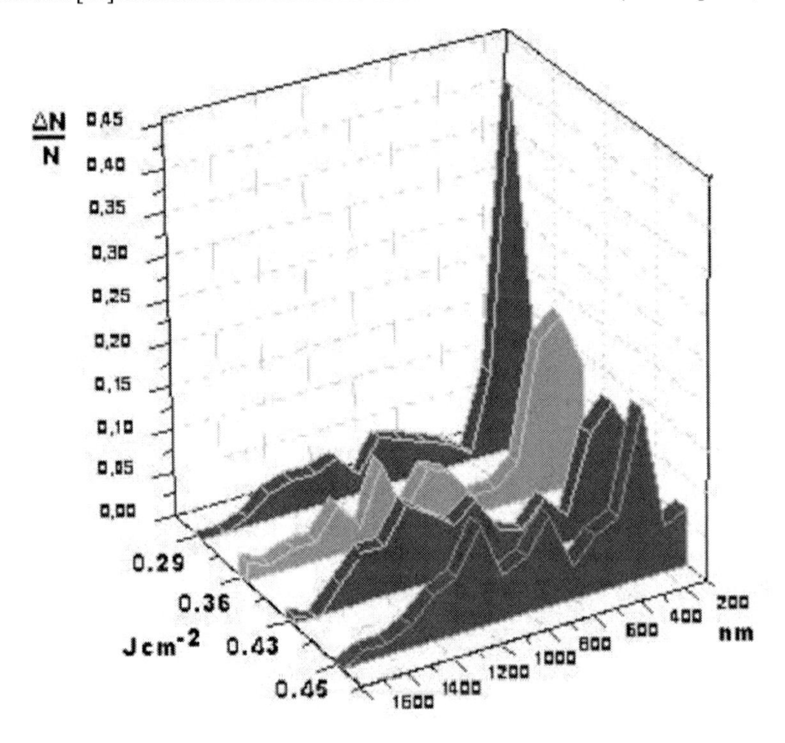

Figure 12. Lateral bimodal size distribution function 0f the nanoparticles ensemble formed on tantalum surface by irradiation under the water with a train of 350 ps laser pulses at different fluencies: F=0.29, 0.36, 0.43 and 0.45 $J*cm^{-2}$ [41]. See Chapter 5).

In the DD interpretation of formation of nanostructures on tantalum in the water, a laser pulse melts the surface layer of the thickness h_m. The subsequent rapid movement of the solidification front is accompanied by the capture of vacancies in this layer. A fraction of captured vacancies is pushed from this layer deeper to the bulk by the normal to the surface strain gradient created by nonuniform laser heating of the material and also by the water vapor pressure. The next pulse in the train again melts the layer h_m eliminating in it vacancies captured by preceding pulse but leaving unaltered the concentration of vacancies beneath this layer. The solidification front movement leads again to the capture of vacancies in the layer h_m, the fraction of which is transferred beneath this layer. The repetition of these processes leads to gradual accumulation of vacancies in the layer of thickness h_v lying beneath the surface layer h_m with the increase of the number of laser pulses. So, as a result of irradiation by a train of laser pulses the surface defect-enriched layer of thickness $h = h_m + h_v$ is created.

We estimate the thickness of the melted layer with the formula

$$h_m = \frac{F(1-R)}{c_v T_m + L_m}, \tag{63}$$

where F is the laser pulse fluency, R is the optical reflection coefficient, c_v is the specific thermal capacity, T_m is the melting temperature and L_m is the latent heat of melting. For tantalum, with $R = 0.78$, $c_v = 2.65\,\text{J/cm}^3\,\text{deg}$, $T_m = 3270\,\text{K}$ and $L_m = 2880.57\,\text{J/cm}^3$ at $F = 0.29\,\text{J/cm}^2$ we have $h_m = 5.5 \cdot 10^{-6}\,\text{cm}$.

For the thickness of buried vacancy-enriched layer we use the formula $h_v = (D_v \theta_v / k_B T)(\partial\xi/\partial z)\tau_{\text{Pr}} N$, where D_v is vacancy diffusion coefficient, $\partial\xi/\partial z$ is strain gradient along the surface normal, τ_{Pr} is the time of existence of normal pressure due to the water vapors, N is the number of laser pulses. At $D_v = 10^{-8}\,\text{cm}^2 * \text{s}^{-1}$, $\partial\xi/\partial z \sim \xi/h_m$, $\xi \sim 10^{-3}$, $\theta \sim 10^2\,\text{eV}$, T~1500K, $\tau_{\text{Pr}} \sim 10^{-7}$ s, $N \sim 10^4$ we have $h_v = 1.5 \cdot 10^{-6}\,\text{cm}$. Thus, the thickness of defect enriched layer is estimated as $h \sim h_m + h_v \sim 7 \cdot 10^{-6}\,\text{cm} \sim 10\,\text{nm}$.

We note that the same estimate h~ 10 nm can be obtained if one assumes that laser hydriding of tantalum (similar to hydriding of titanium [36]) takes place that leads to incorporation of hydrogen (defects) into the molten surface layer of the sample irradiated in the water.

The formation of the DD grating occurs then via the lateral redistribution of defects (vacancies or hydrogen) during the time of existence of high defect mobility, that is during the time of existence of elevated temperature in this layer t_T. The solidification time of the melted layer induced by the action of a pulse is $h_m^{\,2}/\chi$, where χ is the high temperature thermal diffusivity of tantalum ($\chi \sim 10^{-2}$ cm*s^{-1}). Overall time of duration of elevated temperature upon action of the train of N pulses is $t_T = N h_m^{\,2} / \chi \sim 10^{-3}$ s ($h_m \sim 10^{-5}$ cm and $N \sim 10^5$). We assume that in the formula (61) the time $t_0 = t_T$.

Concurrently with the development of the DD instability the etching of the surface molten layer takes place during the action of each laser pulse. It is assumed that after some time spatially uniform etching is replaced by spatially nonuniform etching of the DD structure so that the size of resulting nanoparticles will be determined by the period Λ of the DD grating. Thus, we can use the formula (61) for the size distribution function in the ensemble of nanoparticles.

Figure 13 shows the growth rate (26) and corresponding lateral size distribution function of particles calculated with the help of Eq. (61) at $N_{d0}/N_{cr} = 30$, $N_d(q,0) = \text{const} = 10^{15}$ cm^{-3}, $N_{cr} = 10^{16}$ cm^{-3}, $t_0 = 10^{-3}$ s, and values of parameters characteristic for tantalum: $D_d = 10^{-7}$ cm^2 *s^{-1} (vacancy diffusivity in tantalum at T_m [42]), $\nu = 0.35$, $h = 10^{-5}$ cm , $l_d = 3*10^{-6}$ cm, $L_d = 4.2*10^{-6}$ cm, $\rho c^2 = 10^{12}$ erg*cm^{-3}. We used also the value of the lateral stress estimated with formula $\sigma_{||} = K\ a_f - a_s\ / a_s \sim 10^{10}$ erg*cm^{-3} , where a_f and a_s are, respectively, the lattice constant of film and substrate, K is the elastic modulus ($K \sim \rho c^2$ and $a_f - a_s\ / a_s \sim 10^{-2}$).

It is seen from comparison of figure 11 and figure 12 with figure 13 that the calculated size distribution function has the form that in the main corresponds to the form of the experimental distribution function. It has the intense asymmetric maximum in the range of small (nano) particles and the broad, more symmetric maximum in the range of larger (micrometer size) particles. Asymmetry of the shortwave maximum arises because it is achieved at the wavelength of limiting bending mode $\Lambda_c = 2h$ (see figure 1); bending modes with $q > q_c = \pi / h$ do not exist. This situation is similar to the abrupt cut off of Debay density of phonon states at Brilloun zone boundary at $q = 2\pi / a$, where a is crystal cell size [43]. The locations of both calculated maxima of size distribution function (Figure 13) are close to experimental maxima of the curve corresponding to the lowest fluency F=0.29 J*cm^{-2} (Figure 12).

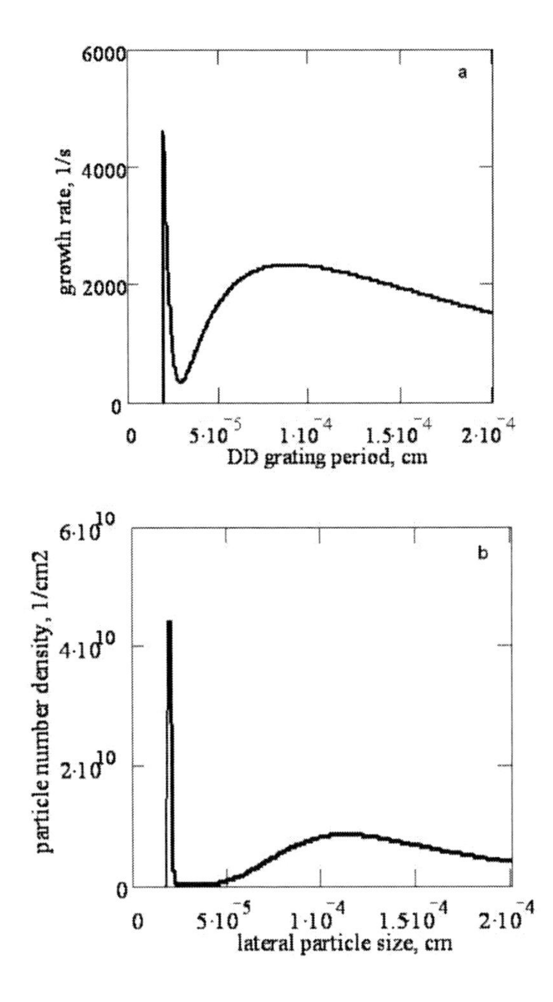

Figure 13. The growth rate of DD gratings (26) as a function of its period (a); the particles surface number density (61) as a function of the lateral particle size (b). Calculated with formulas (26) and (61) at the values of parameters given in the text.

With increase of fluency F the concentration of vacancies captured from the melt N_{d0} is decreased: $N_{d0} \sim \exp(-h_m / v_m \tau)$, where v_m is the velocity of the solidification front and τ is the time of transition of an atom from a displaced (defect) state to a regular position in the crystalline lattice [44]. Using the estimate: $v_m = \chi_m / h_m$, where χ_m is the thermal diffusivity in the melt and taking into account that h_m~F (see (63)), we have $N_{d0} \sim \exp(-h_m^2 / \chi_m \tau) \sim \exp(-F^2 / F_0^2)$, where F_0^2 is a characteristic fluency. Thus, as it seen from (61) and (26), the intensity of the peaks must decrease with increase of fluency (it can be shown by the numerical investigation of (61) that small size peak is more sensitive to the fluency increase). This prediction of the DD theory is in accordance with experiment (see figure 12). At last, since $\Lambda_m \sim h$, (31), and $\Lambda_c = 2h$, (28), and h increases with fluency increase, both maxima must shift to the range of larger particle sizes with fluency increase. This shift, in fact, is observed in experiment (see figure 12). An interesting feature of the experimental size distribution function is modulation (with 1-3 deep minima) of its broad shoulder in the large sizes region (Figure 12). This multimodal size distribution effect needs more studying. From the DD point of view its cause may be nonlinear three DD gratings interactions leading to transformation of the spectrum of Fourier harmonics of the surface relief due to the second harmonic generation and mixing of wave vectors of DD gratings (Secs.3 and 6.3).

As was noted above, the DD theory predicts in general case a gradual transition from the unimodal to the bimodal particle size distribution with increase of defect concentration, i.e. with change of characteristics of laser irradiation. Such transformation of size distribution function is indeed observed in experiment on laser-induced generation of ensemble of nanoparticles (see, for example, [15] and figure 14).

Besides, we note a close similarity of the shapes of bimodal size distribution function and locations and shifts of its maxima in dependence on the thickness of relevant defect-enriched surface layer in the case of laser-induced generation of nanoparticles ensemble (figure 11 and figure 12) and in the case selforganization of nanodots ensemble during molecular-beam epitaxy [45, 46]. This similarity is discussed in more detail in [12] on the basis of DD approach.

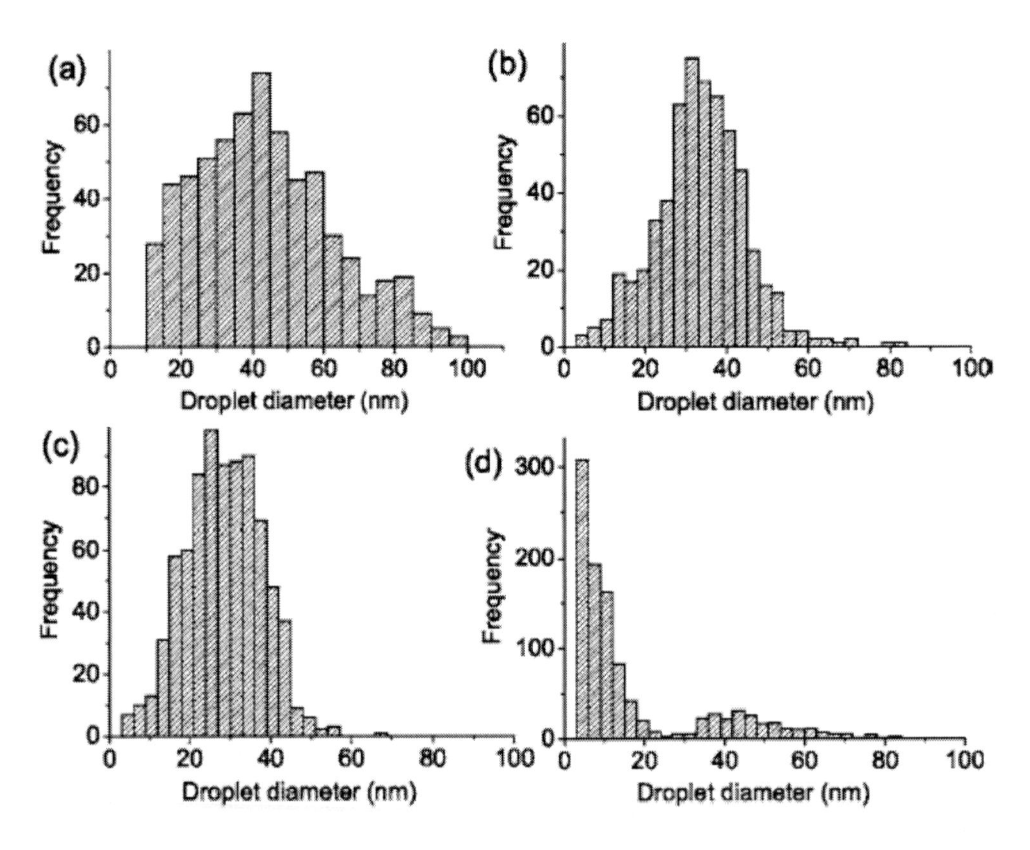

Figure 14. Ag nanoparticle diameter distribution produced by multipulse excimer laser nanostructuring of 5 nm Ag-on-oxide thin film at fluencies of (a) 175 $mJ*cm^{-2}$, (b) 250 $mJ*cm^{-2}$, (c) 400 $mJ*cm^{-2}$, and (d) 600 $mJ*cm^{-2}$ [15].

6.4. Surface Relief Harmonics Generation due to Three DD Gratings Interactions

Going back to the case of Ge (Figure 8) we note, that nonmonochromatic shape of nanorelief given by the profilogram of the surface relief (Figure 8e) was not understood up until now. The clue to the explanation is the observation that the superposition of the first and the second harmonics yields the relief approximately corresponding to the experimental one. Closer correspondence is obtained if also the third harmonic is added to this superposition (Figure 15a). Fourier transformation of experimental profilogram $\zeta\ x$, presented figure 8e, shows that, indeed, the first, second and the third harmonics dominate in the spectrum (Figure 15b) [14]. The generation of third harmonic of surface relief can be explained in terms of the cascade of three DD wave interactions involving SHG: $q_c + q_c = 2q_c$ and wave vectors mixing: $q_c + 2q_c = 3q_c$, considered in Sec.3.

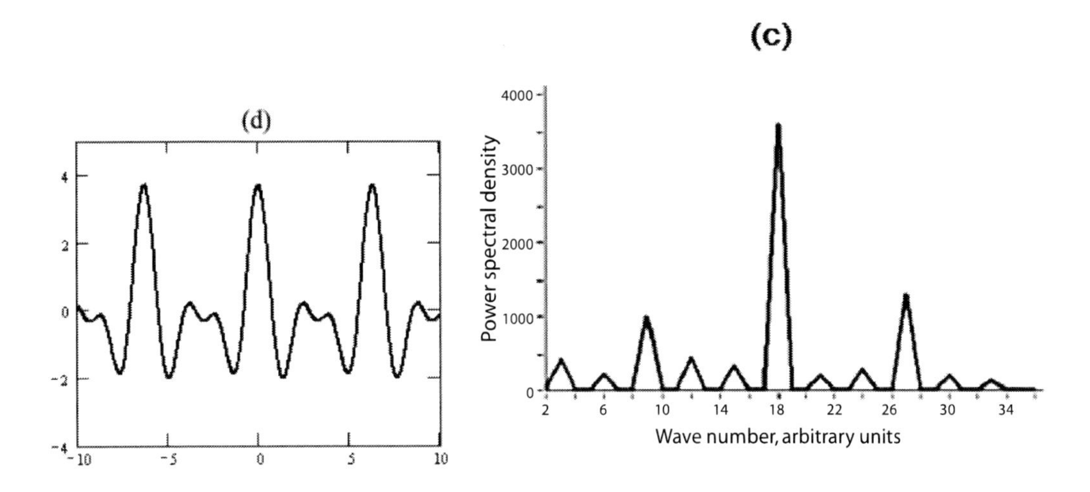

Figure 15. The function $f(x)=\cos(x)+1.7\cos(2x+0.1)+\cos(3x)$ that reproduces well the experimental profilogram $\zeta\ x$ shown in figure 8e under the condition that the large scale modulation of $\zeta\ x$ is eliminated (c); the spectral power $\left|\zeta_q\right|^2$ calculated with experimental profilogram $\zeta\ x$ shown in figure 8e but without large scale modulation [14] (d). Note that the amplitudes of the harmonics in the superposition $f(x)$ correspond to $\left|\zeta_q\right|^2$.

In the work [47] (see also the review [9]) the (100) surface of Si was irradiated by the linear polarized millisecond duration pulse of neodium laser with fluency in vicinity of the melting threshold. The irradiation at normal incidence leads to formation of two-dimensional crystallographically oriented grating of dots with micron-scale periodicity (Figure 10), characteristics of which were described by the theory of surface DD instability [47,9].

At the incidence angle $\theta = 30^{\circ}$, a more complex picture was observed than could be interpreted on the basis of such a simple interpretation (see figure 16).

Large scale (~30 μm) crystallographically oriented blocks were interpreted in [48] as a result of the surface thermo-deformation instability. Here, of interest is the relatively small (~3.5 micrometer) scale relief modulation, which has a complicated shape. Fourier-spectrum of this image presented in figure 16 shows that this relief is formed by superposition of one grating with wave vector parallel to the electric field vector $\mathbf{E}$ of exciting laser radiation the period of which depends on the angle of incidence. This grating is well known ripples which are formed due to interference of exciting and diffracted surface electromagnetic waves [49-51]. Besides, in Fourier-spectrum two pairs of crystallographically oriented reflexes are present along each of two mutually orthogonal [100] type directions. Pairs lying closer to the center correspond to micrometer scale DD (interstitials) gratings with wavevectors $\mathbf{q}_m$ along [100] type directions, formed due to the DD instability [47,9]. Their period is independent of the angle of incidence and the scaling parameter is the length of interstitial diffusion in direction normal to the surface during pulse action: $h = h_{\mathrm{diff}}$ [9]. Pairs lying at a distance from the center exactly two times longer correspond to two mutually orthogonal gratings with wave numbers $2|\mathbf{q}_m|$, that is to SH gratings of surface relief. More close inspection of the surface image shown in figure 16 shows that the DD grating is formed by periodic alternation of two intense dark lines and a grey (less intense) line lying exactly at the center between two dark lines(see figure 17b). Quite a similar picture is obtained as a result of computer modeling of 2D image of the surface with surface relief modulated by the superposition of first and the second harmonics (see figure 17 a).

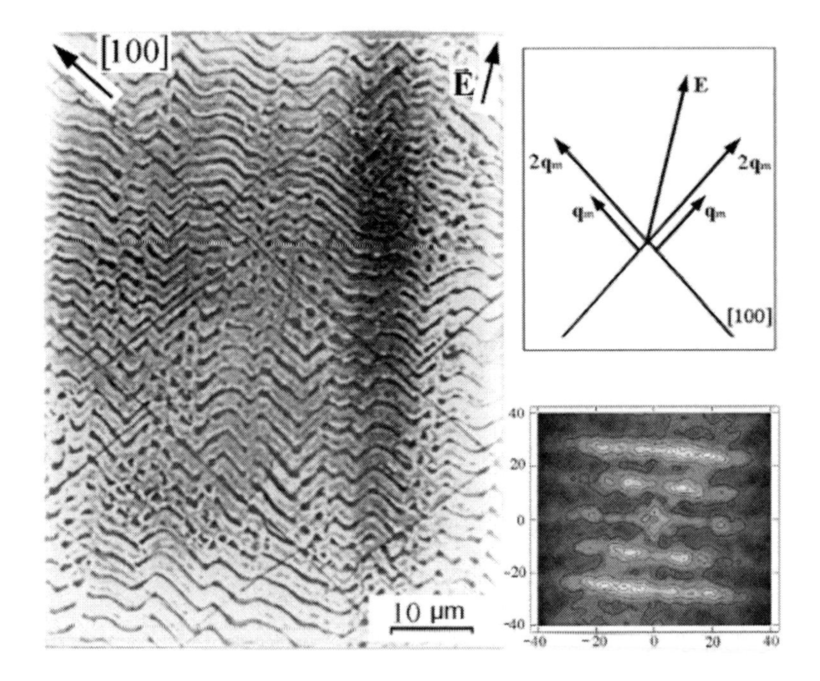

Figure 16. Photograph in optical microscope of Si(100) surface irradiated by millisecond pulse of linear polarized radiation (left) [47]. Fourier-transform of this image (bottom right). Each pair of reflexes lying on the diameter on equal distances from the centre corresponds to a grating of surface relief. The vectors $\mathbf{E}$, $\mathbf{q}_m$ and $2\mathbf{q}_m$ are, respectively, electric field vector of exciting laser radiation, wave vectors of the first and the second harmonics of the surface relief [14].

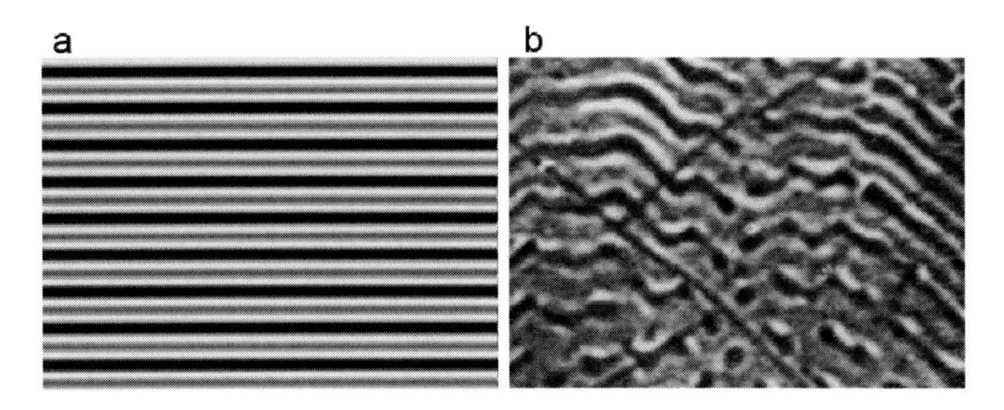

Figure 17. Computer modeled 2D image of surface with relief modulated by the superposition of the first and the second harmonics (a). Part of the image shown in figure 16 of irradiated surface (b). Compare computer generated grating (a) with the experimental DD grating in the upper left corner of (b) [14].

Effects of generation of micron-scale surface relief in Si [47] and nanometer scale relief in Ge [1,2] were observed in quite different irradiation regimes and for the formation of the DD structures in these two cases defects of different types are responsible (interstitials in the case of Si [47] and vacancies in the case of Ge [1,2]). Nevertheless, the DD mechanism is capable of describing from the unified viewpoint basic experimental data, including generation of surface relief harmonics. Adequacy of the DD mechanism in these two cases is supported by the following additive arguments. The characteristic time of formation of DD gratings can be roughly estimated as $t = \Lambda^2 / 4\pi^2 D_d$, so that the ratio of formation times is $t_{Si} / t_{Ge} = (\Lambda_{Si}^2 / \Lambda_{Ge}^2)(D_d^{Ge} / D_d^{Si})$. Experimental conditions in [47] and [1, 2] are such that the left hand side of this relation must be of order of unity. At $\Lambda_{Si} = 3*10^{-4}$ cm , $\Lambda_{Ge} = 5*10^{-5}$ cm, $D_d^{Ge} = 10^{-7}$ cm^2*s^{-1} (vacancies in Ge [31, 32]), and $D_d^{Si} = 10^{-5}$ cm^2*s^{-1} (interstitials in Si [47]) the right hand side is also of order of unity.

Thus, the consideration of this section, shows that the surface relief formed as a result of the DD instability, can have, depending on the exceeding over threshold, either one or two dominant harmonics with wave vectors $\mathbf{q}_m$, (16) and $\mathbf{q}_c$, (14). In both cases, the relief modulation period $\Lambda \sim h$, where *h* is the thickness of defect-enriched subsurface layer [11]. In nonlinear regime, additive harmonics $2q_m$, $3q_m$, $q_c - q_m$ can appear due to three waves DD interactions [14].

The DD mechanism of generation of surface relief harmonics consists in spatial redistribution of defects upon action of the grating of selfconsistent strain on the initial defect grating. For example, in the case of SHG the defect grating $\mathbf{q}_m$ is acted upon by the grating of strain-induced forces with the same wave vector $\mathbf{q}_m$, but shifted in the respect to its phase by $\pi/2$. This leads to a rising of the grating of defect fluxes with wave vector $2\mathbf{q}_m$, that serves as a source in the equation for the defect concentration field, (4) (or, in Fourier representation, in equation (25)). Thus, the nonlinear (quadratic) defect flux in the considered case of generation of the SH of the surface relief is similar to the quadratic polarization (or electric current) in the case of SHG in nonlinear optics.

The occurrence of two maxima of the growth rate and the possibility of nonlinear generation of DD harmonics must be taken into account upon analysis of experimental data on generation of surface relief upon irradiation with energy fluxes. Thus, the occurrence of nanometer and micrometer scales of surface relief modulation is a characteristic feature of processes of generation of nanostructures upon ion and laser beam etching of semiconductors (see review [11]). Three gratings DD interactions, as it is shown in this section (see also [14]), lead to generation of the second harmonic and generation of third harmonic by mixing of wave vectors. One may expect that at still higher levels of surface excitation, nonlinear DD interactions of higher orders will lead to further enrichment of the spectrum of harmonics of the surface relief.

CONCLUSION

Thus a new, DD approach to the problem of the nanostructure self-organization on solid surface is proposed. It is based on a concept of a defect-induced instability involving the quasi-Lamb static deformation waves that emerge in the near-surface layer (film) coupled with the quasi-Rayleigh static deformation waves that emerge in the underlying elastic continuum (substrate). These interrelated surface static deformation waves are analogs of the classical acoustic surface waves (the Lamb wave that propagates in plates and the Rayleigh wave that propagates along the surface of a semi-infinite medium). The generation and stationary maintenance of the new-type coupled static deformation waves in the near-surface layer is due to the self-consistent distribution of defects resulting from DD interaction via the defect deformation potential.

The initial stage in the DD scenario of the surface nanorelief self-organization involving the coupled static Lamb and Rayleigh waves is the formation of the stressed (stretched) surface nanolayer with thickness h that is enriched with mobile point defects. The same scenario that can be called a film on a substrate is also possible in the case of the self-organization of the surface relief microstructures. The extrema of the superpositional surface strain field induced by the continuum of the coupled static Lamb-Rayleigh waves serve as the nanoparticle nucleation centers.

In this work, general expression (24) for the coefficient of the defect-bending coupling of the film is derived with regard to the reaction of the substrate (underlying elastic continuum), which leads to an increase in the effective bending rigidity of the film. It is demonstrated that the substrate reaction can be neglected at a relatively high lateral stress in the film. Thus, we prove the adequacy and indicate the applicability domain of the simplified model of the DD instability for a free film that was introduced in [52], where only the static quasi-Lamb-wave instability is involved. The free film model of DD instability [52] (see also review [9]) was derived from first principles in [53], where also nonlinear computer analysis of it was done.

The establishing of this fact, that the DD instability is described by DD KS equation (Sec. 4) adds a new support to the supposition that the processes of nano- and microstructuring of the surface upon laser and ion beams irradiation are underlied by one and the same DD mechanism [11]. The KS equation with coefficients specific for ion sputtering of the surface is widely used for the description of formation of periodic structures upon ion irradiation of solids [25, 26]. The two-dimensional computer solution of the universal DD KS equation (43) or (45), obtained recently [56], enables one in a unified way to describe the formation of quite similar structures observed under laser and ion beam irradiation (for example, dot structures presented in figure 10 and figure 9).

A distinctive feature of the free film and the film on substrate models is that the lateral size of the nano- and microstructures is proportional to thickness h of the defect-enriched layer. The occurrence of such a universal linear dependence in the experimental results on the generation of the nano- and microstructures of the surface relief upon the laser and ion-beam irradiation of solid surfaces is demonstrated in [11].

In this chapter, we show that two maxima emerge on the curve of the growth rate of the DD instability. This makes it possible to propose a new interpretation of the two experimentally observed scales of the modulation for the quasi-periodic surface relief that results from the laser and ion-beam irradiation of semiconductor surfaces [12]. The existence of the second (micron) scale has been previously interpreted as a consequence of the DD instability of the second (thicker) defect-enriched layer that is formed owing to the defect diffusion from the thin near-surface layer to the bulk medium [11].

The generation of DD harmonics of surface relief due to three DD gratings interactions with wave vectors $\mathbf{q}_1$,$\mathbf{q}_2$ and $\mathbf{q}_3$ is a new aspect of the theory of selforganization of surface DD structures. Previously, three DD gratings interactions were studied in the diffusion approximation under the condition of small length of wave vector of one of the interacting gratings ($|\mathbf{q}_1| << |\mathbf{q}_2|,|\mathbf{q}_3|$. Under certain conditions, ascending diffusion in q-space leads to collapses of distributions of q-vectors in angular and scalar q-space leading to generation of coherent surface DD structures [10, 16].

We developed in this work the new approach to the calculation of the size distribution function for the nanoparticle ensemble that is formed due to the seed DD structure resulting from the surface DD instability. This unconventional approach involving the representation of the distribution function in terms of the instability growth rate is validated by the resulting scenario of a gradual transition from the unimodal nanoparticle size distribution to the bimodal distribution with an increase in the control parameter, which is in agreement with the experiment. The shapes of the resulting distributions and the maxima positions are also in agreement with the experimental data.

In the case of the multipulse laser etching, the material removal at the extrema of the DD field takes place at the maximum (or minimum) rate. The resulting distribution and shape of hills and wells at the surface relief visualizes the latent structure of the extrema in the superpositional DD field on the surface [11]. The formation of the surface DD field with a cellular structure via the development of the surface DD instability can serve as a universal mechanism of nano- and microstructuring of solid surfaces involving the generation of mobile point defects in a thin near-surface layer [54, 11].

REFERENCES

[1] Vinsents, S. V., Zaitsev, V. B., Zoteev, A. V., Plotnikov, G. S. *Semiconductors*, 2002, vol.36, 841.

[2] Vinsents, S.V., Zaitseva, A. V., Plotnikov, G. S. *Semiconductors*, 2003, vol.37, 124.

[3] Baidullaeva, A. A., Bulakh, M. B., Vlasenko, A. I., et al *Semiconductors*, 2004, vol.38, 29. Ibid. 2005, vol.39, 1028.

[4] Tsing, H. H., Finlay, R .J., Wu, C., Deliwala, S., and Mazur, E. *Appl.Phys.Lett.* 1998, vol.73, 1673.

[5] Wu, C., Crouch, C. H., Zhao, L, Carey, J. E, Younkin, R. J., Levinson, A, Mazur, E. *Appl.Phys.Lett.* 2001, vol.78, 1850.

[6] Crouch, C. H., Carey, J. E, Warrender, J. M., Aziz, M. J., and Mazur, E. *Appl.Phys.Lett.* 2004, vol.84, 1850.

[7] Ashkenazi, D. et al, *JLMN-Journal of laser micro/nanoengineering.*, 2006, vol.1, N1, 12.

[8] Dolgaev, S.I., Lavrishev, S.V., Lyalin, A.A, Simakin, A.V., Voronov, V.V., Shafeev, G.A. *Appl.Phys.Lett.*, 2001, vol. 73, 177.

[9] Emel'yanov V.I. *Laser Physics*, 1992, vol.2, 389.

[10] Emel'yanov, V.I. *Quantum Electron* 1999, vol.29, 561.

[11] Emel'yanov, V.I. *Laser Physics* 2008, vol.18, 682.

[12] Emel'yanov, V.I. *Laser Physics* 2008, vol.18, 1435.

[13] Victorov, I.A. *Acoustic Surface Waves in Solids*, Nauka, Moscow, 1981 (in Russian).

[14] Emel'yanov, V.I., Seval'nev, D.M. *Quantum Electron*, 2009, vol.39, N7.

[15] S.J.Henley, J.D.Carey, S.R.P.Silva, *Phys.Rev.* 2005, B72, 195408.

[16] Emel'yanov, V.I., Mikaberidze, A.I. *Phys.Rev.* B72, 235407(pp.1-11), 2005.

[17] Landau, L.D. and Lifshitz, E.M. *Theory of Elasticity* Pergamon, Oxford, 1986.

[18] C.E.Bottani, M.Yakona, J.Phys.;*Condens.Matter* I, 8337, 1989.

[19] 8. B.W.Dodson, J.Y.Tsao, *Appl.Phys.Lett.*,51,1325, 1987

[20] J.J. Yu, J.Y. Zhang, I.W. Boyd, Y.F.Lu, *Appl. Phys.* A 2001, vol.72, 35.

[21] Emel'yanov, V.I., Seval'nev, D.M. *J. Russian Laser Research*, 2009, vol.30, 21.

[22] Kuramoto, Y. Chemical Oscillations, Waves and Turbulence, Springer, Berlin, 1984.

[23] Sivashinsky, G.I. *Ann.Rev.Fluid Mech.*, 1983, vol.15, 179.

[24] Hohenberg, P.C., Shraiman, B.I. *Physica*, (Amsterdam), 1989, vol. 37D, 109.

[25] Bradley, R.M., Harper, J.M.E. *J.Vac.Sci.Technol.A*, 1988, vol. 6, 2390.

[26] Kahng, B, Jeong H., and Barabasi, A.L. *Appl. Phys.Lett.*, 2001, vol. 78, 805.

[27] Bobek, T., Fasko, S., Dekorsy, T., Kurz, H. *Nuc .Instr. and Meth. in Phys. Res. B*, 2001, vol.178,101.

[28] Ruspony, S., Constantini, G., de Mongeot, F.B., Boragno, C., Valbusa, U., *Appl.Phys.Let.*, 1999, vol.75, 3318.

[29] Emel'yanov, V.I., Baidullaeva, A., Vlacenko, A.I., Mozol, P.E. *Quant. Electron.* 2008, vol.38, 245.

[30] Kazakevich, P.V., Simakin, A.V., Shafeev, G.A. *Quant.Electron.* 2005, vol.35, 831.

[31] Emel'yanov, V.I., Vinsents, S.V., Plotnikov, G.S .J. of Surface Investigation. X-ray, Synchrotron and Neutron Technique 2007, vol.1, 667.

[32] Emel'yanov, V.I., Zaitsev, V. B., Plotnikov, G.S. J. of Surface Investigation. X-ray, Synchrotron and Neutron Technique 2008, vol.2, 392.

[33] Kiselev, V.F., Kozlov,S.N., and Zoteev, A.V. *Fundamentals of Physics of Solid Surface*, Moscow Sate University Publ.,1999 (in Russian).

[34] Kudo, A, *Catalysis Surveys from Asia*, 2003, vol.7, 31.

[35] Baery, P., Camprisano, S.U., Foti, G., and Rimini E. *Appl.Phys.Lett.*, 1999, vol.33, 137.

[36] Schwickert, M., Carpene, E., Lieb, K.P., Uhrmacher, M., Schaaf, P., Gibhardt, H. *Physica Scripta*, 2004, vol.T108, 113.

[37] Shen, M.Y., Crouch, C.H., Carey, J.E., and Mazur, E. *Appl.Phys.Lett.*, 2004, vol.85, 5694.

[38] Emel'yanov, V.I. *Quant.Electron.*, 2008, vol. 38, 618.

[39] Gago, R., Vasquez, L., Cuerno, R., Varela, M., Ballesteros, C., Abella, J.M. *Appl.Phys.Lett.*, 2001, vol.78, 3316.

[40] Gago R, Vasquez L, Cuerno R, Varela M, Ballesteros C, Abella JM. *Nanotechnology*, 2002, vol.13, 304.

[41] E.V. Barmina, M. Barberoglou, V. Zorba, A.V. Simakin, E. Stratakis, K. Fotakis, G.A. Shafeev, *Quant.Electron.*, vol. 39., N1, 2009.

[42] Satta, A., Willaime, F., Gironcoli, S. *Phys.Rev.B*, 1999, vol.60, 7001.

[43] Ziman, J.M. *Principles of the theory of solids*, University Press, Cambridge, 1964.

[44] Emel'yanov V.I., Babak D.V., *Appl.Phys.A* 74, p.797-805 (2002).

[45] J.Stangl, V.Holy, G.Bauer, *Rev.Mod.Phys.*, 76 725 (2004).

[46] Chapparo, S.A., Drucker, J., Zhang, Y., et al *Phys.Rev.Lett.* 1999, vol.83, 1199.

[47] Banishev, A.F., Emel'yanov, V.I., and Novikov, M.M. *Laser Physics* 1992, vol.2, 192.

[48] Emel'yanov V.I., Eriomin, K.I. *Quantum Electron*, 2001, vol.31, 154.

[49] Young, J.F., Preston, J.S., van Driel, H.M., and Luscombe, J. *Phys.Rev.B*, 1983, vol. 27, 1141.

[50] Preston, J.S., van Driel, H.M., and Sipe, J.E. *Phys.Rev.B* 1983, vol. 27, 1155.

[51] Akhmanov, S.A., Emel'yanov, V.I., Koroteev, N.I., and Seminogov, V.N. *Sov.Phys.Usp.* 1985, vol.147, 1084.

[52] Emel'yanov V.I. in *Nelieynye Volny* (Lektsii na VII Vses.Shkole po nelineinym volnam, Gor'kii, 1987) (*Nonlinear Waves: Lectures at the Seventh All-Union School on Nonlinear waves,* Gorky,1987) (Moscow, Nauka,1989).

[53] Walgraef, D., Ghoniem, N.M., Lauzeral, J. *Phys.Rev.B*, 1997, 56, 15361.

[54] Emel'yanov V.I. *Microelectronic Eng.,* 2003, 69, 435.

[55] Emel'yanov, V.I. *Laser Physics* 2009, vol.19, 538.

[56] Emel'yanov, V.I., Seval'nev, D.M. (to be published)

INDEX

B

C

D

K

L

M

N

O

P

Q

R

S

U

V

W

Y

Z